Lecture Notes in Computer Science   11158

Commenced Publication in 1973
Founding and Former Series Editors:
Gerhard Goos, Juris Hartmanis, and Jan van Leeuwen

More information about this series at http://www.springer.com/series/7409

Carson Woo · Jiaheng Lu
Zhanhuai Li · Tok Wang Ling
Guoliang Li · Mong Li Lee (Eds.)

Advances in Conceptual Modeling

ER 2018 Workshops Emp-ER, MoBiD,
MREBA, QMMQ, SCME
Xi'an, China, October 22–25, 2018
Proceedings

 Springer

Editors
Carson Woo
University of British Columbia
Vancouver, BC, Canada

Jiaheng Lu
University of Helsinki
Helsinki, Finland

Zhanhuai Li
Northwestern Polytechnical University
Xian, China

Tok Wang Ling
Department of Computer Science
National University of Singapore
Singapore, Singapore

Guoliang Li
Department of Computer Science
and Technology
Tsinghua University
Beijing, Beijing, China

Mong Li Lee
National University of Singapore
Singapore, Singapore

ISSN 0302-9743   ISSN 1611-3349   (electronic)
Lecture Notes in Computer Science
ISBN 978-3-030-01390-5   ISBN 978-3-030-01391-2   (eBook)
https://doi.org/10.1007/978-3-030-01391-2

Library of Congress Control Number: Applied for

LNCS Sublibrary: SL3 – Information Systems and Applications, incl. Internet/Web, and HCI

This Springer imprint is published by the registered company Springer Nature Switzerland AG
The registered company address is: Gewerbestrasse 11, 6330 Cham, Switzerland

Preface

The International Conference on Conceptual Modeling provides a leading international forum for disseminating the latest research on conceptual modeling. It brings together academic and industrial scientists who conduct research on theories of concepts underlying conceptual modeling, methods and tools for developing and communicating conceptual models, techniques for transforming conceptual models into effective implementations, and the impact of conceptual modeling techniques on databases, business strategies, and information systems development, etc. The conference's long history has established itself as the premier research conference in the conceptual modeling area.

Traditionally, the conference includes more specialized or focus sessions such as workshops and prototype demonstrations to create opportunities for a smaller group of researchers working in the same field to meet and exchange ideas. This provides opportunities to discuss new and promising topics or approaches. The 37th International Conference on Conceptual Modeling (ER 2018) continued this strong tradition, producing this companion volume to supplement the main conference proceedings.

This year, papers from the following five workshops are included in this volume:

1. Emp-ER: Empirical Methods in Conceptual Modeling
2. MoBiD: Modeling and Management of Big Data
3. MREBA: Conceptual Modeling in Requirements and Business Analysis
4. QMMQ: Quality of Models and Models of Quality
5. SCME: Conceptual Modeling Education

Each workshop had its own Program Committee and reviewers. We received many high-quality submissions, of which 14 were accepted. Along with the papers, this companion volume includes three keynote abstracts by Tok Wang Ling, João Araujo, and Oscar Pastor from three individual workshops (MoBiD, MREBA, and doctoral symposium), and two papers recommended by the program chairs of the main conference. In addition to the five workshops, we had:

6. An invited conference lunch speaker to discuss a topic that is of general interest and related to conceptual modeling. This year we were honored to have the past president of Taiwan's largest IT R&D organization, Dr. Chintay Shih of Industrial Technology Research Institute (ITRI), who introduced and elaborated on managing IT talents.
7. An interactive workshop CCM (Characterizing the Field of Conceptual Modeling) where there was no paper submissions, but the organizers formed groups during the session to discuss fundamental issues of conceptual modeling.
8. An ER forum for providing a vibrant platform for presenting and discussing novel research, and especially research-in-progress, that addresses any topic related to conceptual modeling. All of these five ER forum papers were recommended by the main conference program chairs.

9. A doctoral symposium (with three PhD students) where current PhD students had the opportunity to receive feedback on their research from an international community of experts and discuss research ideas with them.

10. A prototype demonstration session where participants had the opportunity to visualize and discuss implementation and utilization of new conceptual models. We received a number of high-quality submissions, from which we accepted 13 papers.

11. Four diverse tutorials with each tutorial providing a road map about a subject area related to conceptual modeling. The tutorials covered materials from a variety of different authors. The objective of a tutorial is to cover a subject that has adequate past work but still provides many opportunities for future research.

12. DSBC: Data Science and Blockchain workshop. Because this workshop was a late addition to the program, the papers are not included in this volume.

Each of the ER forum, doctoral symposium, and prototype demonstration presentations have a short version of their corresponding paper, or an extended abstract, in this volume. Information about the invited luncheon speech, interactive workshop CCM (Characterizing the Field of Conceptual Modeling), each of the four tutorials, and the DSBC (Data Science and Blockchain) workshop are also included in this volume.

We wish to thank the chairs of each workshop, the interactive workshop, ER forum, doctoral symposium, prototype demonstration, and tutorials, their respective Program Committee members and reviewers, and most importantly, the authors who submitted their work, all of which contributed to making this part of the conference a success. We hope you will enjoy reading work from this volume.

October 2018

Carson Woo
Jiaheng Lu
Zhanhuai Li
Tok Wang Ling

ER 2018 Conference Organization

Honorary Chair

Shan Wang   Renmin University of China, China

Conference General Co-chairs

Zhanhuai Li   Northwestern Polytechnical University, China
Tok Wang Ling   National University of Singapore, Singapore

Organization Chair

Xuequn Shang   Northwestern Polytechnical University, China

Steering Committee Liaison

Il-Yeol Song   Drexel University, USA

Program Committee Co-chairs

Juan-Carlos Trujillo   University of Alicante, Spain
  Mondéjar
Karen Davis   Miami University, USA
Xiaoyong Du   Renmin University of China, China

Workshop Co-chairs

Carson Woo   University of British Columbia, Canada
Jiaheng Lu   University of Helsinki, Finland

Tutorial Co-chairs

Gillian Dobbie   University of Auckland, New Zealand
Matteo Golfarelli   University of Bologna, Italy

Demo Co-chairs

Qun Chen   Northwestern Polytechnical University, China
Zhifeng Bao   RMIT, Australia

Doctoral Symposium Co-chairs

Xavier Franch   Polytechnic University of Catalonia, Spain
Chaokun Wang   Tsinghua University, China

ER Forum Co-chairs

Heinrich C. Mayr   Alpen-Adria-Universität Klagenfurt, Austria
Oscar Pastor   Universitat Politècnica de Valencia, Spain

Symposium on Conceptual Modeling Education (SCME) 2018 Co-chairs

Isabelle Comyn-Wattiau   ESSEC Business School, France
Hui Ma   Victoria University of Wellington, New Zealand

Treasurer and Registration Chair

Wenjie Liu   Northwestern Polytechnical University, China

Panel Co-chairs

Carlo Batini   University of Milano Bicocca, Italy
Ernest Teniente   Polytechnic University of Catalonia, Spain

Proceedings Co-chairs

Guoliang Li   Tsinghua University, China
Mong Li Lee   National University of Singapore, Singapore

Publicity Co-chairs

Bin Cui   School of EECS, Peking University, China
Moonkun Lee   Chonbuk National University, Republic of Korea
Mengchi Liu   Carleton University, Canada
Selmin Nurcan   University of Paris 1, France

Conference Webmaster

Jialu Hu   Northwestern Polytechnical University, China

ER 2018 Workshop Organization

Characterizing the Field of Conceptual Modeling (CCM) 2018 Co-chairs

Lois M. L. Delcambre   Portland State University, USA
Stephen W. Liddle   Brigham Young University, USA
Oscar Pastor   Universitat Politècnica de Valencia, Spain
Veda C. Storey   Georgia State University, USA

Conceptual Modeling in Requirements and Business Analysis (MREBA) 2018 Co-chairs

Jennifer Horkoff   Chalmers and the University of Gothenburg, Sweden
Renata Guizzardi   Universidade Federal do Espirito Santo, Brazil
Jelena Zdravkovic   Stockholm University, Sweden

Data Science and Blockchain (DSBC) 2018 Co-chairs

Peter Chen   Carnegie Mellon University, USA
Carson Woo   University of British Columbia, Canada

Empirical Methods in Conceptual Modeling (Emp-ER) 2018 Co-chairs

Jennifer Horkoff   Chalmers and the University of Gothenburg, Sweden
Sotirios Liaskos   York University, Canada

Modeling and Management of Big Data (MoBiD) 2018 Co-chairs

Il-Yeol Song   Drexel University, USA
Jesús Peral   University of Alicante, Spain
Alejandro Maté   University of Alicante, Spain

Quality of Models and Models of Quality (QMMQ) 2018 Co-chairs

Samira Si-said Cherfi   Conservatoire National des Arts et Metiers, France
Beatriz Marin   Universidad Diego Portales, Chile
Oscar Pastor   Universitat Politècnica de Valencia, Spain

ER 2018 Workshop Program Committee

DSBC 2018 Program Committee

Arne Solvberg	NTNU, Norway
Rong Chang	IBM Research, USA
Oscar Pastor	Universitat Politècnica de Valencia, Spain
Hasan Yasar	Software Engineering Institute, CMU, USA
Heinrich C. Mayr	Alpen-Adria-Universität Klagenfurt, Austria

Demo 2018 Program Committee

Muhammad Cheema	Monash University, Australia
Farhana Choudhury	RMIT University, Australia
Henning Koehler	Massey University, New Zealand
Yuchen Li	Singapore Management University, Singapore
Wei Lu	Renmin University, China
Ruiming Tang	Huawei, China
Sheng Wang	RMIT University, Australia
Manuel Wimmer	Vienna University of Technology, Austria
Qing Xie	Wuhan University of Technology, China
Jianqiu Xu	NUAA, China
Yi Yu	National Institute of Informatics, Japan
Jelena Zdravkovic	Stockholm University, Sweden
Ping Zhang	Wuhan University, China
Jingbo Zhou	Baidu Inc., China

Doctoral Symposium 2018 Program Committee

Eric Yu	University of Toronto, Canada
Heinrich C. Mayr	Alpen-Adria-Universität Klagenfurt, Austria
Hongzhi Wang	Harbin Institute of Technology, China
Jianliang Xu	Hong Kong Baptist University, SAR China
Peiquan Jin	University of Science and Technology of China, China
Roel Wieringa	University of Twente, The Netherlands

Emp-ER 2018 Program Committee

Silvia Abrahao	Universität Politecnica de Valencia, Spain
Jordi Cabot	Open University of Catalonia, Spain
Marian Daun	Universität Duisburg-Essen, Germany
Marcela Fabiana	University of Castillia, La Mancha, Spain
Miguel Goulao	Universidade NOVA de Lisboa, Portugal

Alicia Grubb	University of Toronto, Canada
Katsiaryna Labunets	Delft University of Technology, The Netherlands
Tong Li	Beijing University of Technology, China
Grischa Liebel	University of Gothenburg, Sweden
Dirk van der Linden	University of Bristol, UK
Feng-Lin Li	Alibaba Inc., China
Lin Liu	Tsinghua University, China
Jan Mendling	Vienna University of Economics and Business, Austria
Alexander Nolte	University of Tartu, Estonia
Xavier Le Pallec	University of Lille 1, France
Oscar Pastor	Universitat Politècnica de Valencia, Spain
Geert Poels	Ghent University, Belgium
Helen Purchase	University of Glasgow, UK
Jan Recker	University of Cologne, Germany
Iris Reinhartz-Berger	University of Haifa, Israel

MoBiD 2018 Program Committee

Yuan An	Drexel University, Philadelphia, USA
Marie-Aude Aufaure	Ecole Centrale Paris, France
Rafael Berlanga	Universität Jaume I, Spain
Sandro Bimonte	Irstea, France
Gennaro Cordasco	University of Salerno, Italy
Dickson Chiu	University of Hong Kong, SAR China
Alfredo Cuzzocrea	University of Calabria, Italy
Stuart Dillon	The University of Waikato, New Zealand
Gill Dobbie	University of Auckland, New Zealand
Jose Luis Fernandez-Aleman	University of Murcia, Spain
John R. Johnson	Pacific Northwest National Lab, USA
Magnus Johnsson	University of Lund, Sweden
Nicholas Multari	Pacific Northest National Laboratory, Richland, USA
Antoni Olive	Polytechnic University of Catalonia, Spain
Jeff Parsons	Memorial University of Newfoundland, Canada
Oscar Pastor	Universitat Politècnica de Valencia, Spain
Nicolas Prat	ESSEC Business School, France
Carlos Rivero	University of Idaho, USA
Pablo Sanchez	University of Cantabria, Spain

MREBA 2018 Program Committee

Okhaide Akhigbe	University of Ottawa, Canada
Claudia Cappelli	Universidade Federal do Estado do Rio de Janeiro, Brazil
Lawrence Chang	University of Texas, USA
Fabiano Dalpiaz	Utrecht University, The Netherlands

Marcela Ruiz	Utrecht University, The Netherlands
Aditya Ghose	University of Wollongong, Australia
Paul Johannesson	Stockholm University, Sweden
Tong Li	Beijing University of Technology, China
Sotirios Liaskos	York University, Canada
Grischa Liebel	Chalmers, Sweden
Lin Liu	Tsinghua University, China
Lidia Lopez	Polytechnic University of Catalonia, Spain
Pericles Loucopoulos	University of Manchester, UK
Joshua Nwokeji	Gannon University, USA
Andreas Opdahl	University of Bergen, Norway
Anna Perini	Fondazione Bruno Kessler, Italy
Jolita Ralyte	University of Geneva, Switzerland
Iris Reinhartz-Berger	University of Haifa, USA
Junko Shirogane	Tokyo Woman's Christian University, Japan
Motoshi Saeki	Tokyo Institute of Technology, Japan
Samira Si-Said Cherfi	Conservatoire National des Arts et Metiers, France
Vitor Souza	Universidade Federal do Esprito Santo, Brazil
Sam Supakkul	Sabre Travel Network, USA
Angelo Susi	Fondazione Bruno Kessler, Italy
Lucineia Thom	Universidade Federal do Rio Grande do Sul, Brazil

QMMQ 2018 Program Committee

Pnina Soffer	University of Haifa, Israel
Asma Sellami	University of Sfax - ISIMS, Tunisia
Jolita Ralyte	University of Geneva, Switzerland
Geert Poels	Ghent University, Belgium
Jeffrey Parsons	Memorial University of Newfoundland, Canada
Jose Ignacio Panach Navarrete	University of Valencia, Spain
Nan Niu	University of Cincinnati, USA
Raimundas Matulevicius	University of Tartu, Estonia
Xuelin Li	University of Texas at Dallas, USA
Julio Cesar Leite	PUC-Rio, Brazil
Zhi Jin	Peking University, China
Cesar Gonzalez-Perez Incipit	Incipit, CSIC, Spain
Sepideh Ghanavati	Texas Tech University, USA
Maya Daneva	University of Twente, The Netherlands
Jose Antonio Cruz-Lemus	University of Castilla-La Mancha, Spain
Cristina Cachero	University of Alicante, Spain
Ladjel Bellatreche	LIAS/ENSMA, France
Faten Atigui	CNAM, France

SCME 2018 Program Committee

Jacky Akoka	CEDRIC-CNAM and IMT-TEM, France
Solomon Antony	Murray State University, USA
Dickson K. W. Chiu	The University of Hong Kong, SAR China
Sebastian Link	The University of Auckland, New Zealand
Marcela Ruiz	Utrecht University, The Netherlands
Monique Snoeck	Katholieke Universiteit Leuven, Belgium

ER FORUM 2018 Program Committee

Raian Ali	Bournemouth University, UK
Nasser Assem	Al Akhawayn University, Ifrane, Morocco
Zhifeng Bao	RMIT, Australia
Luca Cernuzzi	Univ. Católica de Asunción, Paraguay
S. Chittayasothorn	King Mongkut's Inst. of Techn., Thailand
Karen Davis	Miami University, USA
Xiaoyong Du	Renmin University of China, China
Aurona Gerber	University of Pretoria, South Africa
Giovanni Giachetti	Universidad Andres Bello, Chile
Giancarlo Guizzardi	UFES Brazil & Free Univ. Bozen, Italy
Tok Wang Ling	National University of Singapore, Singapore
Beatriz Marín	Universidad Diego Portales, Chile
Heinrich C. Mayr	Alpen-Adria-Universität Klagenfurt, Austria
Zoltán Micsei	Budapest Univ. of Techn. & Eco., Hungary
Oscar Pastor	Universitat Polytécnica de Valencia, Spain
Ricardo Quinteró	Instituto Tecnológico de Culiacán, Mexiko
Daniel Riesco	National University of San Luis, Argentina
Genaina Rodrigues	Univ. of Brasilia, Brazil
Marcela Ruiz	Utrecht University, The Netherlands
Kurt Sandkuhl	University of Rostock, Germany
Flavia Santoro	UNIRIO, Brazil
Juan-Carlos Trujillo Mondéjar	Univ. de Alicante, Spain
Iyad Zikra	Stockholm University, Sweden

Why Mobility of (IT and Other) Talents Matters? (ER 2018 Conference Lunch Talk)

Chintay Shih

National Tsing Hua Univ. & Past President of ITRI, Taiwan

ER 2018 Conference Lunch Speaker, October 23 (Tuesday), 2018

Abstract. For a company or a developing economy, when considering their long-term development strategy, talents shortage is always a top priority issue. Even for an established enterprise or a developed country, how to tap into the resources of human capital is not easy. This talk is about the speaker's 40 years' experience in Taiwan's famous Hsinchu Science Park, participating in and witnessing the development of Taiwan semiconductor industry from nothing to one of the leading centers in the world. Where do the talents came from? How can a company/country attract, keep, and grow talents? and why the mobility of talent between institutions, countries matters? This talk will try to answer some of these questions.

Contents

Doctoral Symposium

ER FORUM 2018

Symposium on Conceptual Modeling Education (SCME) 2018

Empirical Methods in Conceptual Modeling (Emp-ER) 2018

Conceptual Modeling in Requirements and Business Analysis (MREBA) 2018

Modeling and Management of Big Data (MoBiD) 2018

Quality of Models and Models of Quality (QMMQ) 2018

Tutorials 2018

Demo 2018

Preface to Demonstration Track

Demonstration Track at the 37th International Conference on Conceptual Modeling (ER 2018)

Xi'an, China
October 22–25, 2018

Organized by
Zhifeng Bao and Qun Chen

The International Conference on Conceptual Modeling provides a leading international forum for disseminating the latest research on conceptual modeling. It brings together academic and industrial scientists who conduct research on theories of concepts underlying conceptual modeling, methods and tools for developing and communicating conceptual models, techniques for transforming conceptual models into effective implementations, and the impact of conceptual modeling techniques on databases, business strategies and information systems development, etc. The conference's long history has established itself as the premier research conference in the conceptual modeling area.

Traditionally, the conference includes a session for system prototype demonstrations to create opportunities for researchers to demonstrate their proof of concept research ideas. This provides opportunities for participants to visualize and discuss implementation and utilization of new conceptual models, and find further promising research questions from system building. The 37th International Conference on Conceptual Modeling (ER 2018) continues this strong tradition, producing this companion volume to supplement the main conference proceedings.

There were a total of 21 papers submitted to the demonstration track, and we accepted 13 papers, with a 62% acceptance rate, reflecting the high quality of the submission. The topics include conducting entity search over knowledge bases, exploiting conceptual modeling in data crowdsourcing, enterprise data management and data quality control, to name a few.

We wish to thank the seventeen program committee members of the demonstration track from both industry and academia, and the authors who submitted their work, both of which contribute to making this part of the conference a success. We hope the participants will enjoy reading work from this volume.

October 2018

Zhifeng Bao
Qun Chen
Program Co-chairs

Implementation of Bus Value-Added Service Platform via Crowdsourcing Incentive

Yan-sheng Chai[1](✉), Huang-lei Ma[1], Lin-quan Xing[1], Xu Wang[1], and Bo-han Li[1,2](✉)

[1] College of Computer Science and Technology, Nanjing University of Aeronautics and Astronautics, Nanjing 211106, China
1574411019@qq.com, bhli@nuaa.edu.cn
[2] Collaborative Innovation Center of Novel Software Technology and Industrialization, Nanjing 210016, China

Abstract. Sharing economy is prevailing. The network of cars and shared bicycles is convenient for people to travel. We investigate the issue of value-added service based on crowdsourcing for campus shuttles. We can provide diverse services between users by solving matching problems. The service concludes positioning and location services, requesting designating. The efficient incentive mechanisms make the shuttle bus transportation parcel convenient. We use KNN algorithm to establish KD tree to index different parcels nodes. In our app demo, we show how the application execute and how to improve the user experience who involve the orders.

Keywords: Spatio-temporal · KD-tree · Crowdsourcing · Incentive

1 Instruction

The concept of crowdsourcing has entered the public's field of vision starting with the article "The Rise of Crowdsourcing" published by Jeff Howe, a journalist of Wired Magazine in the United States, but no specific definition of crowdsourcing was given at that time. Jeff Howe did a full definition of crowdsourcing in 2008: Crowdsourcing is the process of outsourcing traditional tasks performed by internal employees to an unknown, usually large number of groups of an open manner [1]. However, from the point of view of time, crowdsourcing proposed by the Americans was one year later than the Witchy who was born in China in 2005. In July 2005, Liu Feng first proposed the concepts of "Wisconsin" and "Witchy" in the article "The Predicament and Countermeasures of Search Engines" [2]. Witchy phenomenon and crowdsourcing are similar to the same place, and it refers to an ordinary business model that relies on common mass resources to complete tasks for the enterprise. We apply crowdsourcing theory to study the parcel transportation problem of campus shuttle in two campuses.

© Springer Nature Switzerland AG 2018
C. Woo et al. (Eds.): ER 2018 Workshops, LNCS 11158, pp. 3–6, 2018.
https://doi.org/10.1007/978-3-030-01391-2_1

2   Related Work

The idea of crowdsourcing is the idea of solving problems of group wisdom. Its concrete model is that after the providers of crowdsourcing tasks release their tasks on the Internet, the non-specific groups in the network will undertake and solve them in a voluntary manner, thus it can complete the solution of the distributed problem. Driven by the development of the Internet, this idea of using cluster intelligence to solve distributed problems is realistic in the era of big data [3].

Mavridis et al. studied assignment problems of crowdsourcing. [4] In accordance with the need of crowdsourcing platform of some specific skills to complete tasks, they build fine models of tasks and participants through skills tree, and study efficient models for assigning tasks to participants with similar skills.

As for the incentive mechanism of crowdsourcing, Yang D studied the incentive mechanism to conduct group perception. The two mentioned system models: the supplier award crowdsourcing to share the user mode, and user-centered mode. [5] Our incentive mechanism is user-centered and has certain reference significance.

3   Crowd-Sourced Incentive Value-Added Service Platform

3.1   Definition

Spatio-temporal crowdsourcing task: a spatio-temporal crowd task issued by the requester of the task, usually is defined as the following five-tuple form, denoted as $t = <l, t, r, s, d>$. 'l' denotes the location of the task. 't' is the release time of the task. 'd' is the deadline for the task. 's' is the space scope released by the task, that is, the crowd-sourced participants in this scope have the opportunity to receive the task. 'r' is the reward for completing this task, usually expressed as the price or bonus of the task. [6] Spatio-temporal crowdsourcing: time and space through crowdsourcing, often referred to as mobile Internet devices in real time on spatio-temporal crowdsourcing platform gathering crowdsourcing tasks and crowdsourcing participants, and through the platform of crowdsourcing task scheduling for the distribution and quality control, which makes crowdsourcing participants in the physical world crowdsourcing task and satisfy the process of task constraints. With the framework of our project, we develop the task allocation mechanism based on continuous KNN query (CQ-KNN) algorithm and SEA-CNN algorithm applicable to mobile terminals [7, 8].

3.2   Task Allocation Mechanism

We refer some methods which adopt incentive mechanisms for large-scale participation and task assignment in mobile crowd sensing environment [9, 10].

We Present a Search Algorithm to Ensure a Large Number of Users to Efficiently Query the Nearby Parcel When Accessing the Database:

(1) Take a fixed point as the center, convert the latitude and longitude coordinates of the geode map into the plane artesian coordinates (x, y).

(2) According to the screening conditions, the database task parcel coordinates (x, y) are read, and there are two dimensions (x and y). The KD-tree (k-dimension tree) is established with points as the center:

    A. Calculate the variances in each dimension, choose the one with the larger variances, and treat this dimension as a split dimension.

    B. On the split dimension, sort the split dimension from small to large, select a coordinate in the middle as the split node.

    C. The left and right subtrees is constructed recursively according to the left and right data onto the split node.

(3) Given a participant coordinates a, start from the root node in the KD-tree and search depth first. Calculation of the current node with a Euclidean distance d, use every space node on the characteristics of the space division, decide whether to enter the left sub tree or the right sub tree to find the nearest point and look for the sub tree.

(4) Calculate the distance between the current node and a d'. If d' < d, update the most adjacent points a1 and d, otherwise not updated. Repeat (4), (5) until reach the leaf node. Backtracking, with sample point as the center and d as the radius, draws a circle to check whether there is delivery of the parent node meeting demands. If yes, enter another subtree to search for the nearest point, otherwise this point is the closest.

Query K adjacent parcel, keep with a K node in the linked list, and the maximum distance participants lists nodes, if found nodes in the search process is less than the maximum, the nodes to join the header, delete table node, update the maximum. In order to motivate participants to help each other, we set up a score system to evaluate each task. If having serviced one user, and the value of score is increased by one.

4 Implementation of the Platform

The following screen shots show some achievements of our project:

As shown in Fig. 1, users nearby can take the tasks. As shown in Fig. 2, it indicates releasing a task which needs users to add detailed information. When a user's parcel pass through our bus platform and asks others to help him transport the parcel to a designated place, he can publish the task by issuing task information. When allocating tasks, we can see other people's tasks through our platform and assign tasks to nearby people efficiently based on our task assignment algorithm.

Figure 3 is the process of the delivery, and Fig. 4 is the trajectory of our campus bus taking transport parcels. Users can view their own trajectory at any time to ensure the security of the parcel, and also record data as a necessary reference for our analysis.

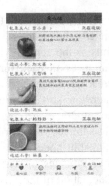

Fig. 1. Users nearby   **Fig. 2.** Task allocation   **Fig. 3.** Status of parcels   **Fig. 4.** Trajectory of parcels

5   Conclusion and Future Work

Crowdsourcing has a very wide application prospect. Our project is to develop an APP combining the crowdsourcing of shuttle bus to deliver parcels so as to improve the added value. If our application works well on the school buses and is widely supported by our staffs and students, we will further promote it at the social vehicle level, including private cars, buses, subways, etc.

References

1. Howe, J.: Crowdsourcing: Why the Power of the Crowd is Driving the Future of Business. www.crowdsourcing.com. Accessed 15 Jan 2014
2. Ye, T., Yu, B.: Analysis of witchy business model. Libr. Inf. Sci. **20**(22), 121–123 (2010)
3. Kleemann, F., Voß, G., Rieder, K.: Un(der)paid innovators: the commercial utilization of consumer work through crowdsourcing. Sci. Technol. Innov. Stud. **4**(2), 5–26 (2008)
4. Mavridis, P., Gross-Amblard, D.: Using hierarchical skills for optimized task assignment in knowledge-intensive crowdsourcing. In: ACM International Conference on World Wide Web, Montreal, pp. 843–853 (2016)
5. Yang, D., Xue, G., Fang, X., et al.: Incentive mechanisms for crowdsensing: crowdsourcing with smartphones. IEEE Trans. Network. **24**(3), 1732–1744 (2016)
6. Tong, Y.X., Yuan, Y., Cheng, Y.R., Chen, L., Wang, G.R.: Survey on spatiotemporal crowdsourced data management techniques. J. Softw. **28**(1), 35–58 (2017)
7. Rubing, L., Qiong, L.: KNN query technology of mobile terminals in highway networks. J. South China Univ. Technol. **40**(1), 138–145 (2012)
8. Xiong, X., Mokbel, M.F., Aref, W.G.: SEA-CNN: scalable processing of continuous k-nearest neighbor queries in spatio-temporal databases. In: International Conference on Data Engineering, Tokyo, pp. 643–654. IEEE (2005)
9. Ogie, R.I.: Adopting incentive mechanisms for large-scale participation in mobile crowdsensing: from literature review to a conceptual framework. Hum. Cent. Comput. Inf. Sci. **6**(1), 24 (2016)
10. Pournajaf, L., Xiong, L., Sunderam, V.: Dynamic data driven crowd sensing task assignment. Procedia Comput. Sci. **29**, 1314–1323 (2014)

SEED V3: Entity-Oriented Exploratory Search in Knowledge Graphs on Tablets

Jun Chen[1,2], Mingrui Shao[1,2], Yueguo Chen[1,2(✉)], and Xiaoyong Du[1,2]

[1] School of Information, Renmin University of China, Beijing, China
chenyueguo@ruc.edu.cn

[2] Key Laboratory of Data Engineering and Knowledge Engineering, MOE, Beijing, China

Abstract. Entity-oriented information access is becoming a key enabler for next-generation information retrieval and exploration systems. Previously, researchers have demonstrated that knowledge graphs allow the exploitation of semantic correlation among entities to improve information access. However, less attention is devoted to user interfaces of tablets for exploring knowledge graphs effectively and efficiently. In this paper, we design and implement a system called SEED to support entity-oriented exploratory search in knowledge graphs on tablets. It utilizes a dataset of hundreds of thousands of film-related entities extracted from DBpedia V3.9, and applies the knowledge embedding derived from a graph embedding model to rank entities and their relevant aspects, as well as explaining the correlation among entities via their links. Moreover, it supports touch-based interactions for formulating queries rapidly.

Keywords: Knowledge Graph · Knowledge presentation
Graph embedding model · Exploratory search · User interface · Tablet

1 Introduction

Entity-oriented information access is becoming a key enabler for next-generation information retrieval and exploration systems. The premise of entity-oriented information access is to better answer users' information needs by returning entities instead of documents [7]. Previously, researchers have demonstrated that knowledge graphs (shorted as KGs) can be applied to improve information access, because they contain high-quality semantic information, which allows us to apply the off-the-shelf semantic correlation for ranking entities and relations, as well as to explicitly show why entities are relevant to each other [3,4].

As illustrated in Fig. 1-c, KGs describe entities in many aspects, each of which is composed of a relation and an entity. For instance, the aspect composed of *Director* and *Robert_Zemeckis* describes that *Forrest_Gump* is directed

This work is supported by the National Science Foundation of China under grants No. 61472426, U1711261, 61432006.

C. Woo et al. (Eds.): ER 2018 Workshops, LNCS 11158, pp. 7–11, 2018.
https://doi.org/10.1007/978-3-030-01391-2_2

by *Robert_Zemeckis*. A particular set of such aspects fosters methods for the understanding of these entities. However, KGs often lack important triples, which affects the effectiveness of methods that simply utilize the explicit links [5]. To address this issue, some knowledge presentation models have been proposed to map entities and relations into the embedding space [2], which provides a principled way to rank entities beyond the explicit links in the existing KGs.

However, less attention is devoted to the user interfaces of tablets for exploring KGs effectively and efficiently. It is faced with two main challenges: (1) conventional user interfaces of tablets are constrained by the performance of virtual keyboard-based and mouse-based interactions, which confronts users with challenges in formulating queries rapidly, as well as providing positive or negative feedbacks with different weights to improve the results; (2) millions of entities are connected by thousands of relations in KGs, which confronts systems with challenges in effectively recommending and presenting relevant information in the limited space. Therefore, it becomes crucial to propose a novel solution that can overcome the limitations of conventional user interfaces of tablets and bring the potential of touch-based interactions to the forefront.

In this paper, we design and implement a system called SEED for entity-oriented exploratory search in KGs on tablets. It utilizes the knowledge embedding derived from a graph embedding model to present and recommend entities and their relevant aspects, as well as explain the correlation among entities via their links. Moreover, it supports users to manipulate entities and aspects directly to formulate queries rapidly.

2   The User Interface of SEED

The user interface of SEED is composed of four areas: (1) **Query Area** (see Fig. 1-a-1 and 1-a-2), (2) **Recommendation Area** (see Fig. 1-a-3 and 1-a-4), (3) **Presentation Area** (see Fig. 1-c), and (4) **Explanation Area** (see Fig. 1-d).

Query Area: To support users to formulate queries rapidly, users can swipe entities or aspects to the left as positive feedbacks (see Fig. 1-a-1) or the right as negative feedbacks (see Fig. 1-a-2). For the entities and aspects in the query area, users can manipulate the bars under them to adjust their weights (i.e., the weight enlarges when moving the bullet on the bar to the right, and vice versa). Users can scroll up in the query area with two fingers to submit a query.

Recommendation and Presentation Area: After submitting a query, the relevant entities and aspects with respect to the query are returned (see Fig. 1-a-3 and a-4). Users can scroll the horizontal or vertical axis to browse the recommended information. When the query is augmented by positive or negative feedbacks, the results are updated accordingly (see Fig. 1-b). When users are interested in a particular entity, they can look up its detailed information by clicking it to foster their understanding towards it (see Fig. 1-c).

Explanation Area: If the mechanism and reasoning of recommendation can be communicated to users in the right way, it can improve acceptance of the

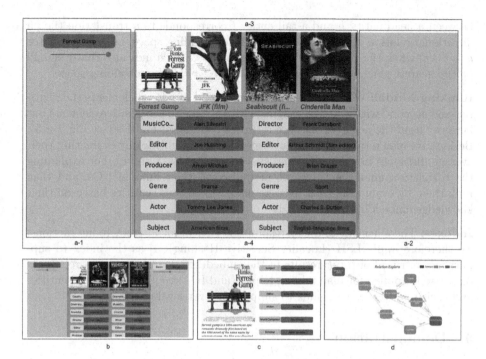

Fig. 1. Figure 1-a shows the main workspace of SEED, which is divided as the query area (a-1 and a-2) and recommendation area (a-3 and a-4). Figure 1-b shows the results augmented by a negative aspect. Figure 1-c shows the relevant aspects of a particular entity by clicking it. Figure 1-d shows the correlation among entities by clicking them together.

prediction and trust towards the system. When users are curious about the correlation among entities in the recommendation area, users can discover the links among entities in KGs by clicking them together, as the explanations to understand why they are recommended (see Fig. 1-d).

3 The Search Engine of SEED

SEED utilizes a dataset of hundreds of thousands of film-related entities extracted from DBpedia V3.9. It applies the knowledge embedding derived from a graph embedding model to rank entities and their relevant aspects. The links among entities in KGs are utilized to explain their correlation. Indexes for embedding and triples are applied to support fast interactive interactions.

Knowledge Embedding: We use a graph embedding model called TransE [2] to generate knowledge embedding, which maps the entities and relations of KGs into a dense, real-valued, and lower-dimensional embedding space, by using the local or global links in the existing KGs. In KGs, the semantic correlation between a head entity h and a tail entity t corresponds to a translation (i.e., a directional relation r) between them in the embedding space, i.e., $h + r \approx t$ (or $h \approx t - r$) when the triple $<h, r, t>$ exists in KGs, which indicates that t should be the nearest neighbor of (or very close to) $h + r$ in the embedding space.

Ranking Model: When given a query composed of several elements (i.e., entities and aspects) with different weights, entities close to the positive elements but far away from the negative ones are returned as relevant entities (see Fig. 1-a-3). Besides, we also return the relevant aspects of these entities by evaluating their average distances in the embedding space (see Fig. 1-a-4 and c). For explaining the correlation among entities, we first discover their links in KGs, and then rank them by evaluating their average distances to these entities based on their knowledge embedding (see Fig. 1-d).

Indexes: The efficiency of query-response is critical for interactive interactions, we therefore apply LSH (Locality Sensitive Hashing) [1] as the index to support fast retrieval of top-k results in the embedding space. Besides, a relational database is applied for maintaining triples, and B$^+$-tree [6] is created on entities and their relations as the index to support fast retrieval of triples.

4   Demonstration

We provide some use cases to foster users' understanding towards SEED. For instance, given an entity *Forrest_Gump* as an initial query, SEED returns the relevant entities and their aspects (see Fig. 1-a-3 and a-4). If users don't like a particular aspect, they can augment the query by swiping it to the right as a negative feedback, and the results are updated accordingly (see Fig. 1-b). If users are interested in the detailed information of a particular entity, they can click it to browse its relevant aspects (see Fig. 1-c). If users are curious about the correlation among entities, users can discover the links among entities in KGs as the explanations by clicking them together (see Fig. 1-d).

5   Conclusions

We design and implement a system for entity-oriented exploratory search in KGs on tablets. We demonstrate how KGs can be exposed for users through the user interface that supports rapid query formulation via touch-based interactions, allows presentation and recommendation of entities and their relevant aspects, as well as explains the correlation among entities via their links. We open new venues for users to turn the tedious typed-query keywords search to entity-oriented exploration, which exploits KGs for an improved information access.

References

1. Andoni, A., Indyk, P.: Near-optimal hashing algorithms for approximate nearest neighbor in high dimensions. Commun. ACM **51**(1), 117–122 (2008)
2. Bordes, A., Usunier, N., García-Durán, A., Weston, J., Yakhnenko, O.: Translating embeddings for modeling multi-relational data. In: NIPS, pp. 2787–2795 (2013)
3. Chen, J., Chen, Y., Du, X., Zhang, X., Zhou, X.: SEED: a system for entity exploration and debugging in large-scale knowledge graphs. In: ICDE, pp. 1350–1353 (2016)
4. Chen, J., Jacucci, G., Chen, Y., Ruotsalo, T.: SEED: entity oriented information search and exploration. In: IUI, pp. 137–140 (2017)
5. Dong, X., et al.: Knowledge vault: a web-scale approach to probabilistic knowledge fusion. In: KDD, pp. 601–610 (2014)
6. Garcia-Molina, H., Ullman, J.D., Widom, J.: Database System Implementation. Prentice-Hall, Upper Saddle River (2000)
7. Meij, E., Balog, K., Odijk, D.: Entity linking and retrieval for semantic search. In: WSDM, pp. 683–684 (2014)

MobiDis: Relationship Discovery of Mobile Users from Spatial-Temporal Trajectories

Xiaoou Ding, Hongzhi Wang[(✉)], Jiaxuan Su, Aoran Xie, Jianzhong Li, and Hong Gao

Harbin Institute of Technology, Harbin 150001, Heilongjiang, China
dingxiaoou@stu.hit.edu.cn
{wangzh,lijzh,honggao}@hit.edu.cn
itx351@gmail.com
admin@xieaoran.com

Abstract. The popularity of smartphones and the advances in location-acquisition technologies witness the development in the research of human mobility. This demo shows a relationship-discovery system of mobile users from their spatio-temporal trajectories. The system first matches all the access device IDs to places of interest (POI) on the map, and then finds out the access device IDs visited by more than one phone frequently or regularly. For these users, a model of historical spatio-temporal trajectories analysis combined with web browsing behavior is proposed to discover the relationship among them. A large-scale real-life mobile data set has been used in constructing the system, the performance of which is evaluated to be effective, efficient and user-friendly.

Keywords: Mobile data · Spatial-temporal trajectories
Relationship discovery

1 Introduction

The popularity of mobile devices and the advances in location-acquisition technologies is a witness to the flowering of mobile-users' behavior analysis. Mobile devices enable people to record the geographical location as well as the web pages they have *visited*. A sequence of such locations changing over time form a *trajectory* [7]. Research on the computing and querying trajectories, such as trajectory pattern mining [2], modelling human mobility trajectories [4] has attracted a wide-spread attention both in academia and industry. Moreover, with the rapid development of mobile social applications, it is convenient, and sometimes necessary for individuals to do their daily activities via their phones.

The study on individuals which has appeared in the same point in time and space in a certain regularity or even occasionally is an important basis for modelling the network structure and behavior regulation. Furthermore, it

© Springer Nature Switzerland AG 2018
C. Woo et al. (Eds.): ER 2018 Workshops, LNCS 11158, pp. 12–16, 2018.
https://doi.org/10.1007/978-3-030-01391-2_3

may also contribute to energy saving and traffic management. Unfortunately, to the best of our knowledge, little existing work has been done to discover user relationships from spatio-temporal trajectory combined with web browsing behavior.

Though challenged, not impossible, the colorful question can be solved by mining coarse trajectories matched with POIs on map as well as web access logs, which motivated the achievement of the system. The contributions of the paper are (1) a system named *MobiDis* is designed and implemented to discover reasonable relationships between mobile phone users, (2) a model of relationship mining is established. We apply the method on large-scale real data from mobile operator. It illustrates that the system is effective, efficient and user-friendly.

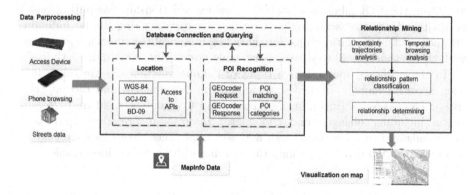

Fig. 1. System architecture

2   System Overview

System Architecture. The system architecture is shown in Fig. 1, consisting of 5 modules, namely data preprocessing, database connection and querying, location module, POI recognition and relationship mining. Original data from mobile operators will be preprocessed both in format and content before computing and querying. All mobile users' behaviour data is anonymized, mobile users are identified by a string (**PhoneID**) encrypted from their own phone number for protecting privacy.

Database connection and querying module provide stable and effective access to internal database and external map information. In location module, it enables transformation among several usual longitude and latitude coordinates. That improves the scalability of the system on different map system by valid access to corresponding APIs. In POI recognition module, all access devices (AD for short) can be located to accurate points in the map, after which they will be matched with POIs. In this way, the area represented by an AD can be obtained definitely with category labels describing its location characteristic.

In relationship mining module, uncertainty trajectories patterns mining and temporal browsing behaviour are taken into account. Several relationships are classified which aims at maximizing the coverage of mobile users. After the influence factors in classification are quantified with different thresholds, the relationship pattern between each pair of users can be found out and determined by the system. Certainly, to make it user-friendly, spatial trajectories are visualized on the map, with the behaviours along the time axis. The relationship discovery result will also shown in the list, correspondingly.

Relationships Discovery. Coarse trajectories generated by mobile users and their historical temporal visit in web pages are two main elements for determining relationships with others. Effective methods and algorithms in trajectory data mining [1,7] are used in frequent trajectories mining among certain POIs, weighting the POI label as a variable. Besides, relationship computing based on trajectories patterns mining [5,6] contributes for clustering similar individuals under different relationships. Furthermore, temporal behaviors are evaluated with the methods in [3] to draw a reasonable portrait of users' daily activities.

In the proposed method, five types of definite relationship are taken into deep account, namely *Relatives* (including family members and distant relatives), *Located-based* (including Neighbors and Community residents), *Professional* (including colleagues and schoolmate), *Interest-based* relationship (including friends and social connections), and *strangers*. These five types are easy to be quantified and can cover a majority of relationships among individuals.

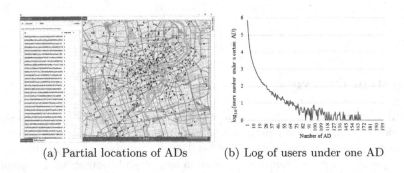

(a) Partial locations of ADs   (b) Log of users under one AD

Fig. 2. Distributions of AD and mobile users

3 Demonstration Scenario

Totally, more than 0.6 million mobile users in a big city in China moving among 1.03 million ADs (shown in Fig. 2(a)) for two months are studied in the system. More than 14 thousand users have trajectories between two or more ADs, which can generate over 0.23 million records half a month on average. Figure 2(b) shows

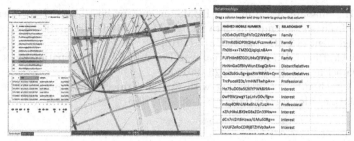

(a) Trajectories and Temporal Reords   (b) Relationship discovery

Fig. 3. System interfaces

logarithmic distributions of users under different ADs. In general, over 6 thousand users have visited more than two ADs frequently, and the largest users number under a certain AD is 2371.

Mobility of Mobile Users. Each spatio-temporal records of mobile users can be shown in the system. We focus on the users appearing in two or more ADs, whose movement among several certain ADs is shown as uncertain trajectories. In addition, the time axis of user' behavior is visualized in the left column, indicating the regulation from the behaviors. Under the scenario, the characteristics in mobility is preliminarily visualized as shown in Fig. 3(a).

Relationship Discovery Under Certain AD. The relationship discovery methods mentioned above is implemented as an important and typical function in the system. For a certain AD, each trajectory connecting with it is shown in the map, various trajectories of different individuals are marked in different colors. The relationship category label of each pair of users under the same AD (at the same time or not) is listed in the right column as shown in Fig. 3(b) after computing in relationship mining module. In the occasional cases that relationship cannot be determined by the system, the pair of mobile-users will be highlighted on the page, system users enable to insert the relationship result to their knowledge into the system.

References

1. Chen, Z., Shen, H.T., Zhou, X.: Discovering popular routes from trajectories. In: IEEE International Conference on Data Engineering, pp. 900–911 (2011)
2. Wu, G., Ding, Y., Li, Y., et al.: Mining spatio-temporal reachable regions over massive trajectory data. In: 33rd IEEE International Conference on Data Engineering, ICDE 2017, San Diego, CA, USA, 19–22 April 2017, pp. 1283–1294 (2017)
3. Jiang, S., Ferreira, J., Gonzalez, M.C.: Discovering urban spatial-temporal structure from human activity patterns. In: ACM SIGKDD International Workshop on Urban Computing, pp. 95–102 (2012)
4. Nguyen, M.N.B., Hasnain, Z., Li, M., et al.: Mining human mobility to quantify performance status. In: 2017 IEEE International Conference on Data Mining Workshops, ICDM Workshops 2017, New Orleans, LA, USA, 18–21 November 2017, pp. 1172–1177 (2017)

5. Pham, H., Shahabi, C., Liu, Y.: EBM: an entropy-based model to infer social strength from spatiotemporal data. In: ACM SIGMOD International Conference on Management of Data, pp. 265–276 (2013)
6. Shahabi, C., Pham, H.: Inferring real-world relationships from spatiotemporal data. IEEE Data Eng. Bull. **38**(2), 14–26 (2015)
7. Zheng, Y., Zhou, X.: Computing with Spatial Trajectories. Springer, New York (2011). https://doi.org/10.1007/978-1-4614-1629-6

Crowd-Type: A Crowdsourcing-Based Tool for Type Completion in Knowledge Bases

Zhaoan Dong[2], Jianhong Tu[2], Ju Fan[2(✉)], Jiaheng Lu[1,2], Xiaoyong Du[2], and Tok Wang Ling[3]

[1] Department of Computer Science,
University of Helsinki, Helsinki, Finland
fanj@ruc.edu.cn
[2] DEKE, MOE and School of Information,
Renmin University of China, Beijing, China
[3] School of Computing, National University of Singapore,
Singapore, Singapore

Abstract. Entity type completion in Knowledge Bases (KBs) is an important and challenging problem. In our recent work, we have proposed a hybrid framework which combines the human intelligence of crowdsourcing with automatic algorithms to address the problem. In this demo, we have implemented the framework in a crowdsourcing-based system, named Crowd-Type, for fine-grained type completion in KBs. In particular, Crowd-Type firstly employs automatic algorithms to select the most representative entities and assigns them to human workers, who will verify the types for assigned entities. Then, the system infers and determines the correct types for all entities utilizing both the results of crowdsourcing and machine-based algorithms. Our system gives a vivid demonstration to show how crowdsourcing significantly improves the performance of automatic type completion algorithms.

1 Introduction

In most KBs, e.g., DBPedia, entities are usually assigned with multiple types, which are organized in a hierarchical ontology [1] for representing semantic concepts of different levels. For example, in DBPedia, *Lionel Messi* has types including "Thing" (*Level=0*), "Person", "Athlete" and "SoccerPlayer". Obviously, fine-grained types (e.g., "SoccerPlayer") are more useful than general ones (e.g., "Person"). However, fine-grained types are more difficult to obtain. Hence, they are more likely to be missing. Existing automatic algorithms have limitations when inferring fine-grained types, while crowdsourcing is better even

This work is supported by National Natural Science Foundation of China (No. 61602488, No. 61632016 and No. 61472427), the Research Funds of Renmin University of China (No. 18XNLG18) and Academy of Finland (No. 310321).

C. Woo et al. (Eds.): ER 2018 Workshops, LNCS 11158, pp. 17–21, 2018.
https://doi.org/10.1007/978-3-030-01391-2_4

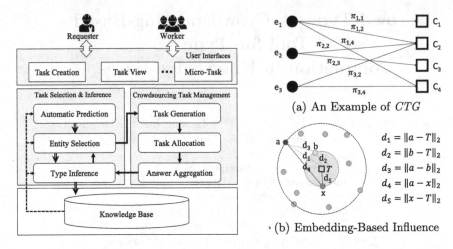

(a) An Example of *CTG*

(b) Embedding-Based Influence

Fig. 1. System architecture   **Fig. 2.** Type Inference method

though some data is missing. Recently, inspired by the studies of machine-crowdsourcing hybrid approaches [5], e.g., web table matching [6] and knowledge extraction [4], we proposed a hybrid framework [3], which combines the human intelligence of crowdsourcing with machine algorithms when to complete missing fine-grained types for entities. We have implemented the hybrid framework in Crowd-Type, a crowdsourcing-based system. This demo paper aims to help the audience better understand the hybrid framework, and to show how crowdsourcing improves the performance of type completion.

2   System Design and Implementation

As shown in Fig. 1, the system architecture of Crowd-Type consists of four parts. We give a brief introduction to the four parts respectively.

2.1   User Interfaces

In this demo, *User Interfaces* mainly contain three pages: `Task Creation` page, `Task View` page and the `Micro-Task` page. We will introduce them in Sect. 3.

2.2   Task Selection and Type Inference

In this part, we have implemented three fundamental algorithms:

Automatic Prediction: Given a set of entities whose types need to be completed, the *Automatic Prediction* algorithm is responsible for predicting possible types for each entity. The target entities together with their candidate types make up a *Candidate Types Graph* (*CTG*). For example, in Fig. 2(a), there are 7 entity-type pairs of 3 entities where $\pi_{1,i}$ denotes the entity-type pair of entity

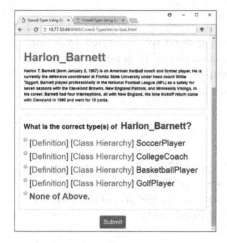
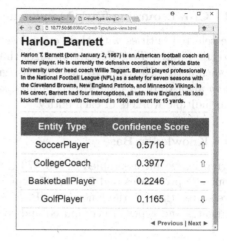

Fig. 3. Micro-task page   **Fig. 4.** Task-view page

e_1 and its possible type C_i ($1 \leq i \leq 3$). Currently, we use SDType, which is reported as the state-of-the-art method [7], to compute a confidence score for each possible entity-type pair. An entity-type pair will be considered as true if its score is greater than the given threshold. However, SDType has limitations as the evidences may be insufficient when inferring fine-grained types, while crowdsourcing is better even though some data is missing.

Entity Selection: Obviously, it would be extremely expensive if all the entity-type pairs are crowdsourced. Therefore, we use an *Entity Selection* algorithm to select the most *"representative"* entities to be verified by human workers. The main idea is, on the one hand, to select those entities which are the most uncertain for machine algorithms to determine their types. On the other hand, those entities that are most helpful to infer others' types, if confirmed by human workers, are easy to be selected.

Type Inference: Based on the *CTG* and crowdsourcing results of selected entities, *Type Inference* algorithm finally determines the score of each entity-type pair. To compute the influences from crowdsourced entities to unselected entities, we have proposed an *Embedding-Based Influence Model* in [3]. The basic idea is that if the entity types are embedded into the same embedding space, entities of the same type are usually not only close to each other, but also to their common types. In Fig. 2b, for example, we use L2-Norm to compute the distance between embeddings. If a is of type T, then b is more likely to be the same type than x, since b is closer to a than x. However, it can not be inferred that the type of a is T, if b is verified by human.

2.3   Crowdsourcing Task Management

The selected entities with their candidate types need to be verified by human workers. Firstly, *Task Generation* algorithm is employed to generate micro-task for each entity. Next, micro-tasks are assigned to human workers by *Task Allocation* algorithm. At last, *Answer Aggregation* algorithm aggregates the answers from multiple workers of each entity.

2.4   Knowledge Base

Currently, we have sampled $2,283,173$ entities, $9,970,687$ predicate instances about 629 predicates and $7,727,665$ type assertions from DBPedia 3.8 [2], and have stored them in the Knowledge Base. Additionally, the statistics of entities, predicates and types are computed and stored in advance.

3   Demonstration Scenarios

The audience can interact with the system in two roles: *Requester* and *Worker*. We will demonstrate three scenarios as follows.

Scenario 1: Task Creation. A requester can create a new type completion task for a set of entities through `Task Creation` page. Because of the limited space, we give the main parameters below instead of a screenshot:
- η: a threshold used by SDType to prune impossible types.
- n: for each micro-task, the number of worker assignments.
- *influence_method*: the method used by Entity Selection algorithm to compute the influences between entities. Two options are available here: *similarity-based* and *embedding-based* [3].
- *agg_method*: Currently, we have two options for aggregating the answers from human workers: *Majority vote* and *EM* algorithm.

Scenario 2: Micro-Task. As shown in Fig. 3, it is very easy for human workers to select the correct types for entity `Harlon_Barnett` with the help of the short description. If necessary, the workers can also click the "Definition" to see the definition of each type and click "Class Hierarchy" for the details of the class structure. However, because the relevant predicates of these types are almost the same, the confidence scores computed by the automatic algorithm are almost the same too. As a result, it is hard for the machine algorithm to determine the correct types solely.

Scenario 3: Task State. The requester can view the state of the tasks through `Task View` page, including not only the progress of the whole type completion task, e.g., how many entities have been completed by human, but also the details of each micro-task. As mentioned above, the Type Inference algorithm updates the scores for each entity-type pair after the crowdsourcing resulted is aggregated. If the confidence score of a type increases, it will be marked with "⇑", otherwise, "⇓", and "−" for no change. Figure 4 shows the score changes of the candidate types of the entity `Harlon_Barnett`. Intuitively, it reflects the effects of crowdsourcing results on the scores.

References

1. DBpedia: Dbpedia ontology. http://wiki.dbpedia.org/services-resources/ontology
2. DBpedia: Dbpedia3.8. https://wiki.dbpedia.org/data-set-38
3. Dong, Z., Fan, J., Lu, J., Du, X., Ling, T.W.: Using crowdsourcing for fine-grained entity type completion in knowledge bases. In: APWeb-WAIM 2018 (Accepted)
4. Dong, Z., Lu, J., Ling, T.W., Fan, J., Chen, Y.: Using hybrid algorithmic-crowdsourcing methods for academic knowledge acquisition. Cluster Comput. 20(4), 3629–3641 (2017)
5. Fan, J., Li, G., Ooi, B.C., Tan, K., Feng, J.: icrowd: an adaptive crowdsourcing framework. In: SIGMOD, pp. 1015–1030 (2015)
6. Fan, J., Lu, M., Ooi, B.C., Tan, W.C., Zhang, M.: A hybrid machine-crowdsourcing system for matching web tables. In: ICDE, pp. 976–987 (2014)
7. Paulheim, H., Bizer, C.: Type inference on noisy RDF data. In: Alani, H., et al. (eds.) ISWC 2013. LNCS, vol. 8218, pp. 510–525. Springer, Heidelberg (2013). https://doi.org/10.1007/978-3-642-41335-3_32

Attribute Value Matching
by Maximizing Benefit

Fengfeng Fan[1,2](✉) and Zhanhuai Li[1,2]

[1] School of Computer Science, Northwestern Polytechnical University, Xi'an, China
fanfengfeng@mail.nwpu.edu.cn
[2] Key Laboratory of Big Data Storage and Management, Ministry of Industry
and Information Technology, Northwestern Polytechnical University, Xi'an, China

Abstract. Attribute value matching (AVM) identifies equivalent values
that refer to the same entities. Traditional approaches ignore the weights
of values in itself. In this demonstration, we present AVM-LB, Attribute
Value Matching with Limited Budget, that preferentially matches the
hot equivalent values such that the maximal benefit to data consistency
can be achieved by limited budget. By defining a `rank` function and
greedily matching the hot equivalent values, the AVM-LB enables users
to interactively explore the achieved benefit to data consistency.

Keywords: Attribute Value Matching · Entity resolution · Hot data
Data cleaning · Big Data

1 Introduction

Due to typographical errors, aliases and abbreviations [1,4], the same real-world
entities may take several distinct representations across data sources, and such
inconsistencies may severely distort the results of data analysis. Hence it is necessary to match and merge those equivalent values by a process called Attribute
Value Matching or AVM [3]. Due to the large data size and limited budget,
it is a very challenging task to identify *all* of underlying equivalents, thus it
is preferred to employ a pay-as-you-go approach [5] to identify the equivalent
attribute values. However, existing approaches ignore the fact that inconsistencies between frequently accessed *hot* attribute values will bring more distortion
to data analysis and matching the *hot* equivalent values will bring more benefit
to data consistencies. In this paper, we propose a demo, denoted by AVM-LB,
which takes the *matching probability* and *data hotness* into consideration, and
interactively explores the achieved benefit by limited budget. To our knowledge,
AVM-LB is the first demonstration that incorporates the *data hotness* into data
cleansing practice. Our contributions can be summarized as follows:

1. AVM-LB provides a `rank` function, which ranks the candidates of value pairs
   for resolving, based on the *matching probability* and *hotness*.

Supported by organization x.

2. Based on the *matching relationship* and the *data hotness*, a benefit metric is devised to quantify the improvement to data consistency.
3. AVM-LB enables users to interactively explore the achieved benefit to data consistencies with limited budget.

2   System Overview

AVM-LB is composed of two components: Benefit Maximization and Performance Evaluation.

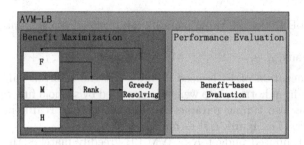

Fig. 1. System overview

2.1   Benefit Maximization

As the Fig. 1 shows, the rank function takes three matrices as input: filter matrix **F**, matching probability matrix **P** and hotness matrix **H**.

Filter Matrix: In AVM-LB, filter matrix **F** maintains the current states between attribute values. A snapshot of **F** is shown in Eq. 1: "1" for matches, "−1" for non-matches, "0" for unknowns, and "*" for links that can be deduced by symmetry.

$$\mathbf{F} = \begin{bmatrix} * & 1 & 0 & 0 \\ * & * & -1 & 0 \\ * & * & * & 1 \\ * & * & * & * \end{bmatrix} \tag{1}$$

As the matching process goes on, more 0-labeled cells will be replaced by either "1" or "−1", depending on the matching results, until no more 0-labels is available or all the budget run out.

Matching Probability: Matching probability matrix **M**, maintains the matching probabilities between attribute values, with $\mathbf{M}[i,j] \leftarrow \mathsf{P}(y_i \cong y_j)$. For simplicity, we approximate the matching probability $\mathsf{P}(y_i \cong y_j)$ by a similarity function $\mathsf{sim}(y_i, y_j)$, which can either be a simple string similarity measurement or some sophisticated metric, e.g., [2].

Hotness: Hotness often reveals the attribute value's importance in data analysis, and it may be a function of the timeliness, occurrences, or access frequencies. For similarity, we estimate the *hotness* of attribute values by their frequencies.

We define the hotness for any attribute value pair $\langle y_i, y_j \rangle$ by Eq. 2:

$$\text{hot}(y_i, y_j) = \text{freq}([y_i]) \cdot \text{freq}([y_j]) \tag{2}$$

where the equivalent class $[y_i]$ denotes the set of attribute values co-referring to the same entity with y_i, and $\text{freq}(\cdot)$ records the frequencies of attribute values. Hotness matrix $\mathbf{H}$, maintains the hotnesses for attribute value pairs, i.e., $\mathbf{H}[i, j] \leftarrow \text{hot}(y_i, y_j)$.

Rank: AVM-LB ranks the value pairs by a integrated scores, which is defined by Eq. 3:

$$\text{rank}(y_i, y_j) = \bar{\mathbf{F}}[i, j] \cdot \mathbf{M}[i, j] \cdot \text{sigmoid}(a \cdot \mathbf{H}[i, j] + b) \tag{3}$$

where $\bar{\mathbf{F}}$, the negation of $\mathbf{F}$, is used to filter out the resolved value pairs, the transformation from hotness into weight is provided by sigmoid function, in which $a \geq 0$ and b as two tuning parameters, and the matrix $\mathbf{Rank}$ maintains the integrated scores, i.e., $\mathbf{Rank}[i, j] \leftarrow \text{rank}(y_i, y_j)$.

Finally with limited budget Ks, AVM-LB greedily matches the equivalents based on the value of $\mathbf{Rank}[i, j]$.

2.2 Performance Evaluation

AVM-LB evaluates the performance by benefit, which is defined by Eq. 4:

$$\text{benefit}(y_i, y_j) = \mathsf{I}(y_i, y_j) \cdot \text{hot}(y_i, y_j) \tag{4}$$

where the indicator function $\mathsf{I}(\cdot)$ will return 1 for $y_i \cong y_j$, and 0 for otherwise. Intuitively, high benefit will bring big improvement to the probability of receiving consistent view of data for random queries.

For demonstrative purpose, we match the Journal values across two public available datasets, DBLP[1] and CiteSeer[2], in which $1,636,497$ and $45,783$ records are analyzed, and $1,666$ and $3,833$ distinct Journal values are extracted respectively. We construct the matching probability matrix $\mathbf{P}$ and hotness matrix $\mathbf{H}$ following the method in [2] and the definition of Eq. 2 respectively.

Figure 2(a) shows the startup user-interface, in which paths for dump file of $\mathbf{P}$ and $\mathbf{H}$ needs to be specified. After setting the valid paths for $\mathbf{P}$ and $\mathbf{H}$, we can interactively explore the achieved benefits by tuning parameters of rank function. For example, Fig. 2(b) shows the accumulated benefit with different budget by different rank function, in which the dashed curve ignores the hotness by setting parameter $a_0 = 0$, while the solid curve fine-tunes the weight of attribute value pairs by setting parameter $a_1 = 0.0001$ and $b_1 = 0$. It can be observed that by tuning parameters, AVM-LB allow us to interactively explore and visualize the benefit to data consistency with limited budget.

---

[1] http://dblp.uni-trier.de/.
[2] https://www.cs.purdue.edu/commugrate/data/citeseer/.

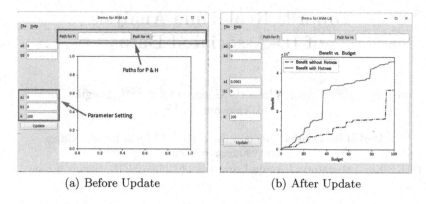

(a) Before Update   (b) After Update

Fig. 2. GUI for AVM-LB

Acknowledgments. The work was supported by the Ministry of Science and Technology of China, National Key Research and Development Program (Project Number 2016YFB1000703), the National Natural Science Foundation of China under No. 61732014 No. 61332006, No. 61472321, No. 61502390 and No. 61672432.

References

1. Batini, C., Scannapieco, M.: Data Quality: Concepts. Methodologies and Techniques. Springer Publishing Company, Incorporated (2010)
2. Fan, F., Li, Z., Wang, Y.: Cohesion based attribute value matching. In: 2017 10th International Congress on Image and Signal Processing, BioMedical Engineering and Informatics (CISP-BMEI), pp. 1–5, October 2017
3. Fan, F., Li, Z., Chen, Q., Chen, L.: Reasoning about attribute value equivalence in relational data. Inf. Syst. **75**, 1–12 (2018)
4. Naumann, F., Herschel, M.: An introduction to duplicate detection. Synth. Lect. Data Manag. **2**(1), 1–87 (2010)
5. Whang, S.E., Marmaros, D., Garcia-Molina, H.: Pay-as-you-go entity resolution. IEEE Trans. Knowl. Data Eng. **25**(5), 1111–1124 (2013)

Towards Real-Time Analysis
of ID-Associated Data

Guodong Jin[1,2], Yixuan Wang[1,2], Xiongpai Qin[1,2(✉)], Yueguo Chen[1,2],
and Xiaoyong Du[1,2]

[1] School of Information, Renmin University of China, Beijing, China
qxp1990@ruc.edu.cn
[2] DEKE Key Laboratory, Renmin University of China, MOE, Beijing, China

Abstract. ID-associated data are sequences of entries, and each entry is
semantically associated with a unique ID. Examples are user IDs in user
behaviour logs of mobile applications and device IDs in sensor records
of self-driving cars. Nowadays, many big data applications generate such
types of ID-associated data at high speed, and most queries over them
are ID-centric (on specific IDs and ranges of time). To generate valuable
insights from such data timely, the system needs to ingest high volumes
of them with low latency, and support real-time analysis over them efficiently. In this paper, we introduce a system prototype designed for this
goal. The system designed a parallel ingestion pipeline and a lightweight
indexing scheme for the fast ingestion and efficient analysis. Besides, a
fiber partitioning method is utilized to achieve dynamic scalability. For
better integration with Hadoop ecosystem, the prototype is implemented
based on open source projects, including Kafka and Presto.

Keywords: Real-time analytics · ID-associated data
Real-time ingestion

1 Introduction

ID-associated data are sequences of entries, and each entry is semantically associated with a unique ID. Examples are user IDs in user behaviour logs of mobile
applications and device IDs in sensor records of self-driving cars. Nowadays,
ID-associated data are ubiquitous in many big data applications. They are generated at high speed, and real-time analysis of them is critical to gain valuable
business insights timely. Take user behaviour analysis in mobile applications as
an example, each time a user clicks inside a mobile application, a log entry with
the user ID recording the user's behaviours is generated automatically and collected by the vendor. The vendor applies real-time analysis over newly collected

This work is supported by Science and Technology Planning Project of Guangdong
under grant No.2015B010131015, 863 key project under grant No.2015AA015307,
and the National Science Foundation of China under grants No.61472426, U1711261,
61432006.

data to identify abnormal user accesses, and study behaviours of a particular group of users over the latest days for better content recommendations.

In many such cases, ID-associated data come at high speed. Typically, most queries over ID-associated data are ID-centric – they retrieve and analyze data of a specified group of IDs over a period of time. To gain valuable insights timely, the system needs to ingest high volumes of data quickly and execute ID-centric queries efficiently. ElasticSearch [5], the distributed search engine, builds inverted indices over data to support real-time searches. However, due to the high latency of indexing, it cannot support data ingestion in real time. SQL-on-Hadoop systems like Spark SQL [2] perform well at batch processing, but they lack the ability of data ingestion. Hive [1] is proven to be a powerful tool for ETL(extract, transform and load) on HDFS. It converts the ETL pipeline into a batch of map reduce jobs, which is slow due to frequent reads and writes of intermediate files.

In this paper, we introduce a system prototype tailored for the real-time analysis of ID-associated data. It integrates ingestion of ID-associated data with ID-centric relational processing. To support this, we designed a parallel ingestion pipeline and a lightweight indexing scheme. The pipeline offers data ingestion with low latency, and a well-designed in-memory columnar store, enabling efficient relational processing over data being loaded. And the indexing scheme helps the system avoid a full scan of the relational table. Combined with the fiber partitioning mechanism, the system provides dynamic scalability. For better integration with Hadoop ecosystem, the prototype is implemented based on an open source messaging system (Apache Kafka [7]), and a popular SQL-on-Hadoop engine (Facebook Presto [8]).

2 System Architecture

As shown in Fig. 1, our system prototype contains six modules (in gray): **Collector**, **Loader**, **Indexer**, **Metadata**, **Connector** and **Coordinator**.

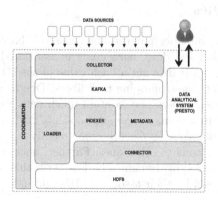

Fig. 1. System architecture

Collector. *Collector* servers as the front-end of the system, which is responsible for collecting data from various sources. Data collected are applied to our fiber partitioning method (details are discussed in Sect. 3), and sent into Kafka with a one-to-one correspondence between fibers and Kafka partitions. Kafka assigns each record with a unique offset to allow re-read of data from a specified offset. We make checkpoints on offsets of loaded data to guarantee data lossless during ingestion.

Loader and Indexer. *Loader* ingests data from fibers assigned by *Coordinator*. It pulls data from Kafka and writes them into HDFS after processing. Generally, data on HDFS are stored in columnar formats such as ORC [3] and Parquet [4] to speed up queries. Data pulled from Kafka are applied with user-defined filters and transformers. Then, data are sorted by their associated IDs and generation time, and appended into a *memory store* (details are discussed in Sect. 3). Finally, the *memory store* flushes data into HDFS as columnar files. In particular, for data in the *memory store*, *Indexer* maintains bloom filters to track existing IDs and records key statistics, such as min and max values of each column (including generation time). When data are flushed into HDFS, these lightweight indices are embedded into files. To efficiently load and index data, a parallel data ingestion pipeline is designed, which is described in Sect. 3.

Metadata, Connector and Coordinator. *Metadata* maintains critical metadata for the system, including definitions of relational tables and columns, user-defined functions, and fiber storage information. When executing queries, reading tasks are pushed down to *Connector* to fetch data needed by query engines. *Connector* is able to fetch data from the *memory store* and HDFS altogether, and make use of existing indices to filter useless files. Thus, analytical systems can be boosted to run queries on both of the most recent data in memory and historical ones on disk. *Coodinator* monitors all running processes in the cluster, and coordinates executions of *Loaders* by dynamically assigning fibers to achieve good scalability and load balance.

3   Key Techniques

Fiber Partitioning. Our fiber partitioning scheme is based on the consistent hashing [6]. In ID-associated data, each entry is associated with a unique ID, and a hash function is applied to get a hash value of the ID. The range of hash values is split into k intervals, mapping to k fibers. Due to the possibly uneven distribution of data, some fibers may contain much data while some may comprise few. During data ingestion, *Coordinator* collects load metrics from *Loaders*. When loads are seriously skewed, *Coordinator* greedily migrates fibers from the node with the heaviest loads to other ones. Further, fibers can be tuned in fine granularity to merge and split dynamically. Based on this partitioning scheme, data are further clustered by their associated IDs and indexed efficiently.

Parallel Ingestion Pipeline. Data pulled from Kafka are maintained as fiber streams separately, and in each stream data are processed with user-defined filters and transformers inside a thread. After filtering and transformation, data are inserted into a *sorted buffer*. Inside the buffer, entries are ordered by their associated IDs and generation time. Once the sorted buffer reaches its max size or user-defined lifetime (elapsed time since the last reset), it appends all data into the *memory store* and resets. The lifetime represents the latency of data ingestion. Data inside the *memory store* is organized as immutable segments and ready to be queried by analytical systems. When the memory store reaches its threshold in size, it flushes segments as columnar files into HDFS. To gain benefits from sequential disk I/O and avoid I/O competitions, a single writer thread is utilized to handle all writing requests.

In-Memory Columnar Storage. *Memory store* organizes data as segments, inside which data are clustered by their associated IDs, and stored in a columnar format to be optimized for cache lines. To reduce memory footprints, lightweight compression schemes such as run length encoding, bit packing, and dictionary encoding are utilized to compress segments. With little overhead, queries can run directly on these lightweight compressed segments. Memory store maintains storage information of each segment, including the storage level (i.e., on-heap, off-heap, on-disk) and location (i.e., object reference, memory address, file path). In-memory segments can be stored on-heap and off-heap based on user configurations. Off-heap storage is recommended by default to avoid GC overheads.

4   Demonstration

The demonstration is set up on a cluster, in which our *Collectors* are deployed along with Kafka, and *Loaders* are distributed with Hadoop and Presto. We design a data generator to generate records based on the result of joining *lineitem* and *orders* table from the TPC-H benchmark. Besides, we add an attribute, called *generation_time*, to identify the generation time of each record. The generator runs in a streaming manner – records are generated and sent to *Collectors* one by one in real time. During the demonstration, a web interface is presented with detailed information of data ingestion, such as throughput of ingestion and loads of each node. Also, users can choose queries from our provided query set or compose some in our web interface. The interface has a query client embedded to issue queries to the Presto and collect results through JDBC. Our provided query set consists of ID-centric queries, which are based on TPC-H queries with time range conditions and designated IDs. The example query[1] analyzes quantity, price and discount of orders made by customer k from *t1* to *t2*. Once a query is submitted, we can track its execution in our web interface. Our system prototype is open source on the Github[2].

---

[1] select sum(quantity), sum(totalprice), min(discount), max(discount), avg(extended price), count(*) from test where custkey = k and *generation_time* > *t1* and *generation_time* < *t2* order by linestatus.

[2] https://github.com/dbiir/paraflow.

References

1. Apache Hive (2011). http://hive.apache.org
2. Spark SQL: relational data processing in spark (2015). http://spark.apache.org/sql/
3. Apache ORC (2018). https://orc.apache.org
4. Apache Parquet (2018). https://parquet.apache.org
5. Gormley, C., Tong, Z.: ElasticSearch: The Definitive Guide: A Distributed Real-Time Search and Analytics Engine. O'Reilly Media Inc, Sebastopol (2015)
6. Karger, D., et al.: Web caching with consistent hashing. Comput. Netw. **31**(11–16), 1203–1213 (1999)
7. Kreps, J., Narkhede, N., Rao, J., et al.: Kafka: a distributed messaging system for log processing. In: Proceedings of the NetDB, pp. 1–7 (2011)
8. Traverso, M.: Presto: interacting with petabytes of data at facebook (2013). Accessed 4 Feb 2014

Phenomenological Ontology Guided Conceptual Modeling for Enterprise Information Systems

Tomas Jonsson[1]([⊠]) and Håkan Enquist[2]

[1] Genicore AB, Göteborg, Sweden
tomas@genicore.se
[2] Department of Applied IT, University of Gothenburg, Gothenburg, Sweden
hakan.enquist@gu.se
http://www.genicore.se
http://www.gu.se/english

Abstract. In model driven Enterprise Information Systems (EIS) conceptual ER models are fundamental. We demonstrate how a phenomenological modeling ontology for conceptual models assists in creating meaningful, auto generated EIS. Application of the phenomenological ontology on conceptual modeling is shown in a graphical, easy-to-use CASE tool, CoreWEB, which support both editing of models and generating user interfaces for consistent manipulation of instance data. A visual and easy-to-grasp demonstration of conceptual modeling with live generation of EIS should be a welcome holistic complement to specialized, focused, in-depth presentations. Audience at all levels find food for thought and inspiration to continue to explore conceptual modeling in general and phenomena oriented approaches with systems generation in particular.

Keywords: Semantic modeling · Ontology · Model driven systems

1 Introduction

The phenomenological conceptual modeling approach is the intellectual tool in focus in this demonstration. It is visualized and explained by using a supporting implementation in CoreWEB, a graphical modeling and execution tool (available with some demo videos at www.ameis.se/cml). The phenomenon ontology is founded on philosophical theory of phenomena as fundamental mind items and defines model language components and runtime system interface components Fig. 1 (Left).

This phenomenological ontology approach for conceptual models has been applied for successful and proven model driven EIS life cycle management, in large scale information systems (1500+ users) [1] over a period of 20 years. Lately, it has also been applied for successful design and maintenance of EIS for less formal organizations e.g. the social network enterprise running Project Lazarus

© Springer Nature Switzerland AG 2018
C. Woo et al. (Eds.): ER 2018 Workshops, LNCS 11158, pp. 31–34, 2018.
https://doi.org/10.1007/978-3-030-01391-2_7

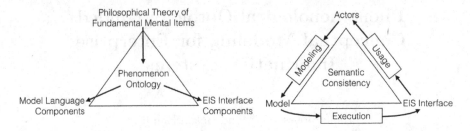

Fig. 1. (Left) The ontology defines modeling language components and EIS interface components. (Right) The ontology guides modeling process in how to map concepts into a model of fundamental mental items, *the phenomena*. The EIS runtime including a user interface is created through model execution, presenting data about *the phenomena*.

(www.projektlazarus.se) a Live Action Role Play (LARP). For ER 2018 this demo contributes specifically to *ontological and cognitive foundations for concept formalization and conceptual modeling*.

2   Phenomenon Ontology Guided Conceptual Modeling

Conceptual models play a fundamental role in life cycle management of model driven EIS. If EIS content is defined completely by the conceptual model, runtime system can be fully generated from the model and semantic consistency, between model and runtime system, can be achieved and maintained [2] hence *modelware becomes the software resource* to manage throughout EIS life cycle.

The EIS life cycle management procedure includes: conceptual modeling with Enterprise actors and domain experts that specify the complete EIS, runtime system generation from model baseline, EIS execution and use by enterprise actors in operations Fig. 1 (Right).

Value and quality of EIS are directly dependent on EIS support for objectives and information tasks of enterprise actors in use situations. Starting in

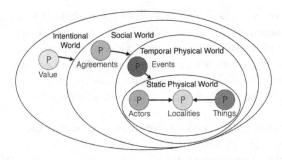

Fig. 2. Ontology foundation with phenomenon kinds, their world awareness contexts, combined with a coloring scheme used in the demo with CoreWEB tool. (arrows read as "relates to")

the enterprise and use of the EIS, conceptual models are created from domain experts' mental models and language as well as from other codified knowledge about the enterprise. If resulting EIS apply the conceptual models exactly, a 'WYSIWYG' relation between use and implementation is achieved, ensuring precision, transparency and maintainability.

This high consistency between minds, conceptual models and EIS require guidance from an ontology of mental models in addition to a modeling language.

The foundation for the phenomenological ontology Fig. 2, is a novel ER modeling approach, different from most referred-to ontologies [3] Bunge Wand Weber (BWW) and Unified Foundation Ontology (UFO) in that modeled entities are phenomena representing fundamental mind items, based on comprehension of existence in a metaphysical sense, Life World (Lebens Welt) and its phenomena categories, as in the philosophical school of phenomenology with philosophers such as Husserl [4], Heidegger [5] and Schutz [6].

The ontology is based on how the mind, when conceptualizing entities, separates different kinds of phenomena, which guides creation and comprehensions of the model. I.e. the concept 'hotel reservation' is a compound of two phenomena, the planned *event* of staying at a hotel and an *agreement* between the booker and the hotel.

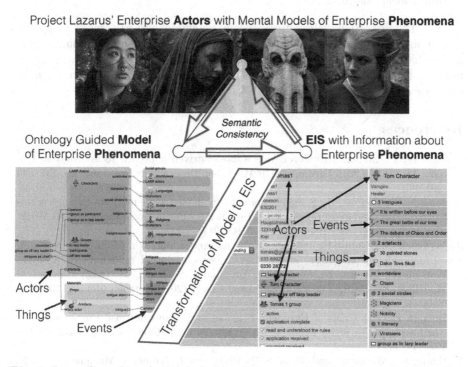

Fig. 3. Demo Case: Live Action Role Play Enterprise, Projekt Lazarus - 2018, with approximately 600 participants. (Top) Lazarus Actors, (Bottom-Left) Phenomena Model, (Bottom-Right) Runtime EIS

Distinguishing phenomena from roles is also part of the ontology, which will reduce definition and data redundancy. I.e. the concepts 'student' and 'teacher' are not phenomena, they are two *roles* that the *actor* phenomena *person* can have.

The color-coding of ontological phenomenon kinds introduced in the ontology graph Fig. 2, is reflected in conceptual model graphs Fig. 3 (Bottom-Left) and in generated runtime user interface Fig. 3 (Bottom-Right), to support semantic consistency throughout the EIS lifecycle.

3 Conceptual Modeling Features Addressed in This Demo

- Understanding the Phenomenological Modeling Ontology Elements, such as kinds of phenomena, kinds of relationships and roles of phenomena and EIS user interface in compliance with phenomenological conceptual models.
- Application of a phenomenon oriented ontology along with modeling rules in interpretation of domain knowledge and conceptual model design choices.
- Graphical editing and organization of model elements in conceptual models for EIS generation and execution of semantically consistent runtime system from conceptual models.

The demonstration utilizes Projekt Lazarus model and EIS to cover application of phenomenological ontology guided conceptual modeling of domain expert knowledge, instant generation of run-time system, interfaces and data management. Color-coded traceability provides intuitive evaluation and enhancement of semantic consistency.

References

1. Jonsson, T., Enquist, H.: CoreEAF - a model driven approach to information systems. In: CAISE 2015 Forum, CEUR Workshop Proceedings, vol. 1367, pp. 137–144 (2015)
2. Jonsson, T., Enquist, H.: Semantic consistency in enterprise models - through seamless modelling and execution support. In: ER 2017 Forum and Demos, CEUR Workshop Proceedings, vol. 1979, pp. 356–359 (2017)
3. Verdonck, M., Gailly, F.: Insights on the use and application of ontology and conceptualmodeling languages in ontology-driven conceptual modeling. In: Comyn-Wattiau, I., Tanaka, K., Song, I.Y., Yamamoto, S., Saeki, M. (eds.) Conceptual Modeling. ER 2016, vol. 9974. Springer, Cham (2016). https://doi.org/10.1007/978-3-319-46397-1
4. Husserl, E.: The Crisis of European Sciences and Transcendental Phenomenology. Northwestern University Press (1970). Trans. D. Carr (Original publ. 1954, written 1934–1937)
5. Heidegger, M.: Being and Time. SCM Press (1962). Trans. J. Macquarrie and E. Robinson (First published 1927)
6. Schutz, A.: The Phenomenology of the Social World. Northwestern University Press (1967). Trans. G. Walsh and F. Lehnert (First published 1932)

Using Clustering Labels to Supervise Mashup Service Classification

Yang Liu$^{(\boxtimes)}$, Lin Li, and Jianwen Xiang

School of Computer and Science, Wuhan University of Technology, Wuhan, China
1716501503@qq.com, cathylilin@whut.edu.cn, xiangjw@gmail.com

Abstract. With the rapid growth of mashup resources, clustering mashup services according to the functions of the mashup services has become an effective way to improve the quality of mashup services management. Clustering is a learning task that classifies individuals or objects into different clusters based on the similarity. The purpose of clustering is to maximize the homogeneity of elements in the same cluster and maximize the heterogeneity of the elements in different clusters. It is a multivariate statistical method for classification. However, compared with the supervised classification, the clustering's ability to categorize is much weaker. Existing methods for mashup services clustering mostly focus on utilizing key features from WSDL documents directly. In this paper, we proposed a method to improve the categorize ability of clustering. That is, applying supervised thought to cluster mashup services. First, taking basic clustering operations on the WSDL documents of mashups to obtain the clustering result for each element. Then, using the WSDL documents as training data, and the clustering results from the first step as pseudo-tags to train a classification learner. Finally, classifying mashups with this classification learner to get the final clustering results.

Keywords: Semi-supervised clustering · Ensemble clustering
Supervised LDA · Pseudo-tag

1 Introduction

In many practical problems of machine learning, it is possible to obtain limited prior knowledge. These prior knowledge can be used to guide the clustering process which will improve the clustering accuracy effectively. Shi [1] et al. used word2vec to obtain prior knowledge to guide the LDA clustering process and improve the clustering accuracy effectively. A single clustering algorithm has the problems of instability, low robustness, poor parallelism, low accuracy, etc. [2]. Clustering integration can improve these issues. Kang [3] et al. proposed a weight-based similarity clustering integration method based on swarm intelligence, and the experimental results proved the effectiveness and performance advantages of ensemble clustering. The ensemble clustering tool based on classification model we proposed can ameliorate the clustering results.

© Springer Nature Switzerland AG 2018
C. Woo et al. (Eds.): ER 2018 Workshops, LNCS 11158, pp. 35–38, 2018.
https://doi.org/10.1007/978-3-030-01391-2_8

In this demonstration, we will present sldaCluster, an ensemble clustering model based on supervised LDA. Users can set the initial clustering model as tfidf or word2vec [4] according to the specific situation. The number of clusters also can be adjusted based on the original data set. Users can also adjust the parameter alpha of the classification model slda [5] to obtain better experimental results. Finally, we will show the comparison between the clustering results of a single clustering algorithm and our proposed clustering method. The demonstration architecture is shown in Fig. 1.

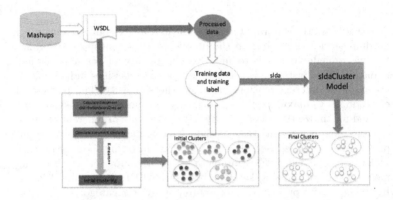

Fig. 1. sldaCluster architecture.

Figure 1 illustrates the overall architecture of sldaCluster. Two kinds of processes will be taken on the source data. The first is to make a simple process to change the source data into the training data which is applicable to the slda model. The second is to cluster the source data to obtain the clustering results. And the training results will be taken as the pseudo-tags to supervise the training of slda model. After the slda learner trained, we can put the processed training data and the presedo-tags into it to get the final clustering results.

2 Demonstration

We will use a mashup data set for experimentation. The data for clustering experiments is the WSDL document of the Mashup service. The experimental environment is Unix, the IDE used is pycharm, and the python version is Python3.5.

2.1 Experimental Procedure

(1) First, download the sldaCluster library from GitHub.
(2) Package and install the library in your environment. The installation details are shown in the README file.

(3) After the installation is complete, the functions in the module can be called and the methods are also shown in the README file.
(4) The tool link is https://github.com/liuyangzy/sldaCluster.git.
(5) The screencast link is https://youtu.be/jBrEHsU3798.

2.2   Comparison and Analysis of the Demonstration Results

In the experiment, the first thing to do is to preprocess the data. Each mashup has a main tag which is going to be used as a category reference for evaluating the clustering results. As categories with less Mashup Services would result in poor clustering performance, we only choose here the top 20 categories. Hence the number of cluster is also set to 20.

The sldaCluster model uses two single models as baseline models, namely tfidf model and word2vec model. The tfidf model used TF-IDF algorithm to calculate text features of each document, and uses cosine method to calculate the similarity between documents. Finally it will use k-means++ algorithm for clustering. The word2vec model uses the word2vec tool to calculate the word vector of each word in the text data. It uses the word vectors of all the words in each document to calculate the document vector. After calculating the similarities between each two documents with the cosine similarity, it will use the k-means++ algorithm to cluster.

We've compared the clustering results of the sldaCluster model with that using a single word2vec clustering method. It is easy to find that the experimental results obtained by using our proposed model have increased by 90.84%, 78.28%, 123.61% and 41.68% respectively under the evaluation index of JC, FMI, F5, Purity. We've also compared the clustering results of the sldaCluster model with that using a single tfidf clustering method. It is easy to find that the experimental results obtained by using our proposed model have increased by 8.67%, 8.31%, 13.53% respectively under the evaluation index of JC, FMI, F5. It means the sldaCluster model can improve the clustering accuracy significantly. The value of entropy has declined, that's to say the sldaCluster is more stable than the single one. The comparing results can be seen in Table 1. All the evaluation results are the best results after adjustment of parameters.

Table 1. Evaluation results of two single clustering models and their sldaCluster model.

Model/Evaluation index	JC	FMI	RI	F1	F5	Purity	Entropy
word2vec	0.0273	0.0557	0.8347	0.0531	0.0415	0.1725	3.8858
w2v_sldaCluster($\alpha = 0.2$)	0.0521	0.0993	0.8181	0.0991	0.0928	0.2444	3.5705
tfidf	0.0715	0.1335	0.8148	0.1334	0.1367	0.3732	2.9619
tfidf_sldaCluster($\alpha = 0.9$)	0.0777	0.1446	0.8580	0.1442	0.1552	0.2821	2.1678

We can have some analysis according to Fig. 2. Taking the word2vec model as the initial clustering algorithm of sldaCluster model will get much better improvement than taking the tfidf model. Through the analysis, it can be known that the difference between the word2vec model and the LDA model is larger relatively. Obviously the model obtained by ensembling the clustering algorithm and the classification model with greater difference is better. This is similar to the requirement that the basic learners are better to have diversity in ensemble learning.

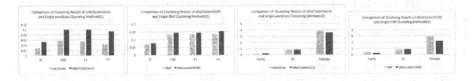

Fig. 2. Comparison of clustering results of sldaCluster and single clustering methods.

Through the comparisons of the above two experiments, two conclusions can be drawn: (1) Compared with a single clustering algorithm, sldaCluster can improve the clustering results significantly and the model is more stable. (2) The smaller the difference between the initial clustering algorithm and the classification model is, the less improvement the clustering can make.

Acknowledgements. This work is supported by the National Social Science Foundation of China (Grant No. 15BGL048), Hubei Province Science and Technology Support Project (Grant No. 2015BAA072), Hubei Provincial Natural Science Foundation of China (Grant No. 2017CFA012). The Fundamental Research Funds for the Central Universities (WUT: 2017II39GX).

References

1. Shi, M., Liu, J., Zhou, D., Tang, M, Cao, B.: WE-LDA: a word embeddings augmented LDA model for web services clustering. In: 24th International Conference on Web Services, Honolulu, HI, USA, pp. 9–16. IEEE (2017)
2. Faceli, K., De Carvalho, A.C., De Souto, M.C.: Multi-objective clustering ensemble. Int. J. Hybrid Intell. Syst. **4**(3), 145–156 (2007)
3. Kang, Q., Liu, S., Zhou, M., Li, S.: A weight-incorporated similarity-based clustering ensemble method based on swarm intelligence. Knowl. Based Syst. **101**, 156–164 (2016)
4. Kusner, M., Sun, Y., Kolkin, N., Weinberger, K.: From word embeddings to document distances. In: 32nd International Conference on Machine Learning, Lille, France, pp. 957–966 (2015)
5. Mcauliffe, J.D., Blei, D.M.: Supervised topic models. In: 21st Advances in Neural Information Processing Systems Conference, Whistler, British Columbia, Canada, pp. 121–128 (2008)

A Multi-Constrained Temporal Path Query System

Jiuchao Shi[1], Guanfeng Liu[2(✉)], Anqi Zhao[1], An Liu[1], Zhixu Li[1],
and Kai Zheng[3]

[1] School of Computer Science and Technology, Soochow University, Soochow, China
[2] Department of Computing, Macquarie University, Sydney, NSW, Australia
guanfeng.liu@mq.edu.au
[3] School of Computer Science and Engineering and Big Data Research Center,
University of Electronic Science and Technology of China, Chengdu, China
zhengkai@uestc.edu.cn

Abstract. The temporal path problem is significant and challenging, where the connections between the vertices are temporal and there can be many attributes on the vertices and edges, such as vehicle speed and the price of a flight. Then in path finding, in addition to the single requirement of the length, or the arrival time, people would like to specify multiple constraints on the attributes to illustrate their requirements in real applications, such as the total cost, the total travel time and the stopover interval of a flight between two cities. In this paper, we devise a system called MCTP to answer the new popular Multi-Constrained Path Queries (MCPQs) in attributed temporal graphs. To the best of our knowledge, this is the first system that supports MCPQs.

Keywords: Path finding · Temporal graph · Multiple constraints

1 Introduction

Path query is popular in many real applications, like the road planning in transportation networks and the route planning on the Internet. The existing methods for the path query in temporal graphs [3] can efficiently deliver a result with considering the path with a single attribute, such as the path with the minimal travel distance or the minimal cost [1,2]. However, there can be many attributes existing on the edges in temporal graphs, forming *attributed temporal graphs*, e.g., the travel time, the cost and the travel distance between two cities by flights. In real applications, people are usually willing to specify the constraints on the attributes to illustrate their requirements in the path query in temporal graphs. The challenge is how to develop an effective and efficient method to solve the Multi-Constrained Path Query (MCPQ) problem that is NP-Complete [4]. We propose a system called MCTP[1], which adopts our novel approximation algorithm and supports four types of important multi-constrained temporal path

---

[1] http://112.74.36.170:8080/MCTP.

© Springer Nature Switzerland AG 2018
C. Woo et al. (Eds.): ER 2018 Workshops, LNCS 11158, pp. 39–43, 2018.
https://doi.org/10.1007/978-3-030-01391-2_9

queries, including *the earliest arrival path, the last departure path, the shortest path* and *the fast path*. To the best of our knowledge, this is the first system that supports the NP-Complete MCPQs in attributed temporal graphs.

2   System Overview

The system architecture of MCTP is shown in Fig. 1, which contains two main layers: *the user interface layer* and *the query processing layer*. The whole sketch of the user interface is shown in Fig. 2, which contains *parameter setting* at the left side and *graph display* at the right side. Besides, four datasets are provided for users to choose from, they are Dblp, Digg, Arxiv and Epinions, each of them has at least 28,000 vertices and 841,000 edges.

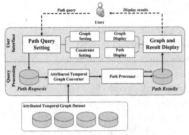

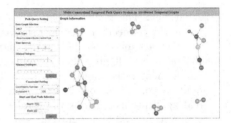

Fig. 1. System architecture   **Fig. 2.** A whole sketch of the user interface

2.1   User Interface

Path Query Setting and *Graph and Result Display* are supported by the web-based user interfaces. *Path Query Setting* is on the left of the UI that contains *Graph Setting* and *Constraint Setting*. These components are designed to help users customize temporal graphs and specify multiple constraints on the attributes respectively. In *Graph Setting*, two drop-down menus and three slide boxes are provided. A user can choose a dataset and a type of temporal path from the first two drop-down menus. An attributed temporal graph structure can be displayed once the setting of the graph is submitted. In *Constraint Setting*, a drop-down menu is provided for users to set the number of constraints and MCTP will create the corresponding number of drop-down menus to set the values of these constraints. After that, users can choose a start vertex, and an end vertex of a temporal path from the selected attributed temporal graph and MCTP will answer the query result once the path query is submitted.

  Graph and Result Display on the right of UI, contains *Graph Display* and *Path Display*. Once the setting of a data graph is submitted, *Graph Display* can display the vertices and edges contained in the graph, where the in-degree and out-degree of each vertex are equal to or greater than the values set by users, and the time points of each edge are in the time interval given by users.

In the display of a temporal graph in *Graph Display*, each vertex is given a digit as its ID, and each edge contains the temporal information of attributes between two vertices. For example, given two vertices v_1 and v_2, supposing there exist two paths p_1, p_2 between them at time point t_1 and t_2 respectively. The two paths have different start time, travel time and attribute values. When users move the mouse over the edge between v_1 and v_2, MCTP will generate a table over it, where the temporal attributes information is displayed. In addition to displaying initial graph structures, The delivered temporal paths will be displayed, and the details of this query process will be presented by *Path Display*.

2.2 Query Processing

In *Query Processing*. MCTP adopts a Two-Pass approximation algorithm that bidirectionally searches the attributed temporal graph with two linear scans from the end vertex v_e to the start vertex v_s and from v_s to v_e respectively to answer MCPQs. In the algorithm, the first *Backward Pass* procedure calculates the temporal objective function values of the *Backward Temporal Paths (BTPs)* from each vertex v_i to v_e, denoted as $p_{v_i,v_e}^B(t)$, to investigate if the *BTP* can satisfy the corresponding multiple constraints. In addition, the aggregated attributes' values of the *BTP* are saved at v_i to be used to investigate the feasibility of a temporal path delivered in the next *Forward Pass* procedure from v_s to v_e for the optimisation of objectives (i.e., the earliest arrival, the lastest departure, the shortest and the fastest optimisation). The details of the temporal path finding methods are discussed in the following Sect. 3.

3 Demonstration

In this section, we outline our demonstration scenario. The demonstration focuses on two key functionalities of MCTP: *Path Query Setting* and *Graph and Result Display*.

The first functionality we provide to the users is the *Path Query Setting* in attributed temporal graphs. This function contains two main sub-functions: *Graph Setting* and *Constraint Setting*. These two functions can help users customise the attributed temporal graph and specify the multiple constraints on the attributes respectively. The *Graph Setting* component allows users to customise the graph based their requirements on path query as the initial graph contains too many vertices and edges. Users may be only interested in the vertices that exist in a certain time interval. For example, users can choose data graphs and choose one type of the path for queries. In addition, users can specify the minimal in-degree, the minimal out-degree and the time interval of the vertices in a data graph. The *Constraint Setting* component is to specify the number and the values of multiple constraints on the attributes. After that, users can select two vertices from the data graph as the start vertex and the end vertex for the path query.

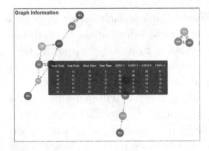

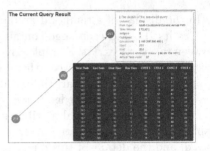

Fig. 3. The display of an attributed temporal graph

Fig. 4. The display of an MCPQ result

3.1 Graph and Result Display

Graph and result display is to display the structure and the associated temporal information for attributed temporal graphs and display the final query results. Once the setting of the graph is submitted, the corresponding attributed temporal graph will be displayed on the right of the UI. Each vertex has a distinct number, and each edge contains the temporal and attributed information between two vertices, such as the start time, the travel time and the value of each attribute. When users move the mouse over the edge, MCTP will generate a table to display the temporal information on the edge. An example is shown in Fig. 3. Once a path query is submitted, MCTP will perform the Two-Pass approximation algorithm, and if there is no feasible temporal path, MCTP returns "no path found" in the middle of the UI. Otherwise, the attributed temporal graph will be displayed. Moreover, the details of this query process are displayed in the form of text at the upper right corner of the *Path Display*. An example is shown in Fig. 4.

4 Conclusion

In this paper, we have proposed an MCTP system to answer MCPQs. Our system and the corresponding methodologies can be used in many temporal graph based applications as the backbone. For example, it can be used in road networks or airline transportation networks for route planning with the constraints of the travel cost, the travel time and the stopover interval of the trip.

References

1. Shah, N., Koutra, D., Zou, T., Gallagher, B., Faloutsos, C.: Timecrunch: interpretable dynamic graph summarization. In: KDD, pp. 1055–1064 (2015)
2. Wang, S., Lin, W., Yang, Y., Xiao, X., Zhou, S.: Efficient route planning on public transportation networks: a labelling approach. In: Proceedings of the 2015 ACM SIGMOD International Conference on Management of Data, Melbourne, Victoria, Australia, 31 May–4 June 2015, pp. 967–982. ACM (2015)

3. Wu, H., Cheng, J., Huang, S., Ke, Y., Lu, Y., Xu, Y.: Path problems in temporal graphs. VLDB **7**(9), 721–732 (2014)
4. Zhang, B., Hao, J., Mouftah, H.T.: Bidirectional multi-constrained routing algorithms. IEEE Trans. Comput. **63**, 2174–2186 (2014)

ACCDS: A Criminal Community Detection System Based on Evolving Social Graphs

Xiaoli Wang⑩, Meihong Wang$^{(\boxtimes)}$, and Jianshan Han

Software School of Xiamen University, Xiamen, China
{xlwang,wangmh}@xmu.edu.cn, 919769245@qq.com

Abstract. This paper presents an intelligent criminal community detection system, called ACCDS, to support various criminal event detection tasks such as drug abuse behavior discovery and illegal pyramid selling organization detection, based on evolving social graphs. The system contains four main components: data collection, community social graph construction, criminal community detection and data visualization. First, the system collects a large amount of e-government data from several real communities. The raw data consist of demographic data, social relations, house visiting records, and sampled criminal records. To protect the privacy, we desensitize the real data using some data processing techniques, and extract the important features for profiling the human behaviors. Second, we use a large static social graph to model the social relations of all residents and a sequence of time-evolving graphs to model the house visiting data for each house owner. With the graph models, we formulate the criminal community detection tasks as the subgraph mining problem, and implement a subgraph detection algorithm based on frequent pattern mining. Finally, the system provides very user-friendly interfaces to visualize the detected results to the corresponding user.

Keywords: Evolving social graphs · Criminal community detection
Subgraph mining · Data visualization

1 Introduction

Nowadays, the community security management system has been widely used in China, and massive resident information has been collected, such as resident demographic data, visiting records, social relationships, criminal records, etc. Many existing works have focused on analyzing such complex data to detect anomaly events. The branch of data mining has wide applications in security, finance, and many others. These methods can be categorized into two groups based on various data models: high-dimensional method (e.g., [1, 2]) and graph based method (e.g., [3–5]). As graph is recently widely used to model real objects, this paper also focuses on the anomaly detection problem based on graph models. Different from existing work concerned with general criminal events, this paper focuses on some urgent real problems such as drug abuse behavior discovery and illegal pyramid selling organization detection. We develop a community data analysis system containing four main components. The components and our main contribution can be summarized as follows.

© Springer Nature Switzerland AG 2018
C. Woo et al. (Eds.): ER 2018 Workshops, LNCS 11158, pp. 44–48, 2018.
https://doi.org/10.1007/978-3-030-01391-2_10

- We collect massive real community data from some cities in China, and use several data cleaning techniques to obtain high-quality data. The sensitive profiles are desensitized before we store the data into the database.
- We use evolving social graphs to model the house visiting data, and employ a subgraph mining algorithm to solve the criminal events detection problem.
- We develop a powerful visualization system to display the evolving social graphs, the warning messages, and the detected criminal communities.

2   System Architecture and Demonstration

Our system is implemented based on J2EE platform and the system architecture is shown in Fig. 1. The system has four main components, including data collection, community social graph construction, data analysis, and result visualization.

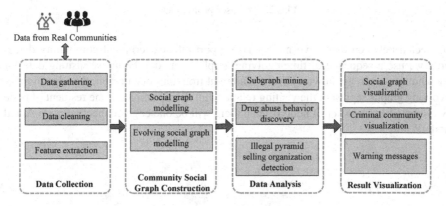

Fig. 1. The system architecture

2.1   Data Collection

We gather data from several real communities in China. The raw data contain massive information, such as resident demographic data, house visiting records, social relationships, and historical criminal records. We do data cleaning first to delete all the noisy data. Then, to protect residents' privacy, we desensitize the demographic data by removing explicit properties and randomly generating some synthetic profiles. Furthermore, we extract the most important features for profiling the human behaviors. The extracted profiles of a resident are shown in our system as in Fig. 2.

2.2   Community Social Graph Construction

We use a large static social graph to model the social relations of all residents, as shown in Fig. 3. In the social graph, each resident is represented as one node. If two residents have relationship such as family relation or friendship, there will be one edge being

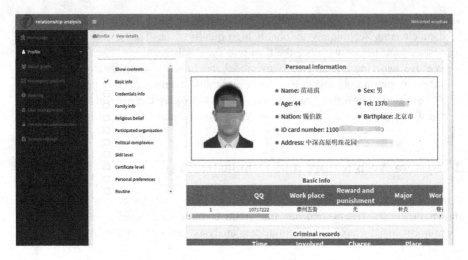

Fig. 2. The user profile page

connected between these two nodes. To support efficient criminal community detection, we use a sequence of time-evolving graphs to model the house visiting data for each house owner, as shown in Fig. 4. Different from the social graph, the edge in time-evolving graphs represent the visiting relationship. For example, if one resident of node A visited a house owner of node B, then there will be one edge connecting node A and node B.

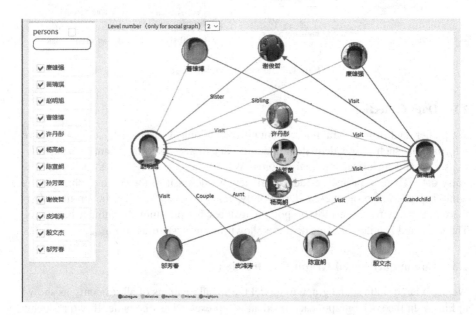

Fig. 3. The community social graph example for several selected residents

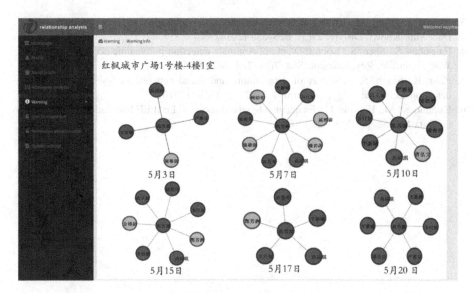

Fig. 4. The evolving visiting graphs for a selected address in May, and a highlighting result of the detected illegal pyramid selling organization (Color figure online)

2.3 Data Analysis and Visualization

Based on the graph models, we formulate the criminal community detection tasks as the subgraph mining problem, and implement a frequent pattern mining algorithm to solve it [6]. The details of the algorithm can be seen in our summited full research paper to a conference. We omit the details here as we will publish a technical report later online. As shown in Fig. 4, the substructures with red nodes and edges are the detected illegal pyramid selling organization. The system contains other pages to show other kinds of detected results. We omit them here for the space constraints.

Acknowledgment. This work is supported by the National Natural Science Foundation of China under Grant No. 61702432, the Fundamental Research Funds for Central Universities of China under Grant No. 20720180070, and the International Cooperation Projects of Fujian Province in China under Grant No. 2018I0016.

References

1. Gupta, M., Gao, J., Aggarwal, C.C., Han, J.: Outlier Detection for Temporal Data. Synthesis Lectures on Data Mining and Knowledge Discovery. Morgan & Claypool Publishers, San Rafael (2014)
2. Mannhardt, F., de Leoni, M., Reijers, H.A., van der Aalst, W.M.P.: Data-driven process discovery - revealing conditional infrequent behavior from event logs. In: Dubois, E., Pohl, K. (eds.) CAiSE 2017. LNCS, vol. 10253, pp. 545–560. Springer, Cham (2017). https://doi.org/10.1007/978-3-319-59536-8_34

3. Akoglu, L., Tong, H., Koutra, D.: Graph based anomaly detection and description: a survey. Data Min. Knowl. Discov. **29**(3), 626–688 (2014)
4. Ranshous, S., Shen, S., Koutra, D., et al.: Anomaly detection in dynamic networks: a survey. Wiley Interdisc. Rev. Comput. Stat. **7**(3), 223–247 (2015)
5. Kaur, R., Singh, S.: A survey of data mining and social network analysis based anomaly detection techniques. Egypt. Inform. J. **17**(2), 199–216 (2016)
6. Kuramochi, M., Karypis, G.: Frequent subgraph discovery. In: IEEE International Conference on Data Mining, pp. 313–320 (2002)

A Prototype for Generating Meaningful Layout of iStar Models

Yunduo Wang, Tong Li[✉], Haoyuan Zhang, Jiayi Sun, Yeming Ni,
and Congkai Geng

Beijing University of Technology, Beijing, China
{15074518,15071109,bumfod,nym16027221,
gengcongkai789123}@emails.bjut.edu.cn
litong@bjut.edu.cn

Abstract. Being able to model and analyze social dependencies is one of the most important feature of iStar modeling language. However, typical layout algorithms (e.g., hierarchical layout and circular layout) are not suitable for laying out iStar models, especially, strategic dependency models. As a result, iStar models are typically constructed and laid out in a manual way, which is time-consuming and tedious, especially when dealing with large-scale models. In this paper, we present a prototype tool which can automatically lay out iStar model in a meaningful way by taking into account the semantics of iStar model elements. In particular, we have developed and integrated an istarml parser with our tool, enabling automatic layout of iStar models that are specified in terms of istarml. We believe our tool would be particularly interesting for people who used to deal with large-scale iStar models.

1 Introduction

iStar modeling language has been proposed for more than two decades, which is an effective approach for social modeling and analysis in the early phase of requirements engineering. Although hundreds of analysis approaches have been proposed based on iStar [4], only few of them concern the layout issue of iStar models. Specifically, jUCMNav[1] has integrated graphviz for layout purpose, which is thus able to automatically visualize and lay out goal models [6]. However, as for the layout of actors and their interrelationship, jUCMNav has implemented it in a rather simple manner, i.e., a matrix-like layout, which does not really contribute to the understandability of iStar models. Similarly, OpenOME also implements a basic graph layout function, but only focuses on intentional elements (e.g., goals and tasks).

The challenge of laying out iStar SD models is that typical layout algorithms, e.g., hierarchical and circular layout algorithms, do no fit iStar syntax well. Consequently, analysts have to manually lay out iStar models, which may

---

[1]http://jucmnav.softwareengineering.ca/jucmnav.

C. Woo et al. (Eds.): ER 2018 Workshops, LNCS 11158, pp. 49–53, 2018.
https://doi.org/10.1007/978-3-030-01391-2_11

cost a significant amount of time when dealing with large-scale scenarios. Moreover, modifications on existing complicated iStar models are also nightmares for
iStar modelers, as inserting/deleting particular actors typically requires holistic
adjustment of existing models.

In this paper, we present a prototype tool which can automatically lay out
iStar models in realtime. The tool implements our previously proposed force-
based layout algorithm [1] with some improvements. In particular, our layout
strategy takes into account the semantics of iStar elements, trying to place them
in a way that is easy and intuitive to understand. For example, actors that have
more interdependencies are placed more closed to each other, resulting in actor
clusters. We believe that people who typically deal with large-scale iStar models
can benefit from using our tool.

Moreover, we have been aware that many modelers have accepted that textual
representation of models are more efficient than graphical diagrams in terms of
editing, storing and versioning models [3]. Also in the iStar community, Grau
et al. propose J-PRiM which does not show the i* elements in a graphical way
but in a tree-form hierarchy [2]. In such a way, as models grow, the scalability
issue can be relieved, and modelers are able to find particular elements in a
faster way. On the whole, we believe our tool can be a useful complement to the
textual representation of iStar models, dynamically (re-)calculating positions for
model elements when they need to be visualized. As such, we have developed and
integrated an istarml parser with our prototype, and thus all iStar models that
are specified in istarml can easily leverage the layout capability of the prototype.

2 Prototype Architecture

The prototype is a browser-based tool, which is implemented using Html, CSS,
and JavaScript. The architecture of the tool is shown in Fig. 1, where our tool
takes istarml as input, and eventually produces position information for each
model element. Although we have not done yet, we plan to deploy the prototype
online as a public model layout service to help researchers in the iStar community.

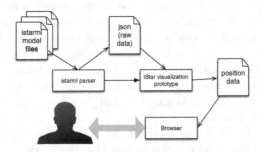

Fig. 1. The architecture of the prototype

The primary feature is to lay out iStar models in realtime. Specifically, our prototype has already implemented the following detailed features:

1. **Meaningful actor distance.** Distances between actors are calculated based on their interdependencies. More interdependencies indicate more closed relations, and thus actors are closed to each other.
2. **Regionalization.** Based on our experiences in dealing with iStar models, it is easier to comprehend a large model when it is laid out in a way that elements are divided in different regions of the canvas. Thus, our algorithm intentionally increase distances among element clusters.
3. **Fixed point.** According to our force-based layout strategy, an element's position will be changed if its connected elements move. Although it is a reasonable setting, sometimes analysts may can have certain elements fixed by a single-click on corresponding elements.

Fig. 2. An automatic layout example

3 Illustrations

Figure 2 shows an automatic layout iStar model, which has not been manually adjusted at all. In particular, actors have been visually clustered based on their interrelationship, and thus each time we can easily comprehend this model region by region. To further adjust the model, analysis can manually fix certain elements according to their preferences. More detailed operations are recorded and have been post online. To facilitate watching the video, we have prepared two versions, a high quality one[2] and a low quality one[3].

4 Conclusions and Future Work

In this paper, we present our prototype tool which use a customized force-based layout algorithm to automatically lay out SD diagram of iStar. Our tool takes into account the semantics of model elements in order to produce more meaningful layout. We believe that the automation of model layout can positively contribute to scalability of iStar modeling language, which has been concerned by more and more modelers [5]. In other words, the tool is suitable for dealing with large-scale iStar models. In addition, our tool complement the textual representation of iStar models, enabling analysts to easily switch between these two types of models in order to maximumly leverage benefits from both types.

As for future work, we plan to first extend our proposal is to deal with automatic layout of the SR view. In addition, we also plan to deploy and publish our tool in order to be practically used by domain experts. Finally, we aim to carry out a series of controlled experiments to evaluate the effectiveness of our tool.

Acknowledgments. This work is supported by National Key R&D Program of China (No. 2017YFC08033007), the National Natural Science of Foundation of China (No. 91546111, 91646201) and Basic Research Funding of Beijing University of Technology (No. 040000546318516).

References

1. Du, X., Li, T., Wang, D.: An automatic layout approach for iStar models (2017)
2. Grau, G., Franch, X., Avila, S.: J-PRiM: A java tool for a process reengineering i* methodology. In: 14th IEEE International Conference on Requirements Engineering, pp. 359–360. IEEE (2006)
3. Gregorics, B., Gregorics, T., Kovács, G.F., Dobreff, A., Dévai, G.: Textual diagram layout language and visualization algorithm. In: 2015 ACM/IEEE 18th International Conference on Model Driven Engineering Languages and Systems (MODELS), pp. 196–205. IEEE (2015)

---

[2] https://www.dropbox.com/s/ezhtkao5szdm3hf/screencast_high_quality.mp4?dl=0.

[3] https://www.dropbox.com/s/bjpn5hwpmz7d6hn/screencast_low_quality.mp4?dl=0.

4. Horkoff, J.: Goal-oriented requirements engineering: an extended systematic mapping study. In: Requirements Engineering, pp. 1–28 (2017)
5. Li, T., Grubb, A.M., Horkoff, J.: Understanding challenges and tradeoffs in iStar tool development. In: 9th iStar Workshop, pp. 49–54 (2016)
6. Rashidi-Tabrizi, R., Mussbacher, G., Amyot, D.: Transforming regulations into performance models in the context of reasoning for outcome-based compliance. In: 2013 Sixth International Workshop on Requirements Engineering and Law (RELAW), pp. 34–43. IEEE (2013)

MDBF: A Tool for Monitoring Database Files

Xiangyu Wei and Jianqiu Xu$^{(\boxtimes)}$ (iD)

Nanjing University of Aeronautics and Astronautics, Nanjing, China
{xywei,jianqiu}@nuaa.edu.cn

Abstract. When complex queries are executed in a database system, several files will be accessed such as relational tables, indexes and profiles. Monitoring database files enables us to better understand the query progress and the storage system. In this demo, we present a tool for monitoring file operations such as when the file is accessed and how many bytes are transferred for read and write. The access information is visually displayed at running time, and also stored as historical data for further analysis and optimization. We provide a graphical interface to report and display the results. MDBF is implemented inside a system kernel with a low overhead. The demonstration is performed by using real trajectory datasets and continuous queries employing three different indexes.

Keywords: Monitor · Database files · Queries · Analysis · Visual

1 Introduction

File monitoring plays an essential role in operating system that constantly watches folders and files. Actions are triggered when files are created or accessed (read/write). Although some tools have been developed, e.g., notify-tools, there exists limited work on this topic in the context of a database system. In principle, a database is a collection of files in a certain format, e.g., relational tables, indexes, metadata and log records. Files reside in secondary storage and will be accessed at query execution time.

Some tools have been developed to monitor a database system [2,5]. A query monitor aims to observe and predict parameters related to the execution of a query. Typical parameters include the cardinality and size of the output of query operators. When execute expensive queries, e.g., *joins*, the I/O communication between main memory and disk usually become a bottleneck. The query monitor should observe database files during the query execution and record statistics such as when the file is accessed and how many bytes are extracted from the file or inserted into the file. The fine-grained information can be further used for the analysis and tuning of the system. Optimizations can be made based on monitoring results, e.g., redesigning the data representation and improving

© Springer Nature Switzerland AG 2018
C. Woo et al. (Eds.): ER 2018 Workshops, LNCS 11158, pp. 54–58, 2018.
https://doi.org/10.1007/978-3-030-01391-2_12

the buffer manager. Monitoring database files will be helpful to understand how queries are executed and the progress of tasks, and find the system bottleneck.

In this paper, we develop a tool called MDBF that monitors database files during the query execution. There are three components: file detector, storage and analysis system, and graphical interface. In the detection phase, a well-known tool Strace is utilized to monitor the operating system and capture all file actions, e.g., read and write blocks of data. We do the filtering by extracting the data related to database file operations and store action statistics in relational tables. The information of accessing files is displayed when the querying is running and also recorded as historical data. A thorough analysis is performed on log data. MDBF is able to monitor all queries incurring database file operations. A graphical interface visualizes the access content to aid our understanding of the query progress. Users can compare the access information of different files in a convenient way. MDBF is developed in an extensible database system SEC-ONDO [4] in C/C++ and Java (JavaSwing and JFreeChart). We demonstrate how to make use of the tool to monitor database files by using real datasets and testing queries, including (i) on-line feedback about database file operations and (ii) historical data analysis and visualization. MDBF incurs a low system overhead and benefits both database users and administrators.

2   MDBF

2.1   An Overview

The monitor consists of three modules: file detector, storage and analysis system and visual interface, as illustrated in Fig. 1. The file detector, serving as the key component, makes use of a tool Strace, which is a diagnostic, debugging and instructional userspace utility. It is used to monitor and tamper with interactions between processes and the Linux kernel, including system calling, signal delivering, and process state changing.

When a query is executed, the tool detects database file operations and generates a monitoring log. The log data is automatically reported to the filter and formatted in a relational table in the system. We are able to display the accessing information at running time and also store them as historical data.

2.2   Data Flow and Log Data

Since Strace captures all system calls in the query evaluation, we are primarily interested in file operations. The data flow received from Strace is filtered to be a certain format and transferred between system modules. A relational table is defined including the operation id, the execution time, the query, the file name, and the number of read/write bytes, as exemplified in Table 1.

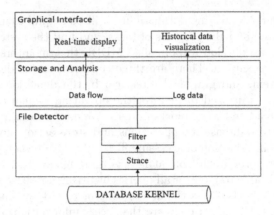

Fig. 1. The framework of MDBF

Table 1. Monitoring table.

Id	Time	Query	Database files	Read(bytes)	Write(bytes)
1	15:39:08	"open database BERLINTEST"	"/home/nuaa/secondo-databases/BERLINTEST/d1.sdb"	512	0
2	15:41:22	"let utrips = Trains feed projec..."	"/home/nuaa/secondo-databases/BERLINTEST/d111.sdb"	0	45
3	15:44:28	"query res10 feed count"	"/home/nuaaxu/secondo-databases/BERLINTEST/d123.sdb"	512	0

3   Demonstration

We demonstrate monitoring databases files by performing queries over real trajectory datasets including 7,915,088 GPS records (50, 000 trips) [1].

Monitoring at Running Time. A list of operators including select and create is executed as test workloads, the tool monitors database files and displays the read/write information, as illustrated in Fig. 2. Note that the data is transferred in block between main memory and disk, and may be buffered before it is flushed into the disk by the system. Users will receive informative feedback of the query progress and watch the execution status.

Historical Data Analysis and Visualization. Continuous distance queries over trajectories are evaluated. The query returns objects within a certain distance to the target at each time point. We employ three indexes (3D R-tree, TB-tree [6] and SETI [3]) to run the query in order to compare the I/O cost of different structures, as illustrated in Fig. 3(a). The monitored data at running time is collected and a graphical interface is developed to visualize the result by histogram and pie chart. The interface displays the number of read and write bytes for data and index files, as shown in Fig. 3(b).

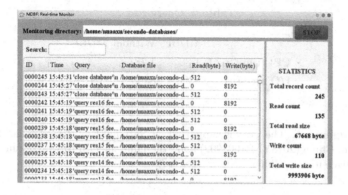

Fig. 2. Real-time data display

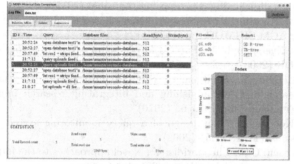

(a) The statistics

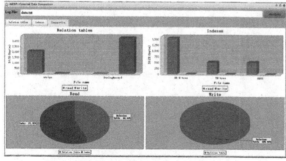

(b) Histogram and pie chart

Fig. 3. Analysis and display of historical data

MDBF Overhead. We record the query cost when MDBF is running and closed. The results, reported in Table 2, show that a low system overhead is incurred. In fact, MDBF requires additional CPU time to do the filtering and data transfer but does not incur any I/O cost.

Table 2. Query cost

	MDBF	Without MDBF
CPU time (sec)	1.86	1.78
I/O accesses	8308	8308

Acknowledgement. The work is supported by National Key Research and Development Plan of China (2018YFB1003902) and "the Fundamental Research Funds for the Central Universities in China", NO. NS2017073.

References

1. https://outreach.didichuxing.com/app-vue/personal (2017)
2. Beeri, C., Eyal, A., Milo, T., Pilberg, A.: Query-based monitoring of BPEL business processes. In: ACM SIGMOD, pp. 1122–1124 (2007)
3. Chakka, V.P., Everspaugh, A., Patel, J.M.: Indexing large trajectory data sets with SETI. In: CIDR (2003)
4. Güting, R.H., Behr, T., Düntgen, C.: SECONDO: a platform for moving objects database research and for publishing and integrating research implementations. IEEE Data Eng. Bull. **33**(2), 56–63 (2010)
5. Mishra, C., Koudas, N.: The design of a query monitoring system. ACM Trans. Database Syst. **34**(1), 1:1–1:51 (2009)
6. Pfoser, D., Jensen, C.S., Theodoridis, Y.: Novel approaches to the indexing of moving object trajectories. In: VLDB, pp. 395–406 (2000)

CusFinder: An Interactive Customer Ranking Query System

Yanghao Zhou, Xiaolin Qin$^{(\boxtimes)}$, Xiaojun Xie, and Xingluo Li

Nanjing University of Aeronautics and Astronautics, Nanjing, China
{zhyreadboy,qinxcs,xiexj,lxldblab}@nuaa.edu.cn

Abstract. Finding a certain number of objects with optimum ranking based on spatial position constraints and given preference of attributes can be essential in numerous scenarios. In this paper, we demonstrate CusFinder, an interactive customer ranking query system to retrieve customers who favour a specific seller more than other people from the perspective of sellers, instead of retrieving sellers for a given customer similar to existing commercial systems. To make the query processing more efficient, a novel indexing is proposed to serve for query engine. Furthermore, we present the result of queries upon multiple visualization views with user-friendly interaction designs.

Keywords: Optimum ranking · Spatial constraints · Multiple views

1 Introduction

As an effective means of retrieving a specified number of objects according to ranking scores, rank-aware query is widely applied in various real-life scenarios. Nevertheless, we have noticed that none of existing work takes into account both sellers' perspective and customer-oriented ranking with spatial constraint, such as the following example.

Example 1.1 Given a collection of restaurants and a set of customers, the manager of one of the restaurants, named *Tom*, is trying to find a small number (denoted by k) of customers who prefer his restaurant compared with other customers. As different customers have different requirements (i.e. taste or price), and customers would not consider those restaurants far away (denoted by σ) from them, *Tom* is supposed to know the rankings of his restaurant for different customers, thereby obtaining those customers who can not only accept the distance between them and *Tom's* restaurant, but also give higher ranking according to their preference for multi-dimension properties.

Moreover, by reviewing the literature and commercial platforms serving similar purpose, we find that few efforts are made on visualizing the query results, which motivates us to develop a visualization-aided system to benefit sellers by identifying potential target customers.

© Springer Nature Switzerland AG 2018
C. Woo et al. (Eds.): ER 2018 Workshops, LNCS 11158, pp. 59–63, 2018.
https://doi.org/10.1007/978-3-030-01391-2_13

Therefore, we propose an interactive customer ranking query system named CusFinder. It distinguishes from the literature in three-fold. First, it supports Reverse Spatial Object k-Ranks Query, a novel query which explores optimal k customers satisfying spatial constraints from the perspective of sellers. Second, CusFinder can answer this query efficiently owing to a hybrid index structure and query processing framework. Finally, it is designed to show the result on map dynamically via an interactive visualization procedure. All these functions are presented based on user-friendly interactions upon multiple visualization views, which heuristically guide users to interact with different visualization results [2].

2   Data Model and Queries

Data Model. We maintain two kinds of data in our system. One is object data, and each object $p = \langle \rho, \phi \rangle$ is a pair of position ρ and a set of associated terms $\phi = \langle t_1, \ldots, t_d \rangle$ which describes the general attributes. The other one is user preference data represented by a triple $u = \langle \rho, \psi, \sigma \rangle$, where $u.\rho$ is the location where u is posted, $u.\psi = \langle w_1, \ldots, w_d \rangle$ is the preference weight (normalized in $[0,1]$ and $\sum_{i=1}^{d} w_i = 1$), and $u.\sigma$ is the maximum distance the user can tolerate between him and any eligible object.

Query Model. We define an object's rank for a given user, namely $rank(u, q)$, by the number of objects with a smaller score than q for u, which is same with the definition in [3]. In this paper, the geographical position of objects and users has been considered by calculating Euclidean distance.

Definition 2.1 **Reverse Spatial Object k-Ranks (RSO-kR) Query.** Given an object set $D = \{p_1, \ldots, p_n\}$, a user set $U = \{u_1, \ldots, u_m\}$, an integer k, a query point q, the RSO-kR will return a set S of k users such that

(1) $\forall u_i \in S, dist(q, u_i) \leq u_i.\sigma$;
(2) $\forall u_i \in S, \forall u_j \in U - S$, if $dist(q, u_j) \leq u_j.\sigma$, then $rank(u_i, q) \leq rank(u_j, q)$.

3   System Architecture

As shown in Fig. 1, CusFinder includes a front-end of visual interface and a back-end of query processing. At the front-end, users could issue queries and view the selected results based on multiple coordinated views. Each new user interaction over the interface will be sent to the back-end, and a search engine is employed to answer the query on top of an indexing module. After the query is processed, results containing target customers will be returned and displayed at the front-end by map view and multidimensional view.

Indexing Module. Different index structures are applied to organize the data. A novel index named *Layer Grid Index* composed by a layer dominant graph I_D [5] and a grid index table I_T is used for seller data, while a histogram is used for customer data.

Search Engine. To conduct filtering efficiently, we propose a progressive strategy, which scans for candidate points layer by layer. Besides, when we are seeking for superior customers, it is a wiser method to roughly estimate ranks with lower and upper bounds of the buckets.

Please refer to our recent research work [4] for more details about indexing design and query algorithm.

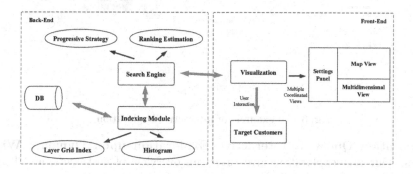

Fig. 1. System architecture

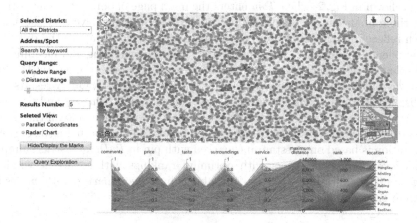

Fig. 2. An overview of the customers

4 Demonstration and Scenarios

We will demonstrate CusFinder based on real-world datasets and scenarios. We crawled restaurant information in ShangHai from Dianping[1], which is one of the biggest lifestyle and group buying websites in China. Each restaurant has descriptive attributes (such as price, taste, surrounding) and location attribute.

Specifically, our demonstration is based on the scenario that *Tom* is trying to find customers who favour his restaurant more than other people, which we have described in *Example* 1.1.

---

[1] http://www.dianping.com.

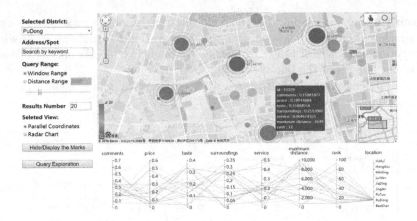

Fig. 3. An example of interactive exploration

Exploration Query 1. Interactive Exploration upon Dynamic Windows. Our first exploration gives users an overview of customers, and helps them understand the distribution and requirements of customers, so that users could retrieve a certain number of customers according to their actual needs.

As shown in Fig. 2, when *Tom* enters the main page of our system, he notices that there is an enormous quantity of colourful points displayed on the map, which reflect the distribution of customers, and the colour implies the different requirements for distance. Considering that it is overwhelming and meaningless to display all customer data in a limited screen with coarser granularity [1], only those customers from current map window and in the RSO-kR Query results will be displayed. All interactions with the map such as zooming in/out and panning will triage a new RSO-kR Query within current window range, and all the result views change simultaneously to keep consistency. As shown in Fig. 3, current exploration is based on a restaurant marked with pin in *PuDong* district, aiming to retrieve twenty customers with optimum rankings within current window range. Finally, 20 results with top-20 rankings have been marked with circles in different sizes and colours.

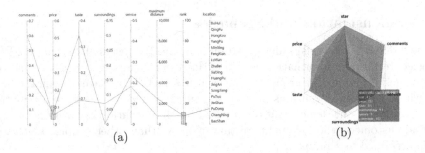

Fig. 4. Detailed comparison: (a) in parallel coordinates; (b) in radar chart.

Exploration Query 2. Detailed Comparison of Selected Customers and Restaurants. After users obtain the filtering results, they want to compare the preference of customers, or discover strengths and weaknesses of their restaurants and competitors. Our second exploration query can help them achieve this goal.

On the basis of interactive exploration, *Tom* intends to know not only the detailed comparison of these selected customers, but also the differences between his restaurants and rivals. With visual assistance of parallel coordinates, he could drag boxes on those coordinate axes indicating properties that he cares about more, thereby observing the corresponding values of these selected customers from each dimension. As shown in Fig. 4(a), although his restaurant is given same ranks by two customers, there are notable differences between the preference. Consequently, he could choose the better one according to his requirements. Besides, *Tom* is able to compare his restaurant with his rivals' restaurants in the radar chart. As shown in Fig. 4(b), he clearly notices that the green coloured restaurant has more favourable comments and lower price compared with others.

Acknowledgement. This research is supported by NSFC 61373015, 61728204, and the Project of State Grid Corporation of China (Storage and Processing of Distributed Parallel Database Based on Big Data Technology).

References

1. Guo, T., Feng, K., Cong, G., Bao, Z.: Efficient selection of geospatial data on maps for interactive and visualized exploration. In: SIGMOD, pp. 567–582. ACM (2018)
2. Wang, S., Li, M., Zhang, Y., Bao, Z., Tedjopurnomo, D.A., Qin, X.: Trip planning by an integrated search paradigm. In: SIGMOD, pp. 1673–1676. ACM (2018)
3. Zhang, Z., Jin, C., Kang, Q.: Reverse k-ranks query. PVLDB **7**(10), 785–796 (2014)
4. Zhou, Y., Qin, X., Xie, X., Guo, C.: A reverse k-ranks query based on clusters. J. Chin. Comput. Syst. (2018, in press)
5. Zou, L., Chen, L.: Dominant graph: an efficient indexing structure to answer top-k queries. In: ICDE, pp. 536–545. IEEE Computer Society (2008)

Doctoral Symposium

Preface to ERDS 2018

Doctoral Symposium (ERDS 2018) at the 37th International Conference on Conceptual Modeling (ER 2018)

Xi'an, China
October 22–25, 2018

Organized by
Xavier Franch and Chaokun Wang

The ER 2018 Doctoral Symposium (ERDS 2018) is hold in Xi'an, China. This symposium gathers together PhD students and academics working in conceptual modelling. ERDS 2018 offers PhD students the opportunity to present and discuss their research and to interact with other researchers, experts in the field of conceptual modeling, who can provide feedback on their research. The symposium is open to any topic covered by the main conference and of strong relevance to the ER community.

We have accepted three submissions as research extended abstracts for the symposium. "An Active Workflow Method for Entity-oriented Data Collection" addresses a practical problem in web search or information retrieval, involving multiple sources, information integration, and entity resolution. "Visual Non-verbal Social Cues Data Modeling" proposes a privacy-preserving data model for the visual non-verbal cues. "Multiple data cleaning on temporal data" proposes a framework for data quality evaluation on imprecise temporal data. The symposium was opened by a keynote given by Prof. Oscar Pastor from Universidad Politécnica de Valencia.

We would like to deeply thank all the authors for submitting their contributions to ERDS 2018 and are pleased to express our appreciation to all Program Committee for their involvement in the evaluation of submitted papers. Without their expertise and dedication, the ERDS 2018 program would not be of such high quality. We would like also thank the co-chairs of the ER conference and the co-chairs of the ER workshops since they have helped us coordinate ERDS 2018 within the ER 2018 conference.

Finally, we are glad to welcome you to Xi'an. We hope you enjoy this doctoral symposium and the benefits from the technical program as well as from sharing your experiences with colleagues from all around the world.

October 2018

Xavier Franch
Chaokun Wang
Doctoral Symposium Co-chairs

Design Science for PhD Research in the Conceptual Modeling Domain

Oscar Pastor

Universitat Politècnica de València, Spain

opastor@dsic.upv.es

Abstract. A sound PhD research work needs to use a solid research methodology background. Design Science is an attractive and useful alternative for facing this problem. Following the last works of Wieringa about this subject, design science is the design and investigation of artifacts in context. Being design and investigation the two main parts of design science, it is fundamental for any research work to identify what kind of research problem is under study: design problems or knowledge questions. Once this identification is done, a design science project iterates over the tasks of a design and engineering cycle, while to answer scientific knowledge questions, an empirical cycle must be used.

This talk will present all these aspects of design science, focusing on what kind of design, engineering or empirical cycles are more appropriate when dealing with conceptual modeling-oriented research works. The problem will be analyzed in the context of a PhD research work, to let PhD students be aware of what benefits can be derived from using it in their works. During the keynote, PhD students will be invited to characterize their PhD research according the ideas that will be presented.

Multiple Data Quality Evaluation and Data Cleaning on Imprecise Temporal Data

Xiaoou Ding[✉]

Harbin Institute of Technology, Harbin 150001, Heilongjiang, China
dingxiaoou@stu.hit.edu.cn

Abstract. With data currency issues draw the attentions of both researchers and engineers, temporal data, which describes real world events with time tags in database, is playing a key role in data warehouse, data mining, and etc. At the same time, 4V features of big data give rise to the difficulties in comprehensive data quality management and data cleaning. On one hand, entity resolution methods are faced with challenges when dealing with temporal data. On another hand, multiple problems existing in data records are hard to be captured and repaired. Motivated by this, we address data quality evaluation and data cleaning issues in imprecise temporal data. This project aims to solve three key problems in temporal data quality improvement and cleaning: (1) Determining currency on imprecise temporal data, (2) Entity resolution on temporal data with incomplete timestamps, and (3) Data quality improvement on consistency and completeness with data currency. The purpose of this paper is to address the problem definitions and discuss the procedure framework and the solutions of improving the effectiveness of temporal data cleaning with multiple errors.

Keywords: Temporal data · Data currency · Multiple data cleaning
Data quality

1 Introduction

Data quality evaluation and data cleaning plays a key role in the whole life circle of big data management today [23]. With the increasing demand in efficiency and currency, the *temporal* feature in data has been recognized as an important issue in data science. However, the quality problems in temporal data are often quite serious and trouble data transaction steps (e.g., acquisition, copy, querying). On one hand, various information systems store data with different formats, which makes entity resolution process necessary. The attribute values are possible to change with the evolution of time. It adds to the difficulties in entity resolution. In practice, certain attribute of records referring to the same entity may change over time. All of these records may be valid and proper for describing a certain entity only at a particular time period. For example, DBLP

© Springer Nature Switzerland AG 2018
C. Woo et al. (Eds.): ER 2018 Workshops, LNCS 11158, pp. 69–75, 2018.
https://doi.org/10.1007/978-3-030-01391-2_14

collects researchers' paper information for many years. The information for the same author about different papers may be different since the author's affiliation, partners or even names may change over time. On another hand, currency, consistency and completeness problems are always costly in multi-source data integration. These problems result in the low reliability of data, and they also add to the confusion in data applications.

Researchers have gone a long way in data quality and data cleaning, particularly in accuracy, consistency, completeness and record de-duplication [3,12,14]. It has been found that currency issues in temporal data also seriously impact data repairing. In addition, these problems are possible to affect each other during repairing, rather than completely isolated [5,14]. However, existing data repairing methods fails to pay attention to the temporal feature in data, without which the performance of them is challenged to break through bottleneck. Motivated by this, we propose multiple data cleaning in temporal data in this project.

2   Problem

We address two main issues in this project, as summarized below.

Problem 1: We first study the entity resolution method on imprecise temporal data. Due to the fact that reliable timestamps are often not available and complete, or even absent in practice, it is necessary to improve the entity resolution method to perform well on imprecise temporal data. We propose the problem definition in Definition 1.

Definition 1. $\mathcal{R} = (A_1, ...A_m)$ *is a relation schema and* $I_R = \{r_1, ..., r_n\}$ *is a set of instances of* $\mathcal{R}$ *which contains* n *records.* I_R *has no precise timestamps.*

*The **Entity Resolution problem on imprecise temporal data** is to cluster the records into clusters, such that records in the same clusters refer to the same entity over time and records in different clusters refer to different entities.*

Problem 2: We study the detecting and repairing approach on incomplete and inconsistent data with incomplete timestamps, to achieve multiple data quality improvement on completeness, consistency and currency. The problem definition is presented in Definition 2.

Definition 2. *Data cleaning on consistency, completeness and currency.* *Given a low-quality data* $\mathcal{D}$, *data quality rules including a set* Φ *of CCs and a set* Σ *of CFDs, and a confidence* σ *for each attributes. Data quality improvement on* $\mathcal{D}$ *with completeness, consistency and currency is to detect the dirty data in* $\mathcal{D}$ *and repair it into a clean one, denoted by* $\mathcal{D}_r$, *where*

(a) $\forall r(r \in \mathcal{D}_r)$ *has a reliable currency order value satisfying the set* Φ *of CCs, denoted by* $(\mathcal{D}_r, \mathcal{D}) \models \Phi$.
(b) $\mathcal{D}_r$ *is consistent referring to the set* Σ *of CFDs, i.e.,* $(\mathcal{D}_r, \mathcal{D}) \models \Sigma$.
(c) *The missing values in* $\mathcal{D}$ *are repaired with the clean ones whose confidence* $> \sigma$ *into* $\mathcal{D}_r$.
(d) *The repair cost* $cost(\mathcal{D}_r, \mathcal{D})$ *is as small as possible.*

3   Proposed Framework and Solution

Solution of Problem 1: The method overview is shown in Fig. 1, which has two main parts: similarity comparison and clustering, respectively. We first define temporal attributes' unstableness, which depicts the evolving probability of attribute values of entities. Accordingly, a dynamic weight scheme is designed to match these records more precisely than the traditional fixed weight scheme.

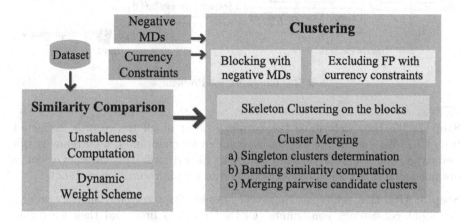

Fig. 1. System architecture for Problem 1

Then, we combine the data currency constraint and matching dependency rules effectively and propose a temporal clustering algorithm that takes data currency into consideration to guarantee result quality.

After that, we develop a rule-conducted uniform framework for resolving temporal records with integration of data quality rules, and design several efficient algorithms to track the problems in the framework. Specifically, we propose algorithms to (1) block records into disjoint blocking and exclude false positives in each block, (2) compute unstableness of attributes and dynamic weight scheme, (3) generate a cluster of each block, and (4) determine initial clusters in the clustering results.

Solution of Problem 2: Since that completeness and consistency are metrics focusing on measuring the quality with features in values, while currency describes the temporal order or the volatility of records in the whole data set. We process consistency and completeness repairing along the currency order, respectively. We outline the 3C data cleaning method in Fig. 2. It includes four main steps.

Step 1: We first construct currency graphs for records with the given CCs, and make conflict resolution in the currency graphs. If conflicts exist, the conflicted CCs and the involved records will be returned. They are supposed to be fixed by domain experts or revised from business process.

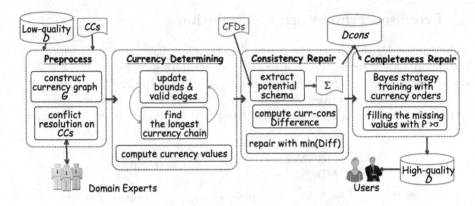

Fig. 2. Framework for Problem 2

Step 2: We then determine the currency order of all the records extracted from CCs. We update the longest currency order chain in each currency graph iteratively, and compute currency values to each record. This currency order is obtained as a direct and unambiguous metric among records on currency.

Step 3: After that, we repair consistency issues with the global currency orders. We input consistency constraints (CFDs for short) first, and then we define a distance metric *Diffcc* to measure the distance between dirty data and clean ones, combining consistency difference with currency orders. That is, we attempt to repair the dirty data with proper values which have the closest current time point.

Step 4: We repair incomplete values with Bayesian strategy because of its obvious advantages in training both discrete and continuous attributes in relational database. We treat currency orders as a weighted feature and train the complete records to fill in the missing values if the filling probability no less than a confidence measure. Up till now, we achieve high-quality data on 3C.

4   Related Work

Study on data quality is extensive for decades in many fields. With the demand for high quality data, many metrics beside accuracy are necessary for quality improvement [27].

Data Quality Evaluation. [3] provides a systematic introduction of data quality methodologies, in which completeness, accuracy, consistency and currency are four important dimensions. As for completeness issues, many kinds of algorithms have been proposed to fill the missing values [10], like statistic-based, probability-based, and learning oriented and etc. As for consistency, different semantic constraints such as FD, CFD, and CIND, have been defined to guide data cleaning under specific circumstance [4,12], where conditional functional dependency (CFD) is a general and effective consistency constraints for querying and inconsistency detection in database [9, 13]. Also, consistency constraints as well as data quality rules discovery problem is proposed and studied in works like [7, 17]. Currency describes to which extent a data set is up-to-date [3]. When various data sources are integrated, timestamps are always neither complete nor uniform. It promotes the study on currency determination without available timestamps. [15] is the first to propose a constraint-based model for data currency reasoning. And several fundamental theoretical problems are discussed in both [12, 15].

Furthermore, as the dimensions are not independent issues in data integration [12], data cleaning approaches have been developed with integrating several data quality dimensions. [14] reports advanced study on critical dimensions and provides a logical framework for a uniform treatment of the issues. [9] propose a framework for quality improvement on both consistency and accuracy. [5] discusses time-related measures with accuracy and completeness, and proposes functions of computing their mutual relationships.

Entity Resolution. (ER) also known as record linkage, duplicate detection, and duplicate identification, aims to take a set of records as input and find out the records referring to the same real-world entities [2, 18, 26, 28]. Researchers have developed multiple similarity metrics for matching duplicate records, including character-based [1, 24], token-based [8] and numeric similarity metrics [19]. [11] summarizes several common categories of ER, including probabilistic matching [25], supervised learning [22], rule-based approaches [19, 20], etc.

With temporal data accumulated rapidly in variety information systems, some methods coped with lexical heterogeneity in ER problem may fail to perform well in resolving temporal records directly due to existence of evolving heterogeneity. [21] studies linking temporal records and propose time decay to capture the evolution of entity value over time. [16, 28] study efficient rule evolution techniques for clustering issues in ER. [6] proposes a fast algorithm to match temporal records.

While advanced techniques in temporal ER problems achieve high efficiency and accuracy, few methods can be well applied to record matching on datasets without timestamps. The reason is that existing methods mostly depend on definite timestamps and under such circumstance, we can only reason a relative currency order with currency constraints. It motivates us to propose a rule-based ER method to address the entity value evolution effectively on imprecise temporal data.

5   Conclusion and Future Work

Now, we are in the midst of cleaning temporal data with multiple errors. We have already propose an entity resolution method in imprecise temporal data. We propose attribute unstableness to capture the entity evolution over time, and apply dynamic weight schema for improving pairwise similarity computation. We apply rules to determine the currency order of records from target attributes, and propose a novel clustering algorithm along with a pruning method to improve the quality of ER.

We have constructed currency order graph with currency constraints, and determined the currency of each record, accordingly. Based on the currency orders deduced from currency constraints, we will construct the complete method of repairing low-quality data with incomplete and inconsistent values, which lacks for available timestamps. Future works includes (1) seeking for more rules to model the evolving trend of the temporal data more accurate and learning efficient methods to find the rules, and (2) design the balance strategy of cleaning cost and effectiveness of the multiple data cleaning.

References

1. UNIMATCH: a record linkage system: users manual. In: Bureau of the Census, Washington DC (1976)
2. Ananthakrishna, R., Chaudhuri, S., Ganti, V.: Eliminating fuzzy duplicates in data warehouses. In: International Conference on Very Large Data Bases, pp. 586–597 (2002)
3. Batini, C., Cappiello, C., Francalanci, C., Maurino, A.: Methodologies for data quality assessment and improvement. ACM Comput. Surv. **41**(3), 16 (2009)
4. Bertiequille, L., Sarma, A.D., Dong, Marian, A., Srivastava, D.: Sailing the information ocean with awareness of currents: discovery and application of source dependence. Computer. Science **26**(8), 1881–3 (2009)
5. Cappiello, C., Francalanci, C., Pernici, B.: Time related factors of data accuracy, completeness, and currency in multi-channel information systems. In: The Conference on Advanced Information Systems Engineering, pp. 145–153 (2008)
6. Chiang, Y.H., Doan, A.H., Naughton, J.F.: Tracking entities in the dynamic world: a fast algorithm for matching temporal records. Proc. VLDB Endow. **7**, 469–480 (2014)
7. Chu, X., Ilyas, I.F., Papotti, P., Ye, Y.: Ruleminer: data quality rules discovery. In: IEEE International Conference on Data Engineering, pp. 1222–1225 (2014)
8. Cohen, W.W.: Integration of heterogeneous databases without common domains using queries based on textual similarity. In: ACM SIGMOD International Conference on Management of Data, pp. 201–212 (1998)
9. Cong, G., Fan, W., Geerts, F., Jia, X., Ma, S.: Improving data quality: consistency and accuracy. In: International Conference on Very Large Data Bases, pp. 315–326 (2007)
10. Deng, T., Fan, W., Geerts, F.: Capturing missing tuples and missing values. In: Twenty-Ninth ACM SIGMOD-SIGACT-SIGART Symposium on Principles of Database Systems, PODS 2010, Indianapolis, Indiana, USA, 6–11 June 2010, pp. 169–178 (2010)
11. Elmagarmid, A.K., Ipeirotis, P.G., Verykios, V.S.: Duplicate record detection: a survey. IEEE Trans. Knowl. Data Eng. **19**(1), 1–16 (2007)
12. Fan, W., Geerts, F.: Foundations of Data Quality Management (2012)
13. Fan, W., Geerts, F., Jia, X.: Conditional dependencies: a principled approach to improving data quality. In: Sexton, A.P. (ed.) BNCOD 2009. LNCS, vol. 5588, pp. 8–20. Springer, Heidelberg (2009). https://doi.org/10.1007/978-3-642-02843-4_4

14. Fan, W., Geerts, F., Ma, S., Tang, N., Yu, W.: Data quality problems beyond consistency and deduplication. In: Tannen, V., Wong, L., Libkin, L., Fan, W., Tan, W.-C., Fourman, M. (eds.) In Search of Elegance in the Theory and Practice of Computation. LNCS, vol. 8000, pp. 237–249. Springer, Heidelberg (2013). https://doi.org/10.1007/978-3-642-41660-6_12

15. Fan, W., Geerts, F., Wijsen, J.: Determining the currency of data. ACM Trans. Database Syst. **37**(4), 71–82 (2012)

16. Fan, W., Jia, X., Li, J., Ma, S.: Reasoning about record matching rules. Proc. VLDB Endow. **2**(1), 407–418 (2009)

17. Fei, C., Miller, R.J.: A unified model for data and constraint repair. In: IEEE International Conference on Data Engineering, pp. 446–457 (2011)

18. Fellegi, I.P., Sunter, A.B.: A theory for record linkage. J. Am. Stat. Assoc. **64**(328), 1183–1210 (1969)

19. Koudas, N., Marathe, A., Srivastava, D.: Flexible string matching against large databases in practice. In: Thirtieth International Conference on Very Large Data Bases, pp. 1078–1086 (2004)

20. Li, L., Li, J., Gao, H.: Rule-based method for entity resolution. IEEE Trans. Knowl. Data Eng. **27**(1), 250–263 (2015)

21. Pei, L.I., Dong, X.L., Maurino, A., Srivastava, D.: Linking temporal records. PVLDB **4**(11), 956–967 (2011)

22. Richman, J., Richman, J.: Learning to match and cluster large high-dimensional data sets for data integration. In: Eighth ACM SIGKDD International Conference on Knowledge Discovery and Data Mining, pp. 475–480 (2002)

23. Sidi, F., Panahy, P.H.S., Affendey, L.S., Jabar, M.A., Ibrahim, H., Mustapha, A.: Data quality: a survey of data quality dimensions. In: International Conference on Information Retrieval and Knowledge Management, pp. 300–304 (2012)

24. Ullmann, J.R.: A binary n-gram technique for automatic correction of substitution, deletion, insertion and reversal errors in words. Comput. J. **20**(2), 141–147 (1977)

25. Verykios, V.S., Moustakides, G.V., Elfeky, M.G.: A bayesian decision model for cost optimal record matching. VLDB J. **12**(1), 28–40 (2003)

26. Verykios, V.S., Elmagarmid, A.K., Houstis, E.N.: Automating the approximate record-matching process. Inf. Sci. **126**(1–4), 83–98 (2002)

27. Wang, R.Y., Strong, D.M.: Beyond accuracy: what data quality means to data consumers. J. Manag. Inf. Syst. **12**(4), 5–33 (1996)

28. Whang, S.E., Garcia-Molina, H.: Entity resolution with evolving rules. Proc. VLDB Endow. **3**(1–2), 1326–1337 (2010)

An Active Workflow Method for Entity-Oriented Data Collection

Gaoyang Guo[✉]

School of Software, Tsinghua University, Beijing 100084, China
ggy16@mails.tsinghua.edu.cn

Abstract. In the era of big data, people are dealing with data all the time. Data collection is the first step and foundation for many other downstream applications. Meanwhile, we observe that data collection is often entity-oriented, i.e., people usually collect data related to a specific entity. In most cases, people achieve entity-oriented data collection by manual query and filtering based on search engines or news applications. However, these methods are not very efficient and effective. In this paper, we consider designing reasonable process rules and integrating artificial intelligence algorithms to help people efficiently and effectively collect the target data related to the specific entity. Concretely, we propose an active workflow method to achieve this goal. The whole workflow method is composed of four processes: task modeling for data collection, Internet data collection, crowdsourcing data collection and multi-source data aggregation.

Keywords: Data collection · Entity-oriented · Workflow

1 Introduction

In our daily life, people are faced with the need to collect data all the time. For example, football fans may want to get the score data for a current football match. Political lovers may want to obtain the latest political news related to a politician. The fans may want to collect the latest news of their favorite stars. These scenarios can be summed up as a single problem: Given a specific entity such as a football match, a politician, or a star, how can we collect valid data about this entity. We call this problem entity-oriented data collection. In fact, people mainly rely on search engines such as Google or news applications such as RSS to solve their own data acquisition requirements by manual query and filtering nowadays. These methods may be low efficiency. Meanwhile, the data collected by these methods may be scattered. In this paper, we propose an active workflow method to solve this problem.

2 Related Work

Web crawler is a common method for data collection from Internet [8]. A web crawler, sometimes called a spider, is an Internet bot that systematically browses

© Springer Nature Switzerland AG 2018
C. Woo et al. (Eds.): ER 2018 Workshops, LNCS 11158, pp. 76–81, 2018.
https://doi.org/10.1007/978-3-030-01391-2_15

the World Wide Web, typically for the purpose of Web indexing [10]. Web search engines and some other sites use web crawling or software to update their web content or indices of sites' web content [2]. Web crawlers copy pages for processing by a search engine which indexes the downloaded pages so that users can search more efficiently. However, web crawler only returns raw web content, which needs users to further dig out valid data. In this paper, our method can perform a series of subsequent data analysis, which outputs the desired data directly. For example, deep learning models such as LSTM networks can be utilized to extract valuable information from texts [6]. We can also try to integrate data from web based on their community structures [7].

Crowdsourcing is a sourcing model in which individuals or organizations obtain goods and services [1,4]. These services include ideas and finances from a large, relatively open and often rapidly-evolving group of Internet users. It divides work between participants to achieve a cumulative result. In this paper, we also utilize the method of crowdsourcing to collect target data from the real world.

A workflow consists of an orchestrated and repeatable pattern of business activity enabled by the systematic organization of resources into processes that transform materials, provide services, or process information [3,5]. It can be depicted as a sequence of operations, the work of a person or a group, the work of an organization of staff, or one or more simple or complex mechanisms. Petri net [9] is a common method of workflow. In this paper, we borrow the idea of workflow to design our method for entity-oriented data collection.

3   Problem Definition

Entity-Oriented Data Collection. Given a specific entity E, sources S and constraints R, collect and aggregate valid data from S which are related to E and satisfy R.

E is our target entity, which could be a person, a place, an event, a thing or an organization. It specifies what information the user wants. S defines the sources where we collect, which could be some seed URLs or some places in the world. R represents the constraints the target entity E should satisfy, which could be some features expressed in texts, images or videos.

4   Method

In this paper, an active workflow method is proposed to solve this problem. The flow char of this method is shown in Fig. 1. As shown in Fig. 1, the whole workflow method is composed of four processes: task modeling for data collection, Internet data collection, crowdsourcing data collection and multi-source data aggregation. Every process follows some specific process rules. With the help of this workflow method, we can achieve the automation of data collection for specific entities. Given an entity-oriented data collection task defined by the target entity E, the sources S and the constraints R, we first input the task into the process of task modeling for data collection. After that, we obtain an EOSQL statement, which is fed into the process of Internet data collection and the crowdsourcing data collection. Then, the scattered valid data from multi sources are aggregated in the process of multi-source data aggregation. Finally, we collect desired valid data which are related to E and satisfy R. The detailed discussion of each process is as follows.

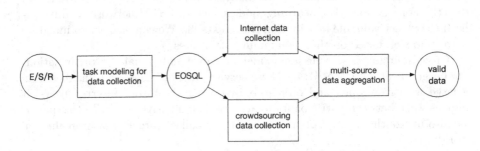

Fig. 1. The active workflow method for entity-oriented data collection

4.1   EOSQL

The first process is task modeling for data collection. In this process, an entity-oriented data collection language called EOSQL is designed to model the data collection task. The design of EOSQL is inspired by SQL. The grammar of EOSQL is similar with SQL, which consists of different types of fields and also supports the clause structure. Users can model the task for entity-oriented data collection by simply using an EOSQL statement following the grammar of EOSQL. The grammatical rules of EOSQL are designed as follows:

$$COLLECT\ [E]\ FROM\ [S]\ WHERE\ [R], \tag{1}$$

where E, S, and R are three fields to model the task for entity-oriented data collection. E indicates the target entity the user is interested in. It specifies what information the following processes should collect. S defines the sources where we collect. For example, S could be some seed URLs which are the entry points of the crawler. Also, S could be some places in the world where the target entity exists. R represents the constraints the target entity E should satisfy. For example, R could be some features expressed in texts, images or videos.

EOSQL plays an important role in the whole workflow method. It is the basis of the following processes since it defines the overall goal for the entity-oriented data collection using a uniform language.

4.2   Distributed Data Crawler

The second process is Internet data collection. In this process, we propose an entity-oriented distributed data crawler, which crawls valid data from Internet related to the specific entity. Based on an EOSQL statement, we can obtain E, S and R. The distributed data crawler then crawls the valid data from S which is related to E and satisfies R. In most cases, the crawler deals with the data collection from Internet.

Our crawler mainly consists of two parts: the preliminary crawling and the further extraction. The preliminary crawling is based on Nutch, which is a highly extensible and scalable open source web crawler software project. This part can be divided into 5 stages: distributed URL injection, distributed URL task allocation, distributed URL queue grabbing, distributed URL web page resolution and distributed database update. The further extraction extracts valid information according to E and R. Based on the preliminary crawling, the further extraction utilizes real-time data processing algorithms to recognize and extract the target entity. For example, we can use information extraction methods such as entity recognition and entity relation extraction to collect valid data from texts according to E and R. We can use computer vision techniques such as face recognition and object detection to extract valid data from images or videos according to E and R.

Based on these two parts, our crawler can achieve the tight coupling of crawling and extraction, which can reduce storage overhead and improve collection efficiency.

4.3   Crowdsourcing Collection Model

The third process is crowdsourcing data collection. In this process, an entity-oriented crowdsourcing collection model is proposed to collect raw data from the real world related to the specific entity based on a crowdsourcing algorithm. Based on an EOSQL statement, we can obtain E, S and R. The crowdsourcing collection model then collect the valid data from S which is related to E and satisfies R. In general, this model deals with the data collection from the real world.

The crowdsourcing collection model mainly relies on human resources. It consists of two steps: the shallow crowdsourcing collection and the deep crowdsourcing collection. In the shallow crowdsourcing, the crowdsourcing participants collect large quantity of uncertain data to be confirmed using collection devices. They filter out useless information to reduce the cost of subsequent collection. In the deep crowdsourcing, the model executes the feature analysis, crowdsourcing task division and crowdsourcing task assignment in turn to further process the uncertain data collected from the first step. Among them, crowdsourcing task assignment is the most challenging since this problem is often NP-hard. To solve the problem of task assignment, we should model it as a NP-hard problem first. It is assumed that there are several crowdsourcing participants and several crowdsourcing tasks. Every participant has a cost and a quality for executing each task. The goal is to find a task assignment scheme to minimize the cost and maximize the quality for executing task. Then we need design an approximate algorithm to solve this optimization problem.

Based on these two steps, our crowdsourcing collection model can collect valid data from the real world with a low cost.

4.4 Multi-source Data Aggregation Model

The last process is multi-source data aggregation. In this process, we propose a multi-source data aggregation model to effectively aggregate data collected from the second process and the third process.

After the second process and the third process, we acquire lots of valid data related to the target entity E both from Internet and the real world. However, these data are from different sources and lack effective organization. Moreover, they may have some quality problems such as data redundancy. Therefore, it is necessary to integrate them all using a uniform data format. The main challenge of this process is to design entity matching algorithm for multi-source and multi-domain data. The data collected from the last two processes may have known patterns or my have unknown patterns. To cope with this scenario, we propose a data blocking strategy in hybrid mode, which not only improves the effect of data blocking by using the known patterns, but also have the universality of handling unknown patterns. On the one hand, we put forward a lazy entity matching strategy based on the query perception technology, which conducts entity matching only for the necessary data and only at the necessary time. This strategy can prevent unnecessary computation cost caused by massive Internet data. On the other hand, we also propose an incremental updating strategy for entity matching to deal with the fact that all data sources are constantly updated. Based on these two strategies, the entity matching algorithm is effective and efficient.

Finally, We aggregate all valid data within a uniform data format utilizing the multi-source data aggregation model. Users can acquire desired data which are related to E and satisfy R from the outputs of the last process.

5   Conclusion

In this paper, we introduce a problem called entity-oriented data collection. An active workflow method is proposed to solve this problem. The active workflow method consists of four processes. We design one model and corresponding process rules for each process. In the future, we will continue to improve our workflow method to conduct entity-oriented data collection.

Acknowledgment. This work was supported in part by the National Key Research and Development Program of China (No. 2017YFC0820402), the Intelligent Manufacturing Comprehensive Standardization and New Pattern Application Project of Ministry of Industry and Information Technology (Experimental validation of key technical standards for trusted services in industrial Internet), and the National Natural Science Foundation of China (No. 61373023).

References

1. Buettner, R.: A systematic literature review of crowdsourcing research from a human resource management perspective. In: Hawaii International Conference on System Sciences, pp. 4609–4618 (2015)
2. Corby, O., Dieng-Kuntz, R., Faron-Zucker, C.: Querying the semantic web with corese search engine. In: Eureopean Conference on Artificial Intelligence, ECAI 2004, Including Prestigious Applicants of Intelligent Systems, PAIS 2004, Valencia, Spain, August, pp. 705–709 (2017)
3. Curcin, V., Ghanem, M., Guo, Y.: The design and implementation of a workflow analysis tool. Philos. Trans. Math. Phys. Eng. Sci. 368(1926), 4193 (2010)
4. Doan, A.H., Ramakrishnan, R., Halevy, A.Y.: Crowdsourcing systems on the world-wide web. Commun. ACM 54(4), 86–96 (2011)
5. Georgakopoulos, D., Hornick, M., Sheth, A.: An overview of workflow management: from process modeling to workflow automation infrastructure. Distrib. Parallel Databases 3(2), 119–153 (1995)
6. Guo, G., Wang, C., Chen, J., Ge, P., Chen, W.: Who is answering whom? Finding "reply-to" relations in group chats with deep bidirectional lstm networks. Clust. Comput. 10, 1–12 (2018)
7. Guo, G., Wang, C., Ying, X.: Which algorithm performs best: algorithm selection for community detection. In: Companion of the The Web Conference, pp. 27–28 (2018)
8. Kobayashi, M., Takeda, K.: Information retrieval on the web. Annu. Rev. Inf. Sci. Technol. 39(1), 33–80 (2005)
9. Murata, T.: Petri nets: properties, analysis and applications. Proc. IEEE 77(4), 541–580 (1977)
10. Shaila, S.G., Vadivel, A.: Architecture specification of rule-based deep web crawler with indexer. Int. J. Knowl. Web Intell. 4(4), 166–186 (2013)

Visual Non-verbal Social Cues
Data Modeling

Mahmoud Qodseya[✉]

Institut de Recherche en Informatique de Toulouse (IRIT),
Université de Toulouse, Toulouse, France
Mahmoud.Qodseya@irit.fr

Abstract. Although many methods have been developed in social signal processing (SSP) field during the last decade, several issues related to data management and scalability are still emerging. As the existing visual non-verbal behavior analysis (VNBA) systems are task-oriented, they do not have comprehensive data models, and they are biased towards particular data acquisition procedures, social cues and analysis methods. In this paper, we propose a data model for the visual non-verbal cues. The proposed model is privacy-preserving in the sense that it grants decoupling social cues extraction phase from analysis one. Furthermore, this decoupling allows to evaluate and perform different combinations of extraction and analysis methods. Apart from the decoupling, our model can facilitate heterogeneous data fusion from different modalities since it facilitates the retrieval of any combination of different modalities and provides deep insight into the relationships among the VNBA systems components.

Keywords: Social signal processing · VNBA systems
Metadata modeling · Visual non-verbal cues

1 Introduction

Observing and understanding human reactions and interactions with other human beings are challenging and interesting research directions for a wide variety of applications from customer satisfaction estimation to social robots modeling. Thus, the main purpose of SSP field is to automatically recognize and understand the human social interactions by analyzing their sensed social cues. These cues are mainly classified into verbal (word, e.g., semantic linguistic content of speech) and non-verbal (wordless and visual) social cues. The verbal cues take into account the spoken information among persons like 'yes/no' response in the answering question context. The non-verbal social cues represent a set of temporal changes in neuromuscular and physiological activities, which send a message about feelings, mental state, personality, and other characteristics of human beings [1].

© Springer Nature Switzerland AG 2018
C. Woo et al. (Eds.): ER 2018 Workshops, LNCS 11158, pp. 82–87, 2018.
https://doi.org/10.1007/978-3-030-01391-2_16

Fig. 1. A general visual non-verbal behavioral analysis schema.

Since crowded places are often noisy, most of the existing methods have been directed toward small group interactions (i.e., meeting, dining). Therefore, various visual-based SSP methods have been introduced in the literature [1–4], which generally can be summarized in a common system schema consisting of five modules, depicted in Fig. 1: (i) data acquisition, (ii) person detection and tracking, (iii) social cues extraction, (iv) contextual information, and (v) social cues analysis.

Different types of sensors and devices like cameras and proximity detectors might be exploited in the data acquisition module to record social interactions. Thus, one or more dedicated computer vision and image processing based (e.g., face detection) techniques could be leveraged for treating the input data to detect and track person(s). The social cues extraction module takes as an input the detected person(s) to extract a feature vector (per person) describing the social cues such as head *pose* (position and orientation). The social cues understanding module deeply analyzes the primitive social cues through modeling temporal dynamics and combining signals extracted from various modalities (e.g., head *pose*, facial expression) at different time scales to provide more useful information and conclusion on the behavioral level of the detected persons. Indeed, this module might optionally leverage additional information describing the context in which the data is captured to provide a precise social behavior prediction and analysis. Finally, the existence of metadata repository decouples the analysis phase from the other components as Fig. 1 shows.

In this paper, we introduce a metadata model for the visual cues as they are the mostly perceived and unconsciously displayed by human among the non-verbal cues. Our data model has many benefits in: (i) fusing heterogeneous data from different sources and modalities via facilitating the retrieval process, (ii) decoupling the social signal extraction phase from the analysis phase, and (iii) preserving privacy by facilitating access control layer integration within the metadata repository [5] to confront the privacy and integrity threats. Therefore, the analysis methods are going to use the extracted metadata instead of the original input data. Moreover, it is easy to extend the designed model for supporting the vocal cues as our model considers *per-frame* metadata associated with the detected persons.

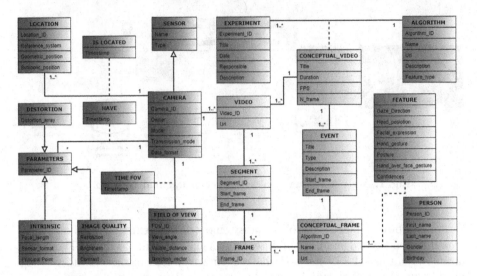

Fig. 2. Generic data model. This generic data model for visual non-verbal social cues shows the relationships that exist between experiment, acquisition, video, and feature groups of entities, which are color-coded as green, orange, yellow, and gray respectively. (Color figure online)

Consequently, this provides a capability to integrate the extracted vocal cues at the *frame-level* similar to *audio-video* synchronization.

Paper Outline. Sect. 2 presents a detailed description of the designed data model. The current state of our research is presented in Sect. 3. The conclusion of our work is summarized in Sect. 4 with some insights for future works.

2 Visual Non-verbal Social Cues Data Model

Visual non-verbal social cues have been received more attention as they are not semantic in nature and often occur unconsciously. The explored visual non-verbal cues include body movements, hand movements, gaze behavior, and signals with higher level of annotation like smiling, facial expressions, hand gestures, hand-over-face gestures, head orientations, and mutual gaze. To handle the high variety of the social visual cues, we design a conceptual data model consisting of experiment, acquisition, video, and feature groups of entities as Fig. 2 shows.

2.1 Experiment Group

Researchers or experts in the VNBA systems are interesting in performing experiments using different configurations of algorithm types and parameters. Thus, the EXPERIMENT class is dedicated to hold such information that includes experiment title, date, responsible person, location, and description. On the other side, researcher could be interested in additional information about the list of algorithms that are used to extract the social cues from a conceptual video. Therefore, the ALGORITHM class is devoted to carry the algorithm's name, URL, description, and feature type (e.g., facial expression, gaze direction) that can be extracted using this algorithm.

2.2 Acquisition Group

Generally, different types of sensors (e.g., camera, GPS, IMU, and microphone) are used for social cues (verbal/non-verbal) acquiring. In the context of VNBA systems, cameras are widely adopted, and thus we cover the camera relevant information within this group in which CAMERA class contains attributes for holding information about the adopted camera(s) in conducting experiments. These attributes include the identity number (e.g., 58395FX), owner (e.g., IRIT), model (e.g., Axis F44 Dual Audio), transmission mode (wired/wireless), and data format (e.g., .mp4) of the camera. Cameras are controlled by *time invariant* parameters at different frequencies, while these parameters include camera intrinsic parameters, location, field of view, distortion, and image quality.

To model these parameters, we propose separated classes for each one of them as follows: (i) INTRINSIC class attributes include camera focal length (Fx,Fy), image sensor format(S), and principal point (Cx,Cy), (ii) LOCATION class contains a system reference as well as the symbolic and geometric (extrinsic camera parameters) position, where the intrinsic and extrinsic camera parameters are used in the computation of the camera projection matrix, (iii) FIELD OF VIEW class contains the attributes (viewable angle, visible distance, and FOV direction) that are used to determine how wide an area of a camera field of view, (iv) DISTORTION class has five attributes that are used for lens distortion correction, and (v) the IMAGE QUALITY class includes common image features such as resolution, brightness, and contrast.

2.3 Video Group

VIDEO, SEGMENT, and FRAME represent a decomposition relationship since a video clip is decomposed into segments which represent sequence of frames. An event is a thing that happens or takes place over a particular time interval (e.g., type of the played music during a time interval). So, a video clip could be decomposed into event(s) that contain(s) a sequence of frames. Although both event and segment represent a part of video, but the event has semantic descriptions (contextual information).

In multiple cameras views scenarios, for each *time-stamp*, we need to process multiple frames (same number of used cameras) together, where same person could appear in more than one frame. Thus, we will have inconsistence feature values related to the same detected person. To overcome this challenge, we introduce both CONCEPTUAL-FRAME and CONCEPTUAL-VIDEO classes. A conceptual frame represents one to N frames (N is the number of the adopted cameras within the experiment) that have a common time stamp and must be analyzed together. A conceptual video represents one to N videos that are recoded within an experiment and must be analyzed together. Therefore, we handle the latter mentioned challenge using the conceptual frame as we will fuse the extracted cues at the *frame-level*.

Since the event is related to particular time interval and location, which are common within the multiple cameras views scenarios, we present an association relationship between EVENT and CONCEPTUAL-VIDEO classes as Fig. 2 shows.

2.4 Features Group

The "FEATURE" class is designed as an association class containing the extracted social cues that are extracted for each detected person inside the conceptual frame. The attributes of feature class include the most common visual non-verbal cues such as gaze direction, head position, facial expression, hand gesture, hand-over-face gesture, and their detection confidences. "PERSON" class contains information (name, age, and birthday) about the experiments' participants.

3 Current Work

Due to the lack of data, we are not able to evaluate the proposed system, so we are implementing an acquisition platform named OVALIE for data recording. Our prototype consists of four cameras that surrounded a table with four chairs, where each person who is sitting on the table his face will appear in two cameras. Furthermore, we have been implemented a software that ables to read multiple videos, detect and track multiple persons based on an upgraded version of OpenFace toolkit [6], and detect the facial expression for each detected person using affectiva SDK[1].

Meanwhile, we implemented a physical model using Mongo database and test (using synthetic data) different quires that can be used for retrieving features for the analysis phase. Currently, we start collecting food images which will be used for YOLO [7] retraining. The customized YOLO deep learning based model could be used for detecting and localizing the different types of food on the table.

4 Conclusion and Future Work

In this paper, we propose a metadata model for the VNBA systems, which is: (i) privacy-preserving, (ii) easily expendable to cover the vocal and verbal cues, (iii) smoothing the data fusion among multiple modalities, (iv) decoupling the social cues extraction phase from the analysis phase, and (v) allowing to evaluate different combinations of the extraction and analysis methods.

As a future direction, we are going to extend our model to include the verbal and non-verbal cues. In addition to that, we intend to collect and annotate a dataset for experimenting and validating our data model.

---

[1] https://developer.affectiva.com/.

References

1. Akhtar, Z., Falk, T.: Visual nonverbal behavior analysis: the path forward. In: IEEE MultiMedia (2017)
2. Vinciarelli, A., et al.: Bridging the gap between social animal and unsocial machine: a survey of social signal processing. IEEE Trans. Affect. Comput. **3**, 69–87 (2012)
3. Cristani, M., Raghavendra, R., Del Bue, A., Murino, V.: Human behavior analysis in video surveillance: a social signal processing perspective. Neurocomputing **100**, 86–97 (2013)
4. Salah, A.A., Pantic, M., Vinciarelli, A.: Recent developments in social signal processing. In: 2011 IEEE International Conference on Systems, Man, and Cybernetics, pp. 380–385, October 2011
5. Kukhun, D.A., Codreanu, D., Manzat, A.-M., Sedes, F.: Towards a pervasive access control within video surveillance systems. In: Cuzzocrea, A., Kittl, C., Simos, D.E., Weippl, E., Xu, L. (eds.) CD-ARES 2013. LNCS, vol. 8127, pp. 289–303. Springer, Heidelberg (2013). https://doi.org/10.1007/978-3-642-40511-2_20
6. Baltrusaitis, T., Zadeh, A., Lim, Y.C., Morency, L.P.: Openface 2.0: facial behavior analysis toolkit. In: 2018 13th IEEE International Conference on Automatic Face & Gesture Recognition (FG 2018), pp. 59–66 (2018)
7. Redmon, J., Farhadi, A.: Yolov3: An incremental improvement. arXiv (2018)

ER FORUM 2018

Preface to ER 2018 FORUM Workshop

Conceptual Modeling: Research in Progress ER 2018 FORUM Workshop at the 37th International Conference on Conceptual Modeling (ER 2018)

Xi'an, China
ctober 22–25, 2018

Organized by
Heinrich C. Mayr and Oscar Pastor

The ER Forum Workshop is a vibrant platform for presenting and discussing novel research that addresses hot topics related to conceptual modelling. Innovation prevails over maturity, in particular regarding emerging new research topics. Consequently, the papers in this chapter address a broad spectrum of interesting approaches: new conceptual patterns for modern accounting information systems, the effectiveness of conceptual modeling languages, the expressiveness of temporal constraints for process models, knowledge graph embedding and view synchronization upon schema evolution.

All submitted papers have been peer reviewed by at least three international Program Committee members. The accepted papers were carefully revised following the reviewers' comments.

We would like to express our gratitude to everyone who helped to make the ER Forum 2018 FORUM Workshop a success. In particular, we thank the authors for their valuable submissions, as well the international Program Committee members for their reviews and helpful comments. The quality of these papers is a tribute to the authors and to the reviewers who guided any necessary improvement. Finally yet importantly, we would like to thank the ER2018 Organizing Committee members for their support.

October 2018

Heinrich C. Mayr
Oscar Pastor
Program Co-chairs

Dependency-Based Query/View Synchronization upon Schema Evolutions

Loredana Caruccio(✉), Giuseppe Polese, and Genoveffa Tortora

Department of Computer Science, University of Salerno,
via Giovanni Paolo II n.132, 84084 Fisciano, SA, Italy
{lcaruccio,gpolese,tortora}@unisa.it

Abstract. Query/view synchronization upon the evolution of a database schema is a critical problem that has drawn the attention of many researchers in the database community. It entails rewriting queries and views to make them continue work on the new schema version. Although several techniques have been proposed for this problem, many issues need yet to be tackled for evolutions concerning the deletion of schema constructs, hence yielding loss of information. In this paper, we propose a new methodology to rewrite queries and views whose definitions are based on information that have been lost during the schema evolution process. The methodology exploits (relaxed) functional dependencies to automatically rewrite queries and views trying to preserve their semantics.

Keywords: Schema evolution · Query rewriting
Functional dependency

1 Introduction

Query/view synchronization (QVS in the following) upon schema evolutions is an extremely critical problem, since it has been estimated that each schema evolution step might affect up to 70% of the queries and views operating on the old schema [12]. The QVS problem has been tackled in several studies, and many approaches and tools have been proposed to support DBAs during the schema evolution process (see [9] for a survey). However, the shortage of proper automated tools has not made the proposed solutions sufficiently practical [1, 11,15], since they often require programmer's work to manually rewrite portions of affected queries and views. Moreover, there are many cases in which it is not possible to correctly synchronize the affected queries and views, especially when the evolution entails loss of information. In fact, all the query/view definitions based on the lost information cannot be rewritten in a simple way, so limiting the possibility to have an automatic synchronization process.

In general, approaches aiming to solve the QVS problem in presence of information loss either prescribe to remove queries and views that cannot be synchronized, or to block the evolution operations preventing the synchronization

© Springer Nature Switzerland AG 2018
C. Woo et al. (Eds.): ER 2018 Workshops, LNCS 11158, pp. 91–105, 2018.
https://doi.org/10.1007/978-3-030-01391-2_17

of some queries or views. Some other approaches prescribe to ask the DBA about the best choice to be made.

In this paper, we propose a methodology aiming to automatically rewrite queries and views relying on the schema constructs that have been deleted during the schema evolution. The methodology exploits semantic correlations of data expressed in terms of relaxed functional dependencies (RFDs) [7], for which automatic tools are now available to extract them from data [6,8]. In particular, the methodology tries to rewrite affected queries and views by seeking possible RFDs between the deleted information and the remaining ones, in a way to preserve or approximate the original semantics of queries and views.

The paper is organized as follows. In Sect. 2 we recall several basic concepts on the theory of Schema Evolution and on the theory of Functional Dependencies. Then, we characterize the whole query/view synchronization problem, and specifically the necessity to manage the loss of information in Sect. 3. In Sect. 4 we present our solution to automatically synchronize queries and views when a loss of information occurs. In Sect. 5 we validate the proposed methodology by presenting the results of several experiments conducted on two public datasets. Related works have been provided in Sect. 6. Finally, conclusions and future research directions are discussed in Sect. 7.

2   Background

In this section we will recall basic notions on the database theory underlying the proposed solution.

2.1   Schema Evolution

Schema evolutions might occur when a system is first released, due to bugs or incomplete functionalities, or to reflect changes in the real world, which might also entail the evolution of the underlying database. Problems related to the schema evolutions have been studied not only for databases, but also for Data Warehouses [14,24] and for Ontologies [23]. In this paper, we focus on the evolution of database schemas.

Definition 1 (Schema evolution). *Let S be a database schema, and $Inst(S)$ be the set of possible instances of S; an evolution of S is the result of one or more changes to the data structures, constraints, or any other artifact of S, entailing modifications to the system catalog.*

Changes might consist of *simple schema modifications*, such as addition, deletion, or renaming of an attribute, of a constraint, or of a relation, and/or *composed schema modifications*, such as join, partition, and decomposition [20]. In what follows, we denote with $S{\rightarrow}S'$ the evolution of the schema S into S', where S and S' are called schema versions.

Example 1. Let us consider the following database schema:

$$S : \text{Doctor}(\underline{\text{idDoctor}}, \text{Name}, \text{Specialization}, \text{Role}, \text{Experience}, \text{Salary}, \text{Level}, \text{Tax}) \quad (1)$$

where the underlined attribute represents the primary key. The schema represents a table of doctors, containing name, specialization, role, experience, annual salary, contractual level, and annual tax. An evolution $S \to S'$ of this schema might concern the removal of the attribute Level, yielding the following new schema version:

$$S' : \text{Doctor}(\underline{\text{idDoctor}}, \text{Name}, \text{Specialization}, \text{Role}, \text{Experience}, \text{Salary}, \text{Tax}) \quad (2)$$

There are two basic strategies to specify a schema evolution $S \to S'$; one strategy describes the operation commands of the procedure to transform S into S', and another first specifies S', and then finds schema correspondences between S and S', which can be represented by means of *mappings*[1] [1–3, 18, 20, 22]. Moreover, there are hybrid strategies mixing the characteristics of the two previous ones. As a consequence, we can have the following three types of schema evolution approaches: *operation-based*, *mapping-based*, and *hybrid*, respectively.

In general, *operation-based* approaches exploit the advantage of knowing a priori how the schema can evolve and the effects of each evolution. On the other hand, *mapping-based* approaches allow us to handle every type of modification [22], but without a complete view of the effects that a single modification will have on the schema.

The impact of schema modifications on the database instances can be characterized through the concept of *information capacity* [17]. The latter specifies whether the set $Inst(S')$ is equivalent to, extends, or reduces $Inst(S)$ [1]. For instance, when an attribute is dropped from a database schema, the corresponding information is lost, hence $Inst(S')$ reduces $Inst(S)$. In this case, we will denote by $NotInst(S, S')$ the set of all instances of S that do not have a corresponding instance in $Inst(S')$.

Schema construct deletions are among the most critical capacity reducing variations, since they entail loss of information that might irremediably invalidate some database components, like queries and application programs.

2.2   Relaxed Functional Dependencies

Let us recall some basic concepts of relational databases.

A relational database schema $\mathcal{R}$ is defined as a collection of relation schemas $(R_1, \ldots, R_n)$, where each R_i is defined over a fixed set of attributes $attr(R_i)$. Each attribute A_k has associated a domain $dom(A_k)$, which can be finite or infinite. A relation instance (or simply a relation) r_i of R_i is a set of tuples such that for each attribute $A_k \in attr(R_i)$, $t[A_k] \in dom(A_k)$, $\forall\, t \in r_i$, where $t[A_k]$

---

[1] A mapping m between two schemas S and S' is a set of assertions of the form $q_S \rightsquigarrow q_{S'}$, where q_S and $q_{S'}$ are queries over S and S', respectively, with the same set of distinct variables, and $\rightsquigarrow \in \{\subseteq, \supseteq, \equiv\}$.

denotes the projection of t onto A_k. A database instance r of $\mathcal{R}$ is a collection of relations $(r_1,\ldots,r_n)$, where r_i is a relation instance of R_i, for $i \in [1,n]$.

In the context of relational databases, data dependencies have been used to define data integrity constraints, aiming to improve the quality of database schemas and to reduce manipulation anomalies. There are several types of data dependencies, including functional, multivalued, and join dependencies. Among these, functional dependencies (FDs) are the most commonly known, mainly due to their use in database normalization processes. Since RFDs extend FDs, let us recall the definition of FD.

Definition 2 (Functional dependency). *A functional dependency (FD) φ, denoted by $X \to Y$, between two sets of attributes $X, Y \subseteq attr(\mathcal{R})$, specifies a constraint on the possible tuples that can form a relation instance r of $\mathcal{R}$: $X \to Y$ iff for every pair of tuples (t_1, t_2) in r, if $t_1[X] = t_2[X]$, then $t_1[Y] = t_2[Y]$. The two sets X and Y are also called Left-Hand-Side (LHS) and Right-Hand-Side (RHS), resp., of φ.*

RFDs extend FDs by relaxing some constraints of their definition. In particular, they might relax on the *attribute comparison* method, and on the fact that the dependency must be satisfied by the entire database.

Relaxing on the attribute comparison method means to adopt an approximate tuple comparison operator, say $\approx$, instead of the "equality" operator. In order to define the type of attribute comparison used within an RFD, we use the concept of *similarity constraint* [10].

Definition 3 (Similarity constraint). *Given an attribute A on a given domain $\mathbb{D}$, let $\phi[A] : \mathbb{D} \times \mathbb{D} \to \mathbb{R}^+$ be a function which evaluates the similarity between value pairs of A.*

As an example, ϕ can be defined in terms of a similarity metric $\approx$, like for instance the edit or Jaro distances [13], such that, given two values $a_1, a_2 \in A$, $a_1 \approx a_2$ is true if a_1 and a_2 are "close" enough w.r.t. a predefined threshold α. We denote the similarity constraint associated to an attribute A as $A_{\leq \alpha}$, which indicates that a pair of values (a_1, a_2) can be considered similar on A if and only if the $\phi[A](a_1, a_2) \leq \alpha$.

Definition 4 (Set of similarity constraints). *Given a set of attributes $X \subseteq attr(R)$ with $X = \{A_1, \ldots, A_k\}$, a set of similarity constraints, denoted as X_Φ, collects the similarity constraints $X_\Phi = \{A_{1 \leq \alpha_1}, \ldots, A_{k \leq \alpha_k}\}$ associated to attributes of X.*

A dependency holding for "almost" all tuples or for a "subset" of them is said to relax on the extent [7]. In case of "almost" all tuples, a *coverage measure* should be specified to quantify the degree of satisfiability of the RFD. Whereas in case of "subset" (*constrained domains* in the following), conditions on the attribute domains should be specified to define the subset of tuples satisfying the RFD.

Definition 5 (Coverage measure). *Given a database instance r of $\mathcal{R}$, and an FD $\varphi : X \to Y$, a coverage measure Ψ on φ quantifies the satisfiability degree of φ on r, $\Psi : dom(X) \times dom(Y) \to \mathbb{R}^+$ measuring the amount of tuple pairs in r satisfying φ.*

As an example, the *confidence measure* introduced in [16] evaluates the maximum number of tuples $r_1 \subseteq r$ for which φ holds in r_1.

Several coverage measures can be used to define the satisfiability degree of an RFD, but usually they return a value normalized on the total number of tuple pairs $\binom{n}{2}$ for n cardinality of $\mathcal{R}$, so producing a value $v \in [0,1]$. In the context of the canonical FDs, this measure evaluates to 1.

Definition 6 (Constrained domain). *Given a relation schema $\mathcal{R}$ with attributes $\{A_1, \ldots, A_k\}$ with attributes $\{A_1, \ldots, A_k\}$ of a given domain $\mathbb{D} = \mathbb{D}_1 \times \mathbb{D}_2 \times \cdots \times \mathbb{D}_k = dom(\mathcal{R})$, let $c_i \; \forall i = 1, dots, k$ be a condition on $\mathbb{D}_i$ the constrained domain $\mathbb{D}_c$ is defined as follows*

$$\mathbb{D}_c = \{t \in dom(\mathcal{R}) | \bigwedge_{i=1}^{k} c_i(t[A_i])\}.$$

Constrained domains enable the definition of "subsets" of tuples on which dependencies apply [5].

Then, a general definition of RFD can be given:

Definition 7 (Relaxed functional dependency). *Let us consider a relational schema $\mathcal{R}$. An RFD ϱ on $\mathcal{R}$ is denoted by*

$$\left[X_{\Phi_1} \xrightarrow{\Psi \geq \varepsilon} Y_{\Phi_2} \right]_{\mathbb{D}_c} \tag{3}$$

where

- $\mathbb{D}_c$ is the constrained domain that filters the tuples on which ϱ applies;
- $X, Y \subseteq attr(R)$, with $X \cap Y = \emptyset$;
- Φ_1 and Φ_2 are sets of similarity constraints on attributes X and Y, respectively;
- Ψ is a coverage measure defined on $\mathbb{D}_c$;
- ε is a threshold.

Given $r \subseteq \mathbb{D}_c$, a relation instance r on R satisfies the RFD ϱ, denoted by $r \models \varrho$, if and only if: $\forall \; (t_1, t_2) \in r$, if Φ_1 is true for each constraint $A_{\leq \alpha} \in \Phi_1$, then *almost always* Φ_2 is true for each constraint $B_{\leq \beta} \in \Phi_2$. Here, *almost always* means that $\Psi(\pi_X(r), \pi_Y(r)) \geq \varepsilon$.

In other words, if $t_1[X]$ and $t_2[X]$ agree with the constraints specified by Φ_1, then $t_1[Y]$ and $t_2[Y]$ agree with the constraints specified by Φ_2 with a degree of certainty (measured by Ψ) greater than ε.

In the following we use RFDs having only one attribute on the RHS; this condition can always be reached by means of the usual transformations of FDs.

Example 2. As an example, for the database schema of **Doctor** shown in (1), it is likely to have the same **Specialization** for doctors having the same **Name** and **PlaceOfBirth**. An FD {Name,PlaceOfBirth} → **Specialization** might hold. However, the names, places and specializations might be stored by using different abbreviations. Thus, the following RFD might hold:

$$\{\text{Name}_\approx, \text{PlaceOfBirth}_\approx\} \longrightarrow \text{Specialization}_\approx$$

where $\approx$ is the string similarity function. On the other hand, the few cases of homonyms for the doctors born in the same place have to be considered. For this reason, the previous RFD should also admit exceptions. This can be modeled by introducing a different coverage measure into the RFD, making it approximate:

$$\{\text{Name}_\approx, \text{PlaceOfBirth}_\approx\} \xrightarrow{\psi(Name, PlaceOfBirth, Specialization) \geq 0.98} \text{Specialization}_\approx$$

3 Problem Description

Let us define the query/view synchronization problem.

Definition 8 (Query/View Synchronization). *Let Q be the set of queries and views defined on a database schema S; upon a schema evolution $S{\to}S'$, the QVS problem consists of finding a transformation τ of Q producing a set Q' of queries and views on S', such that Q' on S' preserves the semantics of Q on S. If such a transformation exists, we say that there exists a synchronization $Q{\to}Q'$, or even that Q' represents a legal rewrite of Q.*

Example 3. Let us consider the view on the database schema S of Example 1:

$$
\begin{aligned}
&\text{CREATE VIEW } \textsf{getDoctorByLevel} \text{ AS}\\
&\text{SELECT } \textsf{Name, Specialization}\\
&\text{FROM } \textsf{Doctor}\\
&\text{WHERE } \textsf{Level} = \text{``}D\text{''}
\end{aligned}
\tag{4}
$$

It extracts the tuples **getDoctorByLevel(Name, Specialization)** for all the stored doctors with a contractual level equal to "D". Clearly, even such a simple view needs be rewritten upon the schema evolution defined in (2).

The preferred automated tools will be those that give the user the illusion of defining queries and views on an older version of the schema even though it has evolved [18].

Example 4. Let us consider the schema S of Example 1. If it evolves into the following schema:

$$
\begin{aligned}
&\textsf{DoctorData}(\underline{\textsf{idDoctor}}, \textsf{Name, Specialization, Role, Experience})\\
&\textsf{DoctorEconomy}(\underline{\textsf{idDoctor}}, \textsf{Salary, Level, Tax})
\end{aligned}
\tag{5}
$$

then it is possible to automatically synchronize all queries and views involving attributes of both **DoctorData** and **DoctorEconomy**, by introducing a *join* between the new two relations.

Unfortunately, it is not always possible to find a synchronization for all queries and views, especially when the modification concerns the deletion of schema constructs instantiated within some queries or views. In the following, we analyze the schema constructs that can affect the results of a query or of a view.

Definition 9. *Given a relation R and a selection condition c $Inst(R, c)$ is the set of instances in R satisfying c [28].*

Definition 10. *Given a schema S, and a query/view Q with selection condition C_Q*

$$s_Q = \{s \in Inst(S, C_Q)\} \tag{6}$$

is the result instance of the query/view Q.

Definition 11. *Let Q be a query/view, and $S \to S'$ an evolution of S, we say that Q can be automatically rewritten upon the deletion of a schema construct if and only if the following properties hold:*

- *There exists an instance $s_i \in Inst(S')$, which corresponds to (is the same of) the result instance of Q when applied on S (result data preserving), i.e. $s_i = s_Q$.*
- *For each selection condition c of Q, there exists a constraint c' such that $s_Q \in Inst(S, c) \cap Inst(S', c')$ (result construction preserving).*

It is worth to noticing that the lack of preservation in result construction does not necessarily produce the lack of preservation in result data. In fact, when $s_Q \notin NotInst(S, S')$, there can exist a condition c_j such that $s_Q \notin Inst(S, c_j) \cap Inst(S', c'_j)$ for each c'_j definable on S'. In other words, this case occurs when the information on the attributes whose values must be outputted by executing Q have not been lost, but there is a condition useful to produce the result instance of Q that cannot be redefined in S'.

4 Methodology

In the literature, there are few methodologies addressing the synchronization of queries and views upon schema evolutions yielding information loss [9]. They propose solutions ranging from the possibility of not including all the information required by Q in the result instance [1], to the possibility of defining a priori some parameters and/or policies on Q that block evolutions yielding loss of information declared as essential for Q [19]. In some other cases, the DBA intervention is required to manage the loss of information [25], or to replace lost data with approximated ones within the result instance [26].

The proposed methodology exploits RFDs [7], aiming to produce result instances equivalent to or approximating those produced on the old schema version. In particular, we focus on schema evolutions yielding attribute deletions, since they can invalidate conditions of queries and views. However, the proposed methodology can be easily extended to the removal of relations, since they can be rewritten in terms of removal of several attributes.

More formally, given a schema evolution $S \to S'$, where $NotInst(S, S') \neq \emptyset$. Let $c_1, \ldots, c_k$ in Q, and let $\bar{c}_1, \ldots, \bar{c}_h$ be the conditions which involve attribute deleted by the evolution $S \to S'$. For each $c \in \{\bar{c}_1, \ldots, \bar{c}_h\}$ we construct the modified condition c' such that $s'_Q \in Inst(S', c')$ and $s'_Q = s_Q$, where $s_Q \in Inst(S, c)$.

In order to construct c', we verify the existence of some RFDs ψ of the form:

$$X_{\Phi_1} \to A_{EQ} \tag{7}$$

where A represents the attribute instantiated in c, and EQ is the equality constraint.

In this way, according to the type of each query/view condition c corrupting Q, it is possible to transform c into c' through the following general formula:

$$c' = \bigvee_{i=1}^{k} (X_1 = x_{1i} \wedge X_2 = x_{2i} \wedge \cdots \wedge X_n = x_{ni}) \tag{8}$$

$\forall X_j \in X$, with $j = 1, \ldots, n$, and $\forall t_i$ such that $t_i[A]\theta y$, where y is a *constant* or an *attribute*, and θ represents one of the possible operators that can be used in c.

In this way, each query/view Q including the condition c can be rewritten into Q' by replacing c with c'.

Example 5. Let us consider the database instance in Table 1, and the view Q in (3).

After the evolution $S \to S'$ defined in (2), if the following RFD φ $\{\mathsf{Role}_\approx, \mathsf{Experience}_\approx\} \to \mathsf{Level}_{EQ}$ holds, then we can automatically transform Q in Q' in the following way:

CREATE VIEW getDoctorByLevel AS
SELECT Name, Specialization
FROM Doctor   (9)
WHERE (Role = *"Specialized"* AND Experience = 2 *years*)
       OR (Role = *"SpecializedDr."* AND Experience = 3 *years*)

producing the same result instance of the original view.

The transformation rule defined in (8) can be specialized based on the condition used in the query/view (see Table 2).

In general, the proposed solution can be applied whenever an RFD is included within the previously defined general form. Moreover, the whole methodology

Table 1. A portion of a doctor database instance.

idDoctor	Name	Specialization	Role	Experience	Salary	Level	Tax
1	George Johnson	Neurology	Junior Dr.	2 years	$118,000	E	$27,140
2	Joe House	Cardiology	Head Physician	10 years	$314,000	B	$94,200
3	Derek Williams	Pediatrician	Specialized Dr.	2 years	$156,000	D	$39,000
4	Henry Jones	Neurology	Specialized	3 years	$158,000	D	$39,500
5	Victor Sanchez	Radiology	Senior Surgeon	5 years	$225,000	C	$63,000
...							

Table 2. The transformation rule (8) specialized for each type of query condition.

Condition Type	Formula
attribute = constant $(A = {}'a')$	$\bigvee_i (X_1 = x_{1i} \wedge X_2 = x_{2i} \wedge \cdots \wedge X_n = x_{ni})$ $\forall t_i$ s.t. $t_i[A] = {}'a'$, and $\forall X_j \in X$ with $j = 1, \ldots, n$
attribute = attribute $(A = B)$	$\bigvee_i (X_1 = x_{1i} \wedge X_2 = x_{2i} \wedge \cdots \wedge X_n = x_{ni})$ $\forall t_i$ s.t. $t_i[A] = t_i[B]$, and $\forall X_j \in X$ with $j = 1, \ldots, n$
attribute θ constant $(A\ \theta\ 'a')$	$\bigvee_i (X_1 = x_{1i} \wedge X_2 = x_{2i} \wedge \cdots \wedge X_n = x_{ni})$ $\forall t_i$ s.t. $t_i[A]\ \theta\ 'a'$, and $\forall X_j \in X$ with $j = 1, \ldots, n$
attribute θ attribute $(A\ \theta\ B)$	$\bigvee_i (X_1 = x_{1i} \wedge X_2 = x_{2i} \wedge \cdots \wedge X_n = x_{ni})$ $\forall t_i$ s.t. $t_i[A]\ \theta\ t_i[B]$, and $\forall X_j \in X$ with $j = 1, \ldots, n$
attribute *in* Query $(A\ in\ Q)$	$\bigvee_i (X_1 = x_{1i} \wedge X_2 = x_{2i} \wedge \cdots \wedge X_n = x_{ni})$ $\forall t_i$ s.t. $t_i[A] = (v_1 \vee, \ldots, \vee v_k)\ \forall v_z \in \{v_1, \ldots, v_k\}$ result value of Q, and $\forall X_j \in X$ with $j = 1, \ldots, n$
attribute *not in* Query $(A\ not\ in\ Q)$	$\bigvee_i (X_1 = x_{1i} \wedge X_2 = x_{2i} \wedge \cdots \wedge X_n = x_{ni})$ $\forall t_i$ s.t. $t_i[A] <> (v_1 \wedge, \ldots, \wedge v_k)\ \forall v_z \in \{v_1, \ldots, v_k\}$ result value of Q, and $\forall X_j \in X$ with $j = 1, \ldots, n$
attribute θ *any* Query $(A\ \theta\ any\ Q)$	$\bigvee_i (X_1 = x_{1i} \wedge X_2 = x_{2i} \wedge \cdots \wedge X_n = x_{ni})$ $\forall t_i$ s.t. $t_i[A]\ \theta\ (v_1 \vee, \ldots, \vee v_k)\ \forall v_z \in \{v_1, \ldots, v_k\}$ result value of Q, and $\forall X_j \in X$ with $j = 1, \ldots, n$
attribute θ *all* Query $(A\ \theta\ all\ Q)$	$\bigvee_i (X_1 = x_{1i} \wedge X_2 = x_{2i} \wedge \cdots \wedge X_n = x_{ni})$ $\forall t_i$ s.t. $t_i[A]\ \theta\ (v_1 \wedge, \ldots, \wedge v_k)\ \forall v_z \in \{v_1, \ldots, v_k\}$ result value of Q, and $\forall X_j \in X$ with $j = 1, \ldots, n$
attribute *between* $'a_1'$ *and* $'a_2'$ $(A\ between\ 'a_1'\ and\ 'a_2')$	$\bigvee_i (X_1 = x_{1i} \wedge X_2 = x_{2i} \wedge \cdots \wedge X_n = x_{ni})$ $\forall t_i$ s.t. $t_i[A] \geq 'a_1' \wedge t_i[A] \leq 'a_2'$, and $\forall X_j \in X$ with $j = 1, \ldots, n$

can be effectively used in practice, due to the fact that many RFDs can be automatically extracted from data [6,21].

Finally, it is fair to note that it is possible to find more than one correct rewritings of queries/views corrupted by an evolution. This is also another issue related to the general QVS problem. To this end, in order to choose the best candidate synchronization, we will select the query/view rewrite maximizing the length of c' in term of $\vee$ sentences (i.e. c' such that the value of k is minimum). In this way, more general rewritings of queries/views will be used.

4.1 Query/View Synchronization with Approximate Results

In case there is no RFD suitable to accomplish the synchronization process, it would be desirable to rewrite queries/views in a way to produce results approximating those of the original query/view version. This can be done due to approximate nature of RFDs.

More formally, in order to transform c into c', so that $s'_Q \in Inst(S', c')$ and $s'_Q \approx s_Q$, we verify the existence of some RFD φ of the form:

$$X_{\Phi_1} \to A_{\Phi_2} \qquad (10)$$

where A represents an attribute instantiated in c that has been deleted during the schema evolution, with Φ_2 a similarity constraint.

Example 6. Let us consider again the database instance in Table 1, and the view Q in (3). After the evolution $S \to S'$ defined in (2), if the RFD φ Experience$_\approx \to$ Level$_\approx$ holds, then we can automatically transform Q in Q' in the following way:

> CREATE VIEW getDoctorByLevel AS
> SELECT Name, Specialization
> FROM Doctor $\qquad (11)$
> WHERE Experience = 2 *years* OR Experience = 3 *years*

producing a result that is not equal, but similar to the one of the original view, as shown in Table 3. It is worth to notice that a level of acceptance for the result approximations can be managed by domain experts, by restricting similarity constraint thresholds of valid RFDs.

Table 3. An approximate version of the view getDoctorByLevel.

idDoctor	Name	Specialization
1	George Johnson	Neurology
3	Derek Williams	Pediatrician
4	Henry Jones	Neurology

5 Evaluation

We made a prototype implementation of the proposed methodology in Java. When synchronizing queries affected by an evolution, the prototype considers all the RFDs that have one of the removed attributes on the RHS and none of the removed attributes on the LHS. The choice of the RFD to be used for the synchronization follows this rule:

Choice the RFD *with an LHS producing the minimum result set of a query constructed as*   SELECT LHS_Attributes FROM r WHERE affected_Condition

where affected_Condition represents the condition instantiating one of the attributes removed during the evolution process, which corresponds to the RHS of the RFD used for the synchronization.

We evaluated the proposed methodology on two different datasets, the *Bridges* and the *Echocardiogram* datasets, drawn from the UC Irvine Machine Learning repository [4]. Statistics on the characteristics of the considered datasets are reported in Table 4.

In order to evaluate the proposed methodology, we defined several queries on each dataset, and randomly removed three attributes from each of them. In particular, there were two queries affected by the attribute removal for the *Bridges* and five for the *Echocardiogram* dataset. We observed the number of attributes on the LHS belonging to the RFD selected for the synchronization, and the growth of the query conditions. In general, the execution time to accomplish the complete synchronization was very small, varying in the range $[0.2, 0.5]$ seconds. Finally, we performed another experimental session, by forcing the selection of RFDs considering approximate matches on the RHS. In this case, we also analyzed the number of false positives introduced during the synchronization process.

Table 4. Statistics on the datasets considered in the evaluation.

Datasets	# Columns	# Rows	# FD	Size [KB]
Bridges	13	108	142	6
Echocardiogram	13	132	538	6

Evaluation results are shown in Fig. 1. We can notice that the growth of query conditions (number of conditions into the synchronized query) does not depend on the LHS cardinality ($|$LHS$|$, number of attributes) of the selected RFD, as shown in Fig. 1(a). However, since the conditions are automatically processed, their size does not affect the human effort in the synchronization process. Moreover, in the case of rewritings with approximate results (Fig. 1(b)), we notice that by using RFDs with similarity constraints on the RHS yields a reduced growth of the query condition, and the LHS cardinalities does not increase. However, as expected, in some cases several false positives are generated. Finally, it is worth to notice that a synchronization process accomplished through an RFD with a higher LHS cardinality increases the risk of possible future synchronizations, since it is highest the probability that one of them could be involved in future schema evolutions. This represents the main limitation of the proposed approach.

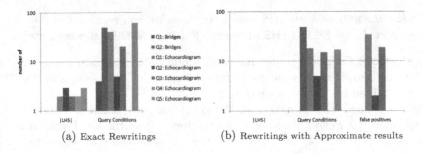

(a) Exact Rewritings   (b) Rewritings with Approximate results

Fig. 1. Evaluation results of the proposed methodology.

6   Related Work

QVS approaches and tools defined in the literature are based on one of the three schema evolution strategies defined in Sect. 2.1: *operation-based, mapping-based,* and *hybrid*. Moreover, the QVS problem raises several issues, ranging from the automation level of the synchronization process to the management of information loss [9]. QVS approaches should provide solutions for all the issues related to the general problem. However, in some cases, solutions useful for some issues might be not appropriate for other cases.

In general, the QVS process permits the synchronization of all the queries/views affected by the evolution of a given schema. In the previous sections we discussed the fact that corrupted queries/views cannot be always synchronized. To this end, the concept of policies guiding the synchronization process can be used [25]. In particular, three types of policies have been defined in the literature: (i) *propagate*, (ii) *block*, and (iii) *prompt*, which prescribe how to handle the portions of the view definitions affected by the schema modification: *propagate* prescribes to apply changes and synchronize Q, *block* forbids changes, and *prompt* prescribes to ask the DBA for the action to be undertaken [11]. Although the latter might appear the most suitable policy, it should not be abused in order to keep the automation level of the QVS process sufficiently high.

Another solution to the QVS problem relies on the concept of *View Evolution Parameters (VEPs)* [29], which enables the possibility to define how to handle the single components of a view during the synchronization process, by specifying a priori whether the component (e.g. an attribute) is replaceable, or whether it is mandatory. In addition, the *View Extent*[2] *Parameter (VE)* [29] associated to a view Q specifies a condition on the extent of a view Q' in order for Q' to be considered an acceptable synchronization of Q. In other words, the VE parameter $\phi \in \{\equiv, \subseteq, \supseteq, \approx\}$, specifies a priori whether the extent of Q' must be equivalent ($\equiv$), be included ($\subseteq$), include ($\supseteq$), or approximate ($\approx$) the extent of Q, in order for Q' to be considered a legal rewrite of Q. In other words, the

---

[2] The view extent is the usually adopted term indicating the result-set of a view statement, i.e. the materialized view.

VE parameter establishes a relationship that must hold between the projections of Q' and Q on their common attributes.

Fault tolerance is the idea underlying the approach [27], in which attempts are made to recover data from the source schema, even with some errors. This has been done by means of default mappings, a formalism based on default logic, which is suitable to express rules allowing exceptions. This solution permits to define approximate versions of queries/views invalidated from the evolution.

Finally, a cooperative approach to querying has been used in the context of logic-based data integration of heterogeneous databases [18]. In particular, this solution exploits heuristics to produce approximate answers, which are submitted to the user or to the DBA for approval. Such a process exploits a dialogue for information seeking, in which participants (user and system) aim at finding an adequate mapping between the query and the modified schema.

All of approaches previously mentioned aim to provide a general solution to the QVS problem. However, they do not propose a specific solution w.r.t. the fact that queries/views might be corrupted even when their result instances can be recovered upon the evolution, and the latter only affects the query/view selection conditions. The proposed methodology aims to isolate this kind of situations, by providing possible rewritings of queries/views based on semantic correlations among data, expressed in terms of RFDs. For this reason, the proposed solution not only can be embedded within more generic solutions, but should also permit to increase the automation level of the QVS process.

7   Conclusion and Future Work

We have proposed a new methodology to automatically rewrite queries/views corrupted as a consequence of schema evolutions causing loss of information on which they were defined. The transformation procedure exploits the semantic correlations among data provided by RFDs, in oder to derive a general transformation formula for the QVS problem. It can be included in every approach or tool aiming to solve the general QVS problem. Moreover, since RFDs can be automatically extracted [6,21], the proposed methodology allows to improve the automation level of the synchronization process.

In the future, we would like to investigate the effects of instance evolutions of previous query/view synchronization processes. In particular, the addition or deletion of some data might change the set of RFDs holding on the dataset, possibly altering parts of previous synchronization processes. Thus, we would like to investigate how to make the query/view synchronization methodology incremental w.r.t. instance updates. To this end, it might be useful to investigate the possibility of exploiting RFDs that relax on the extent [7], that is, RFDs holding only on a subset of the database.

References

1. Bernstein, P.A., Melnik, S.: Model management 2.0: manipulating richer mappings. In: Proceedings of the ACM SIGMOD International Conference on Management of Data (COMAD), pp. 1–12. ACM (2007)
2. Bernstein, P.A., Rahm, E.: Data warehouse scenarios for model management. In: Laender, A.H.F., Liddle, S.W., Storey, V.C. (eds.) ER 2000. LNCS, vol. 1920, pp. 1–15. Springer, Heidelberg (2000). https://doi.org/10.1007/3-540-45393-8_1
3. Bertino, E.: A view mechanism for object-oriented databases. In: Pirotte, A., Delobel, C., Gottlob, G. (eds.) EDBT 1992. LNCS, vol. 580, pp. 136–151. Springer, Heidelberg (1992). https://doi.org/10.1007/BFb0032428
4. Blake, C.L., Merz, C.J.: UCI repository of machine learning databases. http://archive.ics.uci.edu/ml/index.php. Accessed 3 Mar 2018
5. Bohannon, P., Fan, W., Geerts, F., Jia, X., Kementsietsidis, A.: Conditional functional dependencies for data cleaning. In: 2007 IEEE 23rd International Conference on Data Engineering, pp. 746–755. IEEE (2007)
6. Caruccio, L., Deufemia, V., Polese, G.: On the discovery of relaxed functional dependencies. In: Proceedings of the 20th International Database Engineering & Applications Symposium (IDEAS), pp. 53–61 (2016)
7. Caruccio, L., Deufemia, V., Polese, G.: Relaxed functional dependencies - a survey of approaches. IEEE Trans. Knowl. Data Eng. **28**(1), 147–165 (2016)
8. Caruccio, L., Deufemia, V., Polese, G.: Evolutionary mining of relaxed dependencies from big data collections. In: Proceedings of the 7th International Conference on Web Intelligence, Mining and Semantics, WIMS 2017, p. 5 (2017)
9. Caruccio, L., Polese, G., Tortora, G.: Synchronization of queries and views upon schema evolutions: a survey. ACM Trans. Database Syst. (TODS) **41**(2), 9 (2016)
10. Cohen, W., Ravikumar, P., Fienberg, S.: A comparison of string metrics for matching names and records. In: KDD Workshop on Data Cleaning and Object Consolidation, vol. 3, pp. 73–78 (2003)
11. Curino, C.A., Moon, H.J., Zaniolo, C.: Graceful database schema evolution: the prism workbench. Proc. VLDB Endow. **1**(1), 761–772 (2008)
12. Curino, C.A., Tanca, L., Moon, H.J., Zaniolo, C.: Schema evolution in wikipedia: toward a web information system benchmark. In: Proceedings of the 10th International Conference on Enterprise Information Systems (ICEIS), pp. 323–332. Citeseer (2008)
13. Elmagarmid, A.K., Ipeirotis, P.G., Verykios, V.S.: Duplicate record detection: a survey. IEEE Trans. Knowl. Data Eng. **19**(1), 1–16 (2007)
14. Golfarelli, M., Lechtenbörger, J., Rizzi, S., Vossen, G.: Schema versioning in data warehouses: enabling cross-version querying via schema augmentation. Data Knowl. Eng. **59**(2), 435–459 (2006)
15. Hick, J.M., Hainaut, J.L.: Database application evolution: a transformational approach. Data Knowl. Eng. **59**(3), 534–558 (2006)
16. Huhtala, Y., Kärkkäinen, J., Porkka, P., Toivonen, H.: TANE: an efficient algorithm for discovering functional and approximate dependencies. Comput. J. **42**(2), 100–111 (1999)
17. Hull, R.: Relative information capacity of simple relational database schemata. SIAM J. Comput. **15**(3), 856–886 (1986)
18. Lakshmanan, L.V.S., Sadri, F., Subramanian, I.N.: On the logical foundations of schema integration and evolution in heterogeneous database systems. In: Ceri, S., Tanaka, K., Tsur, S. (eds.) DOOD 1993. LNCS, vol. 760, pp. 81–100. Springer, Heidelberg (1993). https://doi.org/10.1007/3-540-57530-8_6

19. Lee, A.J., Nica, A., Rundensteiner, E.A.: The EVE approach: view synchronization in dynamic distributed environments. IEEE Trans. Knowl. Data Eng. **14**(5), 931–954 (2002)
20. Lerner, B.S.: A model for compound type changes encountered in schema evolution. ACM Trans. Database Syst. (TODS) **25**(1), 83–127 (2000)
21. Liu, J., Li, J., Liu, C., Chen, Y.: Discover dependencies from data - a review. IEEE Trans. Knowl. Data Eng. **24**(2), 251–264 (2012)
22. Melnik, S.: Generic Model Management: Concepts and Algorithms. LNCS, vol. 2967. Springer, Heidelberg (2004). https://doi.org/10.1007/b97859
23. Noy, N.F., Klein, M.: Ontology evolution: not the same as schema evolution. Knowl. Inf. Syst. **6**(4), 428–440 (2004)
24. Oueslati, W., Akaichi, J.: A survey on data warehouse evolution. Int. J. Database Manag. Syst. (IJDMS) **2**(4), 11–24 (2010)
25. Papastefanatos, G., Vassiliadis, P., Simitsis, A., Vassiliou, Y.: Policy-regulated management of ETL evolution. In: Spaccapietra, S., Zimányi, E., Song, I.-Y. (eds.) Journal on Data Semantics XIII. LNCS, vol. 5530, pp. 147–177. Springer, Heidelberg (2009). https://doi.org/10.1007/978-3-642-03098-7_6
26. Polese, G., Vacca, M.: A dialogue-based model for the query synchronization problem. In: IEEE 5th International Conference on Intelligent Computer Communication and Processing (ICCP) (2009)
27. Polese, G., Vacca, M.: Notes on view synchronization using default logic. In: Proceedings of 17th Italian Symposium on Advanced Database Systems (SEBD), pp. 253–260 (2009)
28. Poulovassilis, A., McBrien, P.: A general formal framework for schema transformation. Data Knowl. Eng. **28**(1), 47–71 (1998)
29. Rundensteiner, E.A., Lee, A.J., Nica, A.: On preserving views in evolving environments. In: Knowledge Representation meets DataBases (KRDB), CEUR Workshop Proceedings, vol. 8, pp. 13.11–13.11 (1997)

Knowledge Graph Embedding via Relation Paths and Dynamic Mapping Matrix

Shengwu Xiong[1,2], Weitao Huang[1], and Pengfei Duan[1,2(✉)]

[1] School of Computer Science and Technology, Wuhan University of Technology,
Wuhan 430070, China
{swxiong,duanpf}@whut.edu.cn
[2] Hubei Key Laboratory of Transportation Internet of Things, Wuhan 430070, China

Abstract. Knowledge graph embedding aims to embed both entities and relations into a low-dimensional space. Most existing methods of representation learning consider direct relations and some of them consider multiple-step relation paths. Although those methods achieve state-of-the-art performance, they are far from complete. In this paper, a noval path-augmented TransD (PTransD) model is proposed to improve the accuracy of knowledge graph embedding. This model uses two vectors to represent entities and relations. One of them represents the meaning of a(n) entity (relation), the other one is used to construct the dynamic mapping matrix. The PTransD model considers relation paths as translation between entities for representation learning. Experimental results on public dataset show that PTransD achieves significant and consistent improvements on knowledge graph completion.

Keywords: Representation learning · Knowledge graph
Dynamic mapping matrix · Relation path

1 Introduction

Knowledge graphs encode structured information of entities and their rich relations. Although typical knowledge graphs, such as Freebase [18] (Bollacker et al. 2008), WordNet [19] (Miller 1995), Yago [20], usually are large in size and contain thousands of relation types, millions of entities and billions of triples, they are usually far from complete. Knowledge graphs usually contain large-scale structural information by using the form of triples (head entity, relation, tail entity).

Knowledge graph completion is similar to link prediction in social network analysis [11], but traditional approach of link prediction is not capable for knowledge graph completion. Inspired by the translation invariant phenomenon in word vector space [10], Bordes et al. propose the TransE [3] model, which achieves state-of-the-art prediction performance. TransE learns low-dimensional embeddings for each entity and relation. These vector embeddings are donated by the

C. Woo et al. (Eds.): ER 2018 Workshops, LNCS 11158, pp. 106–118, 2018.
https://doi.org/10.1007/978-3-030-01391-2_18

same letter in boldface. The basic idea of TransE is that every relation is regarded as translation between head entity and tail entity in the embedding space. For example, for a triplet (h, r, t), the embedding h is close to the embedding t by adding the embedding r, that is $h + r \approx t$. TransE has outstanding performance in 1-to-1 relations, but it's not good at dealing with 1-to-N, N-to-1 and N-to-N.

In order to overcome the problems of TransE [3] in modeling 1-to-N, N-to-1, N-to-N relations, Researchers propose many methods such as TransH [13], TransR [7] to extend the model and achieve excellent performance. These methods are proposed to assign an entity with different representations when involved in various relations. TransH enables an entity to have distributed representations when involved in different relations. However, TransE and TransH both assume embeddings of entities and relations being in the same space R^k is the semantic space dimension, so it's hard for TransE, TransH to distinguish among the entities which are on the many sides. To deal with the above problem, TransR is proposed to model entities and relations in distinct space, i.e., entity space and multiple relation spaces [4], and performs translation in the corresponding relation space. TransR achieves significant improvements compared to base-lines including TransE and TransH.

In both TransH [13] and TransR [7], all types of entities share the same mapping vectors/matrices. However, different types of entities have different attributes and functions, it is insufficient to let them share the same transform parameters of a relation [14]. And for a given relation, similar entities should have similar mapping matrices and otherwise for dissimilar entities. Furthermore, the mapping process is a transaction between entities and relations that both have various types. TransD [9] names two vectors to represent symbol object (entities and relations). The first one is used to construct mapping matrices, the other one captures the meaning of entity (relation). TransD has less complexity and more flexibility than TransR/CTransR. When learning embeddings of named symbol objects (entities or relations), TransD considers the diversity of them both.

One shortcoming of the above approaches is that they only consider the direct relation between entities. In a knowledge graph, it's known that there are also substantial multiple-step relation paths between entities indicating their semantic relationships. For example, the relation path $h \overset{son}{\rightarrow} e_1 \overset{son}{\rightarrow} t$ indicates the relation grandson between h and t, i.e., $(h, grandson, t)$. Fortunately, the problem is already alleviated effectively by using information of relation paths. PTransE (Guu et al., 2015) [1] tries to extend TransE to model relation paths for representation learning of knowledge graphs and achieves significant and consistent improvements on knowledge graph completion and relation extraction from text.

In this paper, we aim at enhancing TransD [9] with information of relation paths and finally propose a new variant named PTransD. The model is evaluated on a typical large-scale knowledge graph Freebase (Bollacker et al., 2008). In this paper, we adopt a dataset extracted from Freebase, i.e., FB15K. Experimental results show that modeling relation paths provide a good supplement for representation learning of knowledge graphs.

2 Our Method

In Sect. 1, we introduce the advantages and disadvantages of existing methods, including TransE [3], TansH [13] and TransR/CTransR [7]. In this section, PTransD model will be introduced in detail.

2.1 Based Method: TransD

Model TransR/CTransR [7] segments triples of a specific relation r into several groups and learns a vector representation for each group. However, entities also have various properties. In both TransH [13] and TransR/CTransR [7], all types of entities share the same mapping vectors/matrices. It is insufficient to show different attributes and functions of different types of entities. And for a given relation similar entities should have similar mapping matrices and otherwise for dissimilar entities. TransD [9] considers different types of both entities and relations, to encode knowledge graphs into embedding vectors via dynamic mapping matrices produced by projection vectors.

In TransD [9] model, each named symbol object (entities and relations) is represented by two vectors. The first one captures the meaning of entities (relations), the other one is used to construct mapping matrices. For example, given a triplet (h, r, t), its vector are h, h_p, r, r_p, t, t_p, where subscript p marks the projection vectors, $h, h_p, t, t_p \in R^n$ and $r, r_p \in R^m$. For each triplet (h, r, t), TransD [9] set two mapping matrices $M_{rh}, M_{rt} \in R^{m \times n}$ to project entities from entity space to relation space. They are defined as fellow:

$$M_{rh} = r_p h_p^\top + I^{m \times n} \tag{1}$$

$$M_{rt} = r_p t_p^\top + I^{m \times n} \tag{2}$$

where $I^{m \times n}$ is an identity matrix of $m \times n$, $h_p^\top$ and $t_p^\top$ are the transpose matrices of h_p and t_p. The mapping matrices M_{rh} and M_{rt} are determined by both entities and relations, and this kind of operation makes the two projection vectors interact sufficiently because each element of them can meet every entry comes from another vector. With the mapping matrices, the projection vectors were defined as follows:

$$h_\perp = M_{rh}h, t_\perp = M_{rt}t \tag{3}$$

The score function is:

$$f_r(h, t) = -||h_\perp + r - t_\perp||_2^2 \tag{4}$$

In experiments, TransD [9] enforce constrains as $||h||_2 \leqslant 1, ||t||_2 \leqslant 1, ||r||_2 \leqslant 1, ||h_\perp||_2 \leqslant 1, ||t_\perp||_2 \leqslant 1$. To train the model, TransD [9] generate negative samples and use the margin-based rank loss. The overall function of the TransD model is:

$$L = \sum_{(h,r,t) \in S} \sum_{(h',r',t') \in S^-} [\gamma + f_r(h', r', t') - f_r(h, r, t)]_+ \tag{5}$$

where $[x]_+ = max(0, x)$, S and S^- denote golden triples and negative triples. S^- is constructed as follows:

$$S^- = (h', r, t) \cup (h, r', t) \cup (h, r, t') \tag{6}$$

And γ is the margin separating golden triples and negative triples. (h, r, t) and (h', r', t') denote a golden triplet and a corresponding negative triplet, respectively. The process of minimizing the above objective is carried out with stochastic gradient descent (SGD) [12]. In order to speed up the entity and relation embeddings with the results of TransE [3] and initiate all the transfer matrices with identity matrices.

2.2 Path-Based TransD: PTransD

Compared to TransE [3], PTransE [1] achieves consistent and significant improvements on knowledge graph completion and relation extraction from text. Obviously, it is effectively to improve the ability of knowledge representation by considering multiple-step relation paths. Therefore, we propose a path-based TransD [9] model (denoted as PTransD) (Fig. 1).

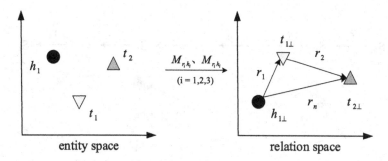

Fig. 1. An illustration of the PTransD model. $M_{r_i h_i}$ and $M_{r_i t_i}$ are mapping matrices of i-th ($i = 1, 2, ..., $ n) head entity h_i and tail entity t_i, PTransD set two mapping matrices to project entities from entity space to relation space.

Multi-step relation paths between entities indicating their semantic relationships. The relation paths reflect complicated inference patterns among in knowledge graphs. However, on the one hand, not all relation paths are meaningful and reliable for learning. For example, a typical relation path $h \xrightarrow{Friend} e_1 \xrightarrow{profession} t$, but actually it does not indicate any semantic relationship between h and t. In experiments, it will bring negative influence. Therefore, in the model we use path-constraint resource allocation algorithm to measure the reliability of relation paths and we select the reliable relation paths for representation learning. To evaluate the reliability of a path, we use a path-constraint resource algorithm (PCRA) just like PTransE [1] model. The basic idea is that assuming a certain amount of resource is associated with the head entity h, and will flow following the given path p. We use the resource amount that eventually flows to the tail entity t to measure the reliability of the path p.

Formally, for a path triple (h, r, t), we compute the resource amount flowing from h to t given the path $p = (r_1, ..., r_l)$. Starting from h and following the relation path p, the following path can be represented by $S_0 \xrightarrow{r_1} S_1 \xrightarrow{r_2} ... \xrightarrow{r_l} S_l$, where $h = S_0$ and $t \in S_l$. For any entity m of S_i (i.e., $m \in S_i$), its direct predecessors along relation r_i in S_{i-1} is donated as $S_{i-1}(\cdot, m)$, The resource flowing to m is defined as (Fig. 2)

$$R_p(m) = \sum_{n \in S_{i-1}(\cdot, m)} \frac{1}{|S_i(n, \cdot)|} R_p(n) \tag{7}$$

where $S_i(n, \cdot)$ is the direct successors of $n \in S_{i-1}$ following the relation r_i, and $R_p(n)$ is the resource obtained from the entity n. For each relation path p, we set the initial resource in h as $R_p(h) = 1$. By performing resource allocation recursively from h through the path p, the tail entity t eventually obtains the resource $R_p(t)$ which indicates how much information of the head entity h can be well translated. We use $R_p(t)$ to measure the reliability of the path p given (h, t), i.e., $R(p|h, t) = R_p(t)$. It can be simply described as Fig. 3.

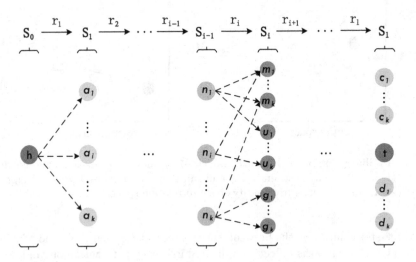

Fig. 2. An illustration of PCRA [17].

On the other hand, it is necessary to represent relation paths in the low-dimensional space. It is obvious that the semantic meaning of a relation path depends on all relations in this path. Given a relation path $p = (r_1, ..., r_n)$, we will define and learn a binary operation function ($\circ$) to obtain the path embedding p by recursively composing multiple relations, i.e. $p = (r_1 \circ ... \circ r_n)$. We add the

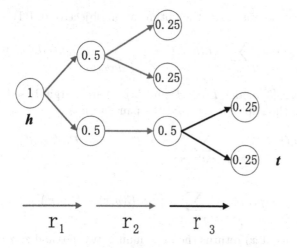

Fig. 3. An simple illustration of PCRA.

embedding of these primitive relations, given by

$$p = r_1 + r_2 + ... + r_n \tag{8}$$

For that in PTransE [1] model, addition operation outperforms other composition operations including multiplication operation and recurrent neural network.

In PTransD model, for each triple (h, r, t), we define the energy function as

$$G(h, r, t) = E(h, r, t) + E(h, P, t) \tag{9}$$

where $E(h, r, t)$ models correlations between relations and entities with direct relations. $E(h, P, t)$ models the inference correlations between relations with multiple-step relation path triples, which is defined as

$$E(h, P, t) = \frac{1}{Z} \sum_{p \in P(h,t)} R(p|h, t) E(h, p, t) \tag{10}$$

where $R(p|h, t)$ indicates the reliability of the relation path p given the entity pair (h, t), $Z = \sum_{p \in P(h,t)} R(p|h, t)$ is a normalization factor, and $E(h, p, t)$ is the energy function of triple (h, p, t).

For a multiple-step relation path triple (h, r, t), we could have followed TransE [3] and define the energy function as $E(h, p, t) = ||h + p - t||$. However since we have minimized $||h + r - t||$ with the direct relation triple (h, r, t) to make sure $r \approx t - h$, we may directly define the energy function of (h, p, t) as

$$E(h, p, t) = ||p - (t - h)|| = ||p - r|| = E(p, r) \tag{11}$$

which is expected to be a low score where the multiple-step relation path p is consistent with the direct relation r, and high otherwise, without using entity embeddings.

In conclusion, we formalize the optimization objective of PTransD as

$$L(S) = \sum_{(h,r,t)\in S} [L(h,r,t) + \frac{1}{Z}\sum_{p\in P(h,t)} R(p|h,t)L(p,r)] \tag{12}$$

Following TransE [3] model, $L(h,r,t)$ and $L(p,r)$ are margin-based loss functions with respect to the triple (h,r,t) and the pair (p,r):

$$L(h,r,t) = \sum_{(h',r',t')\in S^-} [\gamma + E(h,r,t) - E(h',r',t')]_+ \tag{13}$$

and

$$L(p,r) = \sum_{(h',r',t')\in s^-} [\gamma + E(p,r) - E(p,r')]_+ \tag{14}$$

where $[x]_+ = max(0,x)$ returns the maximum between 0 and x, γ is the margin, S is the set of valid triples existing in a knowledge graph and S^- is the set of invalid triples. The objective will favor lower scores for valid triples as compared with invalid triples. The invalid triples set with respect to (h,r,t) is defined as $S^- = (h',r,t) \cup (h,r',t) \cup (h,r,t')$. That is, the set of invalid triples is compose of the original valid triple (h,r,t) with one of three components replaced.

For a entity pair (h,t), we consider its direct relation and multiple-step relation paths. The score function of PTransD is defined as

$$G(h,r,t) = ||h_\perp + r - t_\perp||_{L_1/L_2}$$
$$+ \frac{1}{Z}\sum_{p\in P(h,t)} R(p|h,t)||p-r||_{L_1/L_2} \tag{15}$$

and the score function is further defined as

$$G(h,r,t) = E(h,r,t) + E(h,P,t)$$
$$= ||(r_p h_p^\top + I^{m*n})h + r - (r_p t_p^\top + I^{m*n})t||_{L_1/L_2}$$
$$= \frac{1}{Z}\sum_{p\in P(h,t)} R(p|h,t)||p-r||_{L_1/L_2} \tag{16}$$

where h_p, r_p, t_p are projection vectors which are used to construct dynamic mapping matrices. $A^\top$ is the transport matrix of A. We use I^{m*n} to denote the identity matrix of size m × n. $R(p|h,t)$ indicates the reliability of the relation path p given the entity pair (h,t), $\sum_{p\in P(h,t)} R(p|h,t)$ is a normalization factor.

2.3   Training Method and Implementation Details

Existing knowledge graphs only contain correct triples. We get the set of invalid triples by replacing one of three components of original valid triple (h,r,t). When corrupting the triple, we follow (Wang et al. 2014) [13] and assign different probabilities for head/tail entity replacement. For those 1-to-N, N-to-1 and

N-to-N relations, by giving more chance to replace "one" side, the chance of generating false-negative instances will be reduced. In experiments, we initialize entity vectors and relation vectors by the training result of TransE [3] to avoid overfitting.

For optimization, we use stochastic gradient descent (SGD) to minimize the loss function. We also enforce constraints on the norms of the embeddings h, r, t. That is, we set $||h||_2 \leqslant 1, ||t||_2 \leqslant 1, ||r||_2 \leqslant 1, ||h_p||_2 \leqslant 1, ||r_p||_2 \leqslant 1, ||t_p||_2 \leqslant 1, ||M_{rh}h||_2 \leqslant 1, ||r_\perp||_2 \leqslant 1, ||M_{rt}t||_2 \leqslant 1$. The head entity vector and the tail entity vector are projected from the entity semantic space R^n to the relation semantic space R^m through the mapping matrices M_{rh} and M_{rt}. It obvious that projected vectors rely on entities and relations.

In addition, there are some implementation details that will influence the performance of representation learning. In this paper, we add reverse relations for each relation in knowledge graph and We just consider 2-step relation paths existed in knowledge graph.

3   Experiments and Analysis

3.1   Data Set

Freebase is a huge and growing knowledge graph of a large number of the world facts, there are currently around 1, 2 billion triples and more than 80 million entities. In this paper, we adopt the subset of Freebase: FB15K (Bordes et al. 2014). Table 1 lists statistics and triple types of the FB15K dataset.

Table 1. Statistics of dataset FB15K.

Dataset	Ent	Rel	Train	Test	Valid	1-to-1	1-to-N	N-to-1	N-to-N
FB15K	14,951	1,345	483,142	59,071	50,000	26.2 (%)	22.7 (%)	29.3 (%)	22.8 (%)

In experiments, we use approach in TransH (Wang et al. 2014) [13] to replace the head and tail entity when corrupting the triple, which depends on the mapping property of the relations.

For comparison, we select all methods in (Lin et al., 2015) [1] as our baselines and use their reported results directly since the evaluation dataset is identical. The learning process of PTransD is carried out using stochastic gradient descent (SGD). To avoid overfitting, we initialize entity and relation embeddings with results of TransE [3], and initialize relation matrices as identity matrices.

3.2   Link Prediction

Link prediction [8] is to predict the missing h or t for a golden triple (h, r, t). In this task, we remove the head entity or tail entity and then replace it with all the entities of the dictionary in turn for each triple in test set. We first compute

Table 2. Evaluation results on entity prediction.

Metric	Mean Rank		Hits@10(%)	
	Raw	Filter	Raw	Filter
RESCAL [15]	828	683	28.4	44.1
SE [6]	273	162	28.8	39.8
SME(linear) [2]	274	154	30.7	40.8
SME(bilinear) [2]	284	158	31.3	41.3
LFM [16]	283	164	26.0	33.1
TransE [3]	243	125	34.9	47.1
TransH [13]	212	87	45.7	64.4
TransR [7]	198	77	48.2	68.7
TransE(our)	235	136	51.6	71.4
TransD [9]	221	78	51.9	77.4
PTransE(ADD,2-step) [1]	200	54	51.8	83.4
PTransE(MUL,2-step) [1]	216	67	47.4	77.7
PTransE(RNN,2-step) [1]	242	92	50.6	82.2
PTransE(ADD,3-step) [1]	207	58	51.4	84.6
PTransD(ADD,2-step)	**147.32**	**21.04**	**54.97**	**92.58**

the scores of corrupted triples and then rank them in descending order, the rank of correct entity is finally store.

For evaluation, we use two ranking measures: the mean of those predicted ranks and the Hits@10, i.e. the proportion of correct ranked in the top 10. We call the original setting "Raw". Consider that a corrupted triple may also exist in knowledge graphs, the corrupted triple should be regarded as a correct triple. To avoid such a misleading behavior, we propose to remove from the list of corrupted triples all the triples that appear either in the training, validation or test set. We call this evaluation setting "Filter". In this paper, we provide mean rank and Hits@10 of the two settings.

For experiments of PTransD, we select learning rate λ among $\{0.1, 0.01, 0.001\}$, the dimensions of entity embedding m and relation embedding n among $\{20, 50, 100\}$, and the margin γ among $\{1, 2, 4\}$. The best configuration is determined according to the mean rank in validation set. The optimal configurations are $\lambda = 0.001$, the margin $\gamma = 1$, $m = n = 50$ and taking L_1 as dissimilarity. For the FB15K dataset, we traverse to training for 1000 rounds.

Evaluation results of entity prediction are show in Table 2. For that the addition operation and 2-step paths outperforms other composition operations in both mean rank and Hits@10, we just consider 2-step relation paths and use addition operation to represent the relation path.

From Table 2 we observe that: (1) PTransE [1] gets surprisingly low mean rank. It indicates that PTransD handles complicated internal correlations of entities and relations in knowledge graphs better than other methods. (2) In Hits@10, PTransD significantly and consistently outperforms other baselines including PTransE [1] and TransD [9].

As defined in (Bordes et al. 2013), relations in knowledge graphs can be divided into various types according to their mapping properties such as 1-to-1, 1-to-N, N-to-1, N-to-N. In order to further observe the performance of PTransD and other methods facing different complex relations, Table 4 shows that detailed results by mapping properties of relations on FB15K.

Table 3. Evaluation results by mapping properties of relations (%)

Tasks	Prediction head entities (Hits@10)				Prediction tail entities (Hits@10)			
Relation category	1-to-1	1-to-N	N-to-1	N-to-N	1-to-1	1-to-N	N-to-1	N-to-N
SE [6]	35.6	62.6	17.2	37.5	34.9	14.6	68.3	41.3
SME(linear) [2]	35.1	53.7	19.0	40.3	32.7	14.9	61.6	43.3
SME(bilinear) [2]	30.9	69.6	19.9	38.6	28.2	13.1	76.0	41.8
TransE [3]	43.7	65.7	18.2	47.2	43.7	19.7	66.7	50.0
TransH [13]	66.8	87.6	28.7	64.5	65.5	39.8	83.3	67.2
TransR [7]	78.8	89.2	34.1	69.2	79.2	37.4	90.4	72.1
TransD [9]	86.1	95.5	39.8	78.5	85.4	50.6	**94.4**	81.2
PTransE(ADD,2-step) [1]	91.0	92.8	60.9	83.8	91.2	74.0	88.9	86.4
PTransE(MUL,2-step) [1]	89.0	86.8	57.6	79.8	87.8	71.4	72.2	80.4
PTransE(RNN,2-step) [1]	88.9	84.0	56.3	84.5	88.8	68.4	81.5	86.7
PTransE(ADD,3-step) [1]	90.1	92.0	58.7	86.1	90.7	70.7	87.5	88.7
PTransD(ADD,2-step)	**91.79**	**96.51**	**82.48**	**92.47**	**91.79**	**94.88**	89.58	**94.31**

From Table 3 we can conclude that, where the head entities are predicted: (1) PTransD significantly outperforms other baselines including TransD [9] and PTransE [1] when dealing with the 1-to-1, N-to-1 and N-to-N relations types. (2) The performance of PTransD when dealing with the 1-to-N relations type outperform than other baselines. When the tail entities are predicted: (1) Compared with other models, PTransD is a more fine-grained model which considers the 1-to-1 and 1-to-N types of relations. (2) PTransD has poor ability of knowledge representation when dealing with N-to-1 type of relations, and it is inferior to TransD [9] model. (3) PTransD model performs well in dealing with N-to-N type of relations. It is observed that, on some mapping types of relations, PTransD achieves significant improvement as compared with other models. However, the performance of PTransD is not very well with respect to some types of relations [6].

3.3   Relation Prediction

Relation prediction aims to predict the relation of two entities, which is an important information source to enrich knowledge graphs. We use FB15K as our dataset. We use the score function to rank the candidate relations in relation prediction. Evaluation results are showed in Table 4, where we report Hits@1 instead of Hits@10 for comparison, because Hits@10 for both TransE [3] and PTranE [1] exceeds 95%. Similar to Hits@10, Hits@1 represents the proportion of correct ranked in the number one.

The best configurations of PTransD for relation prediction is $\lambda = 0.001$, $\gamma = 1$, $m = n = 50$ and take L_1 as dissimilarity.

Table 4. Evaluation results on relation prediction (%)

Metric	Mean Rank		Hits@1	
	Raw	Filter	Raw	Filter
TransE [3]	2.8	2.5	65.1	84.3
PTransE(ADD,2-step) [1]	1.7	1.2	69.5	93.5
PTransE(MUL,2-step) [1]	2.5	2.0	66.3	89.0
PTransE(RNN,2-step) [1]	1.9	1.4	68.3	93.2
PTransE(ADD,3-step) [1]	1.8	2.5	68.5	94.0
PTransD(ADD,2-step)	**1.5**	**1.1**	**70.83**	**94.76**

From Table 4, we can observe that: (1) PTransD (in mean rank respect) is lower than the results of TransE [3] and PTransE [1] models under both raw and filter settings. (2) In Hits@1, PTransD obtains 0.2 to 0.5% points higher than other models. That is, PTransD handles complicated internal correlations of entities and relations in knowledge graphs better than other methods. Obviously, PTransD significantly and consistently outperforms other baselines, which indicates that embedding by relation paths and dynamic mapping matrix provide a good supplement for representation learning.

4　Conclusions and Future Work

This paper proposes a noval representation learning method for knowledge graphs named PTransD, which embeds both entities and relations into a low-dimensional space for their completion [5]. In PTransD, to take advantage of relation paths, we consider multiple-step relation paths between entities and use the addition operation to represent relations paths for optimization; We divide entities and relations into different semantics spaces and construct mapping matrix dynamically. Experimental results on FB15K dataset show that PTransD outperforms TransE [3], TransD [9] and PTransE [1] on two tasks including link prediction and relation prediction.

As shown in Tables 2 and 3, evaluation results on some types of relations (e.g. 1-to-N) preforms not very well. One possible reason is that the way to replace the head and tail entity provides negative effect to the evaluation results. In the future, we will explore a new approach to replace the head and tail entity.

Besides, we will extend PTransD to better deal with complicated scenarios of knowledge graphs.

Acknowledgments. This work was partially supported by National Key R&D Program of China (No. 2016YFD0101903), National Natural Science Foundation of China (No. 61702386, 61672398), Major Technical Innovation Program of Hubei Province (No. 2017AAA122), Key Natural Science Foundation of Hubei Province of China (No. 2017CFA012), Applied Fundamental Research of Wuhan (No. 20160101010004), Fundamental Research Funds for the Central Universities (WUT:2018IVB047) and Excellent Dissertation Cultivation Funds of Wuhan University of Technology (2017-YS-061).

References

1. Lin, Y., Liu, Z., Luan, H., Sun, M., Rao, S., Liu, S.: Modeling relation paths for representation learning of knowledge graphs. In: The Conference on Empirical Methods in Natural Language Processing (EMNLP 2015) (2015)
2. Bordes, A., Glorot, X., Weston, J., et al.: Joint learning of words and meaning representations for open-text semantic parsing. In: AISTATS 2012, vol. 22, pp. 127–135 (2012)
3. Bordes, A., Usunier, N., Garcia-Duran, A., et al.: Translating embeddings for modeling multi-relational data. In: Advances in Neural Information Processing Systems, pp. 2787–2795 (2013)
4. Bengio, Y., Courville, A., Vincent, P.: Representation learning: a review and new perspectives. IEEE Trans. Pattern Anal. Mach. Intell. **35**(8), 1798–1828 (2013)
5. Xie, R., Liu, Z., Jia, J., et al.: Representation learning of knowledge graphs with entity descriptions. In: AAAI 2016, pp. 2659–2665 (2016)
6. Bordes, A., Weston, J., Collobert, R., et al.: Learning structured embeddings of knowledge graphs. In: Conference on Artificial Intelligence (2011). (EPFL-CONF-192344)
7. Lin, Y., Liu, Z., Sun, M., Liu, Y., Zhu, X.: Learning entity and relation embeddings for knowledge graph completion. In: The 29th AAAI Conference on Artificial Intelligence (AAAI 2015) (2015)
8. Yang, B., Yih, W., He, X., et al.: Embedding entities and relations for learning and inference in knowledge bases. arXiv preprint arXiv:1412.6575 (2014)
9. Ji, G., He, S., Xu, L., et al.: Knowledge graph embedding via dynamic mapping matrix. In: ACL, vol. 1, pp. 687–696 (2015)
10. Mikolov, T., Chen, K., Corrado, G., et al.: Efficient estimation of word representations in vector space. arXiv preprint arXiv:1301.3781 (2013)
11. Yang, C., Liu, Z., Zhao, D., et al.: Network representation learning with rich text information. In: IJCAI 2015, pp. 2111–2117 (2015)
12. Bottou, L.: Large-scale machine learning with stochastic gradient descent. In: Lechevallier, Y., Saporta, G. (eds.) Proceedings of COMPSTAT'2010, pp. 177–186. Physica-Verlag, Heidelberg (2010). https://doi.org/10.1007/978-3-7908-2604-3_16
13. Wang, Z., Zhang, J., Feng, J., Chen, Z.: Knowledge graph embedding by translating on hyperplanes. In: Proceedings of AAAI, pp. 1112–1119 (2014)
14. Fan, M., Zhou, Q., Chang, E., et al.: Transition-based knowledge graph embedding with relational mapping properties. In: PACLIC 2014, pp. 328–337 (2014)
15. Egger, P., Pfaffermayr, M.: The proper panel econometric specification of the gravity equation: a three-way model with bilateral interaction effects. Empir. Econ. **28**(3), 571–580 (2003)

16. Jenatton, R., Roux, N.L., Bordes, A., et al.: A latent factor model for highly multi-relational data. In: Advances in Neural Information Processing Systems, pp. 3167–3175 (2012)
17. Duan, P., Wang, Y., Xiong, S., Mao, J.: Space projection and relation path based representation learning for construction of geography knowledge graph. In: China Conference on Knowledge Graph and Semantic Computing, CCKS 2016 (2016)
18. Bollacker, K., Evans, C., Paritosh, P., et al.: Freebase: a collaboratively created graph database for structuring human knowledge. In: Proceedings of the 2008 ACM SIGMOD International Conference on Management of Data, pp. 1247–1250. ACM (2008)
19. Miller, G.A.: WordNet: a lexical database for English. Commun. ACM **38**(11), 39–41 (1995)
20. Suchanek, F.M., Kasneci, G., Weikum, G.: Yago: a core of semantic knowledge. In: Proceedings of the 16th International Conference on World Wide Web, pp. 697–706. ACM (2007)

Expressiveness of Temporal Constraints
for Process Models

Johann Eder[✉], Marco Franceschetti, Julius Köpke, and Anja Oberrauner

Department of Informatics-Systems, Alpen-Adria-Universität Klagenfurt,
Klagenfurt, Austria
{johann.eder,marco.franceschetti,julius.koepke}@aau.at,
aoberrau@edu.aau.at
https://www.aau.at/isys

Abstract. Temporal constraints are an important aspect for the modeling of processes. We analyze the expressiveness of different features for the representation of temporally constrained processes. In particular we are able to show that the provision of non-contingency for activities and of references to start events of activities are redundant. We provide transformations for mapping process models employing the richer sets of features to process models using a reduced set of features. We show the equivalence of these process models and discuss the advantages of reducing redundancies in the representation of temporal constraints.

1 Introduction

Process models have been successfully introduced for modeling dynamic phenomena in many areas like business, production, health care, etc. In many of these application areas of process technology temporal issues are crucial aspects, which motivated a substantial body of research to master these requirements: modeling of temporal aspects such as durations, deadlines, and other temporal constraints, formulating different notions of correctness of process models with temporal constraints, checking the temporal correctness of process definitions, computing execution schedules for processes and supporting adherence to temporal constraints at runtime with proactive time management (see [4,8,12] for overviews).

The work on time patterns for business processes and workflows [15,16] brought a necessary consolidation to the formulation of temporal constraints for business processes by collecting and unifying various existing notions of representing temporal aspects of business processes and formally defining the semantics of all these constructs. It was not the aim of these papers to present a minimal basis of temporal constraints which are able to express all the other constraints.

In this paper we contribute to the question of identifying redundancies in temporal constraint pattern languages. Note that it is not the intention of this work to remove some of the patterns, as they are very well suited in aiding

© Springer Nature Switzerland AG 2018
C. Woo et al. (Eds.): ER 2018 Workshops, LNCS 11158, pp. 119–133, 2018.
https://doi.org/10.1007/978-3-030-01391-2_19

modelers in expressing temporal aspects. However, for researchers developing and proving algorithms for, e.g., checking controllability of process models, or for implementors of tools dealing with all these patterns, it is worthwhile to know, whether it is possible to express all these patterns with a reduced set of more basic patterns.

In this paper we focus on two features which are contained in many constraint languages (e.g. [8]: (1) non-contingent duration of activities (contained in pattern *TP2* [Durations] of [14,16] and formally defined in [6]) in addition to contingent duration of activities and (2) constraints on start events of activities (contained in the patterns *TP1* [time lags between activities] and *TP3* [time lags between events] of [16]) in addition to constraints o end events. We will show that both are redundant and can be expressed by other patterns. And we will give some examples for the benefits of a smaller set of basic features.

The remainder of the paper is organized as follows: in Sect. 2 we introduce a quite lean generic process model with temporal constraints on this model. We formally define the (temporal) semantics of temporally constrained process models by specifying which scenarios (runs, traces) conform with the specified constraints. In Sect. 3 we define transformations mapping process models using the extended pattern set to process models using only the basic set of patterns (without non-contingent activities and without references to start events in constraints). We define mappings between the schedules of these models. We also give examples of the transformations. In Sect. 4 we state the main theorem that the original and the transformed process models are equivalent and sketch the proof. Section 5 provides some discussions why we regard the investigation of redundancies in constraint formulation patterns worthwhile. Section 6 then discusses related work and in Sect. 7 we draw some conclusions.

2   Process Models and Temporal Constraints

We consider here a generic process model represented as an acyclic process graph composed of nodes and edges, where nodes are the activities and control structures of the model and the edges express precedence constraints between the nodes. Such a process model generalizes in particular the well-known basic workflow control patterns [19] such as sequence, XOR-splits and -joins, AND-splits and -joins, etc.

Temporal aspects are represented by durations and time points in form of natural numbers, where all durations have to be greater or equal 0. Time points are represented as distance to a time origin (here usually the time point of the start of a process). Activity instances start at a certain time point (time point for the start event of an activity instance) and end at a certain time point (end event); their distance is the duration of an activity instance. A duration can be *contingent*, i.e. the actual duration of an activity between the minimum and maximum duration can only be observed but not controlled, or *non-contingent*, which means that the actual duration between a minimum and a maximum duration can be influenced by a controller.

2.1 Running Example

In Fig. 1 we present, as our running example, a BPMN fragment of a temporally constrained process model from the medical domain, where various patterns for temporal constraints and non-contingent activities are modeled. First (activity A), a patient is admitted to the hospital for treatment. The admission procedure is known to last between 0 and 1 h. Subsequently, the patient is hospitalized and taken to her room (activity H), which requires 1 to 2 h. Then, two activities M and C are performed in parallel. M (taking 1 to 3 h) is performing a MRT scan of the patient, and C is growing a cell culture from a patient's sample. Growing the cell culture is a task which takes at least 24 h and at most 60 h, but depending on factors such as urgency or availability of technicians, can be controlled in its duration, provided that it is within the allowed interval. C has therefore a non-contingent duration. In the following figures we denote non-contingent durations in *italics*. A decision on the type of intervention is taken in activity D, taking 1 to 4 h. Then, depending on the decision, either a non-invasive, long lasting treatment is done in activity T (72 to 144 h), or a surgery is performed in activity S (24 to 48 h including post-surgical rehabilitation). Finally, the patient is released (activity R), which requires 0 to 1 h. Three upper-bound constraints are stated in the process:

1. *ubc(H.s, T.s, 72)*: hospital policies enforce the treatment to start no more than 72 h after the hospitalization has started.
2. *ubc(M.e, D.s, 2)*: in order for a doctor to take a decision on the type of intervention with up-to-date information, the MRT scan must have been completed no more than 2 h before the start of the decision-making task. Notably, the presence of a parallel task with long duration (activity C) together with the upper-bound constraint forces a delay in the execution of activity M.
3. *ubc(A.e, R.e, 240)*: hospital policies enforce the completion of the process after no more than 10 days (240 h) after patient admission.

After formally introducing our generic process model, we show how it can be used to express the hospitalization process using a reduced set of constraint patterns and only contingent activities.

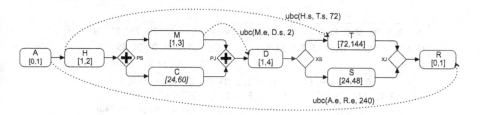

Fig. 1. BPMN fragment of a medical process with temporal constraints.

2.2   Process Model

We now formally define our process model including the formulation of temporal constraints.

Definition 1 (Process Model). *A process model P is a tuple (N, E, C), where:*

- *N is a set of nodes.*
- *$E \subseteq N \times N$ is a set of directed edges.*
- *$N^e = \{n^s, n^e | n \in N\}$ is a set of start and end events of the nodes in N.*
- *C is a set of temporal constraints consisting of*
  - *duration constraints for each $n \in N : d(n, n.d_{min}, n.d_{max}, n.contingent)$;*
  - *upper-bound constraints: $ubc(a, b, \delta)$, where $a, b \in N^e, \delta \in \mathbb{N}$.*
- *$N^c = \{n | n \in N, n.contingent = true\}$. $N^{nc} = N - N^c$.*

Duration constraints associate each node with an interval $[n.d_{min}, n.d_{max}]$ expressing its minimum and maximum duration. The actual runtime duration of node execution falls in this interval. The boolean property $n.contingent$ expresses whether the node and its duration is considered contingent or non-contingent. N^c is the set of all contingent nodes, N^{nc} the set of all non-contingent nodes. Contingent activities have a fixed minimum and maximum duration, but their actual duration at runtime cannot be controlled; instead, the duration of a contingent activity can only be observed after the activity was executed. An example of a contingent activity is a bank money transfer, for which some time between one and four days is required but there is no way to control it. Non-contingent activities have a defined minimum and maximum duration as well, but the actual duration can be controlled by the process execution system.

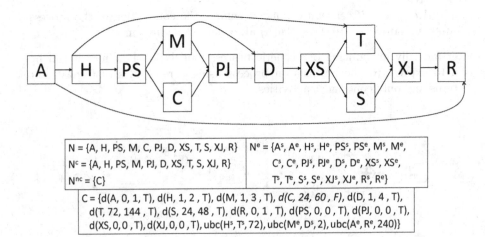

N = {A, H, PS, M, C, PJ, D, XS, T, S, XJ, R}	Nᵉ = {Aˢ, Aᵉ, Hˢ, Hᵉ, PSˢ, PSᵉ, Mˢ, Mᵉ,
Nᶜ = {A, H, PS, M, PJ, D, XS, T, S, XJ, R}	Cˢ, Cᵉ, PJˢ, PJᵉ, Dˢ, Dᵉ, XSˢ, XSᵉ,
Nⁿᶜ = {C}	Tˢ, Tᵉ, Sˢ, Sᵉ, XJˢ, XJᵉ, Rˢ, Rᵉ}

C = {d(A, 0, 1, T), d(H, 1, 2 , T), d(M, 1, 3 , T), *d(C, 24, 60 , F)*, d(D, 1, 4 , T),
d(T, 72, 144 , T), d(S, 24, 48 , T), d(R, 0, 1 , T), d(PS, 0, 0 , T), d(PJ, 0, 0 , T),
d(XS, 0, 0 , T), d(XJ, 0, 0 , T), ubc(Hˢ, Tˢ, 72), ubc(Mᵉ, Dˢ, 2), ubc(Aᵉ, Rᵉ, 240)}

Fig. 2. Process model for the example of Fig. 1.

Each node n in the process model is associated with 2 events: the start (n^s) and the end (n^e) event of the execution of the node. Temporal constraints can then be defined on these events. Upper-bound constraints express maximum allowed durations between events. An upper-bound constraint ($f^s, g^e, 12$) means that the time-span between the start of node f and the end of node g must not be longer than 12 time units.

Figure 2 shows the representation of the hospitalization process of Fig. 1 according to our formalism. For simplicity, we have visualized the sets of nodes and edges in the form of a graph.

2.3   Semantics

We define the semantics of a temporally constrained process model by defining which executions of the process model are valid. This has 2 aspects: (1) is the sequence of events (or of node-executions) admissible according to the process model, and (2) are all temporal constraints satisfied? In this paper we focus on question (2). An execution (also called run, possible trace, or instance type) of a process model is a subgraph of the process graph. Which subgraphs of the process graph are correct executions - question (1) - depends on the types of the nodes (e.g. AND-split, XOR-split, etc.) and the control semantics. To remain as general as possible in our considerations, in this paper we regard this as externally defined and do not reason about particular properties.

We define a scenario as a run with time-stamps for the start and end events of the nodes. Then we define an execution as (temporally) correct, if the scenario fulfills all temporal constraints (valid scenario).

Definition 2 (Scenario). $\bar{S}(P) = (N, E)$ *is a scenario for a process model* $P(N', E', C)$, *iff* $N \subseteq N', E \subseteq E', E \subseteq N \times N$ *and each* $n^s, n^e \in N^e$ *is associated with a time-stamp* $n^s.t, n^e.t \in \mathbb{N}$, *the time points of the start and end events of node* n *in the process instance.*

The definition of scenarios and valid scenarios is generic in the sense that a decision which subgraphs of the process graph are admissible remains generic, i.e. it depends on the specialization of the generic process model defining the types of provided (control) nodes and their specific semantics.

Definition 3 (Valid Scenario). *A scenario* $\bar{S}(P) = (N, E)$ *of a process model* $P(N', E', C)$ *is valid, iff*

1. $\forall n \in N : 0 \le n^s.t$
2. $\forall (m, n) \in E : m^e.t \le n^s.t$
3. $\forall d(n, d_{min}, d_{max}, n.contingent) \in C : d_{min} \le n^e.t - n^s.t \le d_{max}$
4. $\forall ubc(s, d, \delta) \in C : d.t \le s.t + \delta$

Following Definition 3, it is easy to verify that the scenario depicted in Fig. 3 constitutes a valid scenario for the process of our running example.

A process model is *satisfiable*, iff it has a valid scenario. The notion of satisfiability is frequently used in model checking approaches for validating process

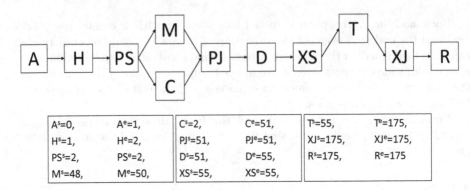

A^s=0,	A^e=1,	C^s=2,	C^e=51,	T^s=55,	T^e=175,
H^s=1,	H^e=2,	PJ^s=51,	PJ^e=51,	XJ^s=175,	XJ^e=175,
PS^s=2,	PS^e=2,	D^s=51,	D^e=55,	R^s=175,	R^e=175
M^s=48,	M^e=50,	XS^s=55,	XS^e=55,		

Fig. 3. Scenario for the process example of Fig. 2.

models. It has the shortcoming, that the correct execution might depend on external properties (like the actual duration of contingent activities) which cannot be influenced by the process controller [6,21]. We therefore strive usually for the different forms of controllability which provide better guarantees for a correct execution.

2.4 Controllability

Controllability [6,20] is defined as a characteristic of a process model which guarantees that it is within the capabilities of the process execution system (the controller) to guarantee for an execution of a process instance satisfying all temporal constraints, irrespective of influences from the environment - such as the actual duration of contingent activities.

A schedule defines when each node has to start its execution and an interval when the execution has to finish.

Definition 4 (Schedule). *A schedule S for a process $P(N,E,C)$ associates each $n \in N$ with $n.s$ the start time for node n, and the end time interval as $[n.e_e, n.e_l]$ with $n.e_e$ the earliest end time and $n.e_l$ the latest end time for node n. We call $(n, n.s, n.e_e, n.e_l) \in S$ a schedule entry.*

The property of controllability of a process requires that there is a schedule for the process, such that all scenarios are valid for which the time-stamps of the scenarios are taken from the respective intervals of this schedule. We will call a schedule with such a property in the following "controllability schedule".

Definition 5 (Controllability). *A temporally constrained process $P(N,E,C)$ is controllable, iff it has a schedule S, such that all scenarios $\bar{S}(P)$ are valid, iff $\forall n \in N : n.s = n^s.t \leq n.e_e \leq n^e.t \leq n.e_l$.*

We now define conditions for schedules to be correct in such a way that the existence of a correct schedule is a necessary and sufficient condition for controllability.

Definition 6 (Correct Schedule). *A schedule S for a process $P(N, E, C)$ is correct, iff for all nodes $n, m \in N$, for all $(n, m) \in E$, for all $a, b \in N$:*

1. $n.s \geq 0$
2. $n.e_l \leq m.s$ *(successor m starts after termination of n)*
3. $n.e_e \geq n.s + n.d_{min}$
4. $n.e_l \leq n.s + n.d_{max}$
5. $n.e_e \leq n.e_l$
6. $n.e_e = n.s + n.d_{min}$, *if $n.contingent=true$*
7. $n.e_l = n.s + n.d_{max}$, *if $n.contingent=true$*
8. $\forall ubc(a^s, b^s, \delta) \in C : b.s \leq a.s + \delta$
9. $\forall ubc(a^e, b^s, \delta) \in C : b.s \leq a.e_e + \delta$
10. $\forall ubc(a^s, b^e, \delta) \in C : b.e_l \leq a.s + \delta$
11. $\forall ubc(a^e, b^e, \delta) \in C : b.e_l \leq a.e_e + \delta$

Note that for contingent nodes the end time interval is a function of the start time, while for non-contingent nodes the end time interval can be chosen within the limits defined by the maximum and minimum duration. It is easy to see that the schedule for the running example process in Fig. 7 (top-left) is correct.

Lemma 1. *A process is controllable, iff it has a correct schedule.*

Proof. That a correct schedule fulfills all requirements of a controllability schedule follows from the definitions. In the other direction: each controllability schedule is correct. □

3 Transformations

In this section we define transformations which replace all non-contingent nodes in the process graph and all references to start events in the upper-bound constraints. With these transformations we show that non-contingent nodes and references to start events in upper-bound constraints are redundant in the formalisms for temporally constrained process models. We show the equivalence of the transformed processes to their original ones in the next section.

3.1 Transformation of Non-contingent Nodes

As shown in Fig. 4, we replace each non-contingent node n with 2 nodes n^f (the front node) and n^t (the tail node) connected with an edge (n^f, n^t). Both nodes are contingent, the front node has minimum and maximum duration 0, and the tail node has the minimum duration of n as both minimum and maximum durations. All incoming edges at n are redirected to n^f, all edges originating from n are transferred to originate from n^t.

In all upper-bound constraints, references to the start event of n are replaced with the end event of the front node, and all references to the end event of n are replaced with the end event of the tail node. An additional upper-bound constraint is included which limits the time-span between the end event of the front node and the end event of the tail node to the maximum duration of the non-contingent node.

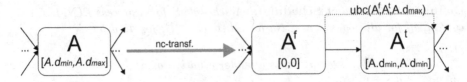

Fig. 4. Transformation of non-contingent nodes to contingent nodes.

Definition 7 (nc-transformation). *Let $P(N, E, C)$ be a process model, let N^{nc} be the set of non-contingent nodes in N. Let the substitution $\theta(x) = (n^f)^e$, if $x = n^s, n \in N^{nc}; (n^t)^e, if x = n^e, n \in N^{nc}$; or x, otherwise. Then the nc-transformation $T^1(P) = P'(N', E', C')$ is defined as follows:*

1. $N' = N^c \cup \{n^f, n^t | n \in N^{nc}\}$
2. $E' = \{(n, m) | n, m \in N^c, (n, m) \in E\}$
   $\cup\{(n, m^f) | n \in N^c, m \in N^{nc}, (n, m) \in E\}$
   $\cup\{(n^t, m) | n \in N^{nc}, m \in N^c, (n, m) \in E\}$
   $\cup\{(n^t, m^f) | n, m \in N^{nc}, (n, m) \in E\}$
   $\cup\{(n^f, n^t) | n \in N^{nc}\}$
3. $C' = \{d(n, d1, d2, true) | d(n, d1, d2, true) \in C\}$
   $\cup\{d(n^f, 0, 0, true), d(n^t, d_{min}, d_{min}, true) | d(n, d_{min}, d_{max}, false) \in C\}$
   $\cup\{ubc(a, b, \delta) | ubc(x, y, \delta) \in C, a = \theta(x), b = \theta(y)\}$
   $\cup\{ubc((n^f)^e, (n^t)^e, n.d_{max}) | n \in N^{nc}\}$

3.2   Transformation of Upper-Bound Constraints on Start Events

In this sub-section we propose a transformation which replaces all references to start events of activities in upper-bound constraints with the corresponding end events. The basic idea is shown in Fig. 5, which reports the three relevant cases involving start events, while the formulas are defined in Definition 8.

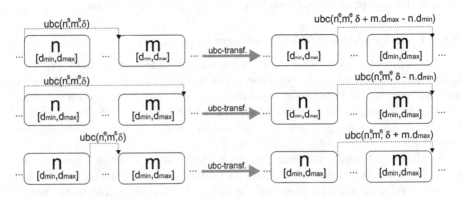

Fig. 5. Transformation of upper-bound constraints on start events.

Definition 8 (ubc-transformation). *Let $P(N, E, C)$ be a process graph.*
$T^2(P) = P'(N, E, C')$ *is defined as:*
$C' = \{t(ubc(n, m, \delta)) | ubc(n, m, \delta) \in C\}$, *where for all $n, m \in N$:*
$t(ubc(n^s, m^s, \delta)) = ubc(n^e, m^e, \delta + m.d_{max} - n.d_{min})$
$t(ubc(n^s, m^e, \delta)) = ubc(n^e, m^e, \delta - n.d_{min})$
$t(ubc(n^e, m^s, \delta)) = ubc(n^e, m^e, \delta + m.d_{max})$
$t(ubc(n^e, m^e, \delta)) = ubc(n^e, m^e, \delta)$

We now combine the replacement of the non-contingent nodes and of references to start events into a single transformation.

Definition 9 (Transformation). *The combined transformation $T(P)$ of T^1 and T^2 is defined as: $T(P) = T^2(T^1(P))$.*

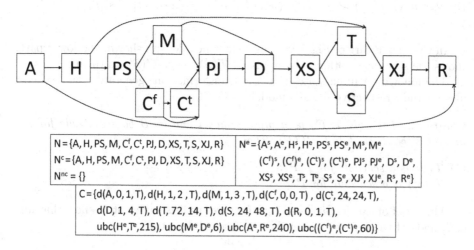

N = {A, H, PS, M, C^f, C^t, PJ, D, XS, T, S, XJ, R}	N^e = {A^s, A^e, H^s, H^e, PS^s, PS^e, M^s, M^e,
N^c = {A, H, PS, M, C^f, C^t, PJ, D, XS, T, S, XJ, R}	(C^f)^s, (C^f)^e, (C^t)^s, (C^t)^e, PJ^s, PJ^e, D^s, D^e,
N^{nc} = {}	XS^s, XS^e, T^s, T^e, S^s, S^e, XJ^s, XJ^e, R^s, R^e}

C = {d(A, 0, 1, T), d(H, 1, 2 , T), d(M, 1, 3 , T), d(C^f, 0, 0, T) , d(C^t, 24, 24, T),
d(D, 1, 4, T), d(T, 72, 14, T), d(S, 24, 48, T), d(R, 0, 1, T),
ubc(H^e,T^e,215), ubc(M^e,D^e,6), ubc(A^e,R^e,240), ubc((C^f)^e,(C^t)^e,60)}

Fig. 6. Transformed process model for the medical example of Fig. 1.

Figure 6 shows the resulting process from the application of the combined transformation on our running example process.

Now we are ready to define transformations on schedules. $T(S)$ is a transformation which transforms a schedule for an original process model to a schedule for the transformed process model; $T'(S')$ is the inverse transformation.

Definition 10 (Schedule-transformations). $T(S)$ *is defined as follows. Let S be a schedule for a process model $P(N, E, C)$.*
$S' = T(S) = \{(n, n.s, n.e_e, n.e_l) | (n, n.s, n.e_e, n.e_l) \in S, n \in N^c\}$
$\cup \{(n^f, n.s, n.s, n.s), (n^t, n.e_e - n.d_{min}, n.e_e, n.e_e) | (n, n.s, n.e_e, n.e_l) \in S, n \in N^{nc}\}$

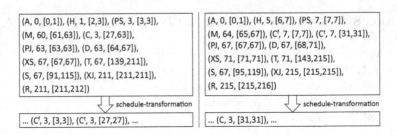

Fig. 7. Two different schedules. On the left: for the example process with non-contingent activities before (top) and after (bottom) applying $\mathcal{T}(S)$. On the right: for the process without non-contingent activities, before and after $\mathcal{T}'(S')$.

$\mathcal{T}'(S')$ is defined as follows. Let S' be a schedule for $\mathcal{T}(P)$.
$$S'' = \mathcal{T}'(S') = \{(n, n.s, n.e_e, n.e_l) | (n, n.s, n.e_e, n.e_l) \in S', n \in N^c\} \cup$$
$$\{n, n^f.s, n^t.e_e, n^t.e_l) | n \in N^{nc}, (n^f, n^f.s, n^f.e_e, n^f.e_l), (n^t, n^t.s, n^t.e_e, n^t.e_l) \in S'\}$$

In the upper part of Fig. 7 two different schedules are shown; the lower part shows the schedules obtained by applying $\mathcal{T}(S)$, resp. $\mathcal{T}'(S')$ to them.

The following lemma states that the transformations generate actual schedules for their respective process models.

Lemma 2. *Let $P(N, E, C)$ be a process model and let S be a schedule for P, and let S' be a schedule for $\mathcal{T}(P)$. Then:*

(1) $\mathcal{T}(S)$ *is a schedule for $\mathcal{T}(P)$;*
(2) $\mathcal{T}'(S')$ *is a schedule for P.*

The proof of this lemma is straightforward, as it is sufficient to show that for each node of the process model there is a schedule entry.

4   Equivalence

In this section we show that a process model and its transformation are equivalent. In particular, we show that if the process model is controllable, so is its transformed process model and vice versa.

We need the following assumptions about admissible runs: if we replace all contingent nodes in the run in the same way as we replace them in the process model then we derive an admissible run of the transformed process model. Formally, we assume: iff (N', E') is a run for $P(N, E, C)$ and $(N'', E'', C'') = \mathcal{T}^1(N', E', C)$ then (N'', E'') is a run for $\mathcal{T}^1(P)$. It is easy to see that this assumption holds for all workflow nets.

Theorem 1 (control-equiv). *Let P be a process model and $\mathcal{T}(P)$ the transformed process model. P is controllable, iff $\mathcal{T}(P)$ is controllable.*

Proof. Proof-sketch: We show: if S is a correct schedule for P then $\mathcal{T}(S)$ is a correct schedule for $\mathcal{T}(P)$ and if S' is a correct schedule for $\mathcal{T}(P)$ then $\mathcal{T}'(S')$ is a correct schedule for P. □

The proof is not difficult but somewhat lengthy. The theorem can be proved straightforward from the definition of correct schedules and the definition of the transformations but requires a considerable amount of term transformations using in particular transitivity of inequalities. Details of the proof can be found in [18].

5 Discussion

In this section we discuss some of the possible advantages and consequences of the findings presented above.

- Academic curiosity: it is a well-established tradition in computer science research to search for a minimum basis of some formalisms. So, part of our motivation was to see whether certain features in temporal constraint languages for processes were actually redundant. We do not argue that they should be removed from these languages, as they frequently make it easier for designers to express temporal constraints in process models as they are seen in the application domain. However, knowing about these redundancies allows for transformations which reduce the complexity of internal representations.
- The reduction of the number of features leads to more compact languages or meta-models for the representation of processes. This makes it easier to reason about properties of process models (such as controllability here), and significantly reduces the size of definitions, algorithms and proofs. For an example, after the transformation we introduced here, the definition of a correct schedule can be reduced to 3 clauses instead of 11. Below we show how the definitions of the notions of process, schedule, and correct schedule can be simplified if we do not provide non-contingent activities and references to start events in our process model.
- The elimination of non-contingent nodes has the disadvantage of increasing the number of nodes, but has the advantage that for example schedules can be expressed much more compact, since it suffices to state one of the parameters start time, earliest end time, or latest end time. The other parameters can then be calculated.
- When all upper-bound constraints can be expressed on the end events of nodes then the mapping of process models to temporal constraint networks can be reduced to mapping only the end events and not the start events to variables of the temporal constraint network. Many algorithms for these networks have a higher than linear complexity [2] (some have even exponential complexity) in the number of variables. To halve the number of variables reduces the runtime of these algorithms enormously.

Here we show how the definitions of process model, of schedule, and of correct schedule can be drastically shortened, if non-contingent activities and references to start events are dismissed.

Definition 11 (Process Model-2). *A process model P is a tuple (N, E, C), where:*

- *N is a set of nodes.*
- *$E \subseteq N \times N$ is a set of edges.*
- *C is a set of temporal constraints consisting of*
  - *duration constraints for each $n \in N$: $d(n, n.d_{min}, n.d_{max})$;*
  - *upper-bound constraints: $ubc(a, b, \delta)$, where $a, b \in N, \delta \in \mathbb{N}$.*

A schedule defines when each node has to start execution and an interval when the execution has to finish. Since the end-time interval for contingent activities is a function of the start time it is sufficient to state the start time for defining a schedule.

Definition 12 (Schedule-2). *A schedule S for a process $P(N, E, C)$ associates each $n \in N$ with $n.s$ the start time for node n. We call $(n, n.s) \in S$ a schedule entry.*

We now define conditions for schedules to be correct in such a way that the existence of a correct schedule is a necessary and sufficient condition for controllability. The most dramatic reduction is the definition of a correct schedule where now 3 clauses are sufficient compared to the original 11.

Definition 13 (Correct Schedule-2). *S is a correct schedule for a process $P(N, E, C)$, iff for all nodes $n, m \in N$, for all $(n, m) \in E$, for all $a, b \in N$:*

1. *$n.s \geq 0$*
2. *$n.s + n.d_{max} \leq m.s$ (successor m starts after termination of n)*
3. *$\forall ubc(a, b, \delta) \in C : b.s + b.d_{max} \leq a.s + a.d_{min} + \delta$*

Comparing these definitions with the original ones shows the reduction of redundant features for temporally constrained process models leads to significantly simpler definitions and algorithms.

6   Related Work

Time management in business processes and workflows has been matter of studies for the last decades. Overviews on the related works can be found in [4,8,12]. Temporally constrained business processes have been considered in [1,11,17] for the analysis of their temporal qualities, by the application of different techniques such as network analysis or constraint analysis. These works focus on determining whether the temporal constraints in a process model can be fulfilled and if deadlines can be met. Here, however, we concentrate on the modeling aspects and in particular we strive for a simplified modeling language without losing expressive power for activity durations and temporal constraints.

This work simplifies the representation of temporal constraints expressible from the workflow time patterns identified in [3,15]. Temporal constraints are classified in 10 time patterns, which can refer to any combination of start/end events of activities. Our proposed transformations give modelers flexibility to define temporal constraints on any combination of start/end events (thus in accordance to the patterns), yet simplifying their representation in the underlying model for easier reasoning, as discussed in Sect. 5.

Distinguishing between contingent and non-contingent activities, and assessing temporal qualities in the presence of uncertain durations dates back to works on temporal reasoning for planning in various research domains such as Artificial Intelligence [20,21]. The application of Temporal Constraint Networks [9] for representing business processes and reasoning on their temporal qualities has been studied extensively in recent years [6,8] with several refinements on the notion of controllability, resulting in more advanced algorithms for its check [2,5,7]. Other works [10,13] considered different formalisms for representing timed workflows and determining correct schedules for execution of activities also in the case of cross-organizational cooperations.

7   Conclusions

The ambition of this work was to investigate redundancies in patterns for expressing temporal aspects of process models. We focused on two particular features of temporal constraints: non-contingent activities, where the controller can influence the actual duration of the activity execution within a minimum and a maximum duration, and references to start events in activities in formulating upper-bound constraints (max-gaps) between events of the process execution.

We proposed some transformations which transform process models with the richer set of features to a process model using fewer features, and we were able to show that we do not lose information, i.e. the original model and the transformed process model are equivalent.

By defining the notions of process model, schedule and correct schedule using the reduced set of features we were able to show that these reductions are highly significant. Moreover, we discussed that encodings of process models in temporal constraint networks can be essentially halved. We therefore conclude that the strive for less redundancy in models as investigated in this paper is worthwhile and does not only satisfy academic curiosity but also delivers significant improvements for practice.

References

1. Bettini, C., Wang, X., Jajodia, S.: Temporal reasoning in workflow systems. Distrib. Parallel Databases **11**(3), 269–306 (2002)
2. Cairo, M., Rizzi, R.: Dynamic controllability made simple. In: 24th International Symposium on Temporal Representation and Reasoning (2017)
3. Cheikhrouhou, S., Kallel, S., Guermouche, N., Jmaiel, M.: Toward a time-centric modeling of business processes in BPMN 2.0. In: Proceedings of International Conference on Information Integration and Web-based Applications and Services, p. 154. ACM (2013)
4. Cheikhrouhou, S., Kallel, S., Guermouche, N., Jmaiel, M.: The temporal perspective in business process modeling: a survey and research challenges. Serv. Oriented Comput. Appl. **9**(1), 75–85 (2015)
5. Combi, C., Hunsberger, L., Posenato, R.: An algorithm for checking the dynamic controllability of a conditional simple temporal network with uncertainty-revisited. In: Filipe, J., Fred, A. (eds.) ICAART 2013. CCIS, vol. 449, pp. 314–331. Springer, Heidelberg (2014). https://doi.org/10.1007/978-3-662-44440-5_19
6. Combi, C., Posenato, R.: Controllability in temporal conceptual workflow schemata. In: Dayal, U., Eder, J., Koehler, J., Reijers, H.A. (eds.) BPM 2009. LNCS, vol. 5701, pp. 64–79. Springer, Heidelberg (2009). https://doi.org/10.1007/978-3-642-03848-8_6
7. Combi, C., Posenato, R.: Towards temporal controllabilities for workflow schemata. In: 2010 17th International Symposium on Temporal Representation and Reasoning (TIME), pp. 129–136. IEEE (2010)
8. Combi, C., Pozzi, G.: Temporal conceptual modelling of workflows. In: Song, I.-Y., Liddle, S.W., Ling, T.-W., Scheuermann, P. (eds.) ER 2003. LNCS, vol. 2813, pp. 59–76. Springer, Heidelberg (2003). https://doi.org/10.1007/978-3-540-39648-2_8
9. Dechter, R., Meiri, I., Pearl, J.: Temporal constraint networks. Artif. Intell. **49**(1–3), 61–95 (1991)
10. Eder, J., Gruber, W., Panagos, E.: Temporal modeling of workflows with conditional execution paths. In: Ibrahim, M., Küng, J., Revell, N. (eds.) DEXA 2000. LNCS, vol. 1873, pp. 243–253. Springer, Heidelberg (2000). https://doi.org/10.1007/3-540-44469-6_23
11. Eder, J., Panagos, E., Rabinovich, M.: Time constraints in workflow systems. In: Jarke, M., Oberweis, A. (eds.) CAiSE 1999. LNCS, vol. 1626, pp. 286–300. Springer, Heidelberg (1999). https://doi.org/10.1007/3-540-48738-7_22
12. Eder, J., Panagos, E., Rabinovich, M.: Workflow time management revisited. In: Bubenko, J., Krogstie, J., Pastor, O., Pernici, B., Rolland, C., Søvberg, A. (eds.) Seminal Contributions to Information Systems Engineering, pp. 207–213. Springer, Heidelberg (2013). https://doi.org/10.1007/978-3-642-36926-1_16
13. Eder, J., Pichler, H., Tahamtan, A.: Probabilistic time management of choreographies. In: Ardagna, D., Mecella, M., Yang, J. (eds.) BPM 2008. LNBIP, vol. 17, pp. 443–454. Springer, Heidelberg (2009). https://doi.org/10.1007/978-3-642-00328-8_45
14. Lanz, A., Posenato, R., Combi, C., Reichert, M.: Controlling time-awareness in modularized processes. In: Schmidt, R., Guédria, W., Bider, I., Guerreiro, S. (eds.) BPMDS/EMMSAD -2016. LNBIP, vol. 248, pp. 157–172. Springer, Cham (2016). https://doi.org/10.1007/978-3-319-39429-9_11
15. Lanz, A., Reichert, M., Weber, B.: Process time patterns: a formal foundation. Inf. Syst. **57**, 38–68 (2016)

16. Lanz, A., Weber, B., Reichert, M.: Workflow time patterns for process-aware information systems. In: Bider, I., et al. (eds.) BPMDS/EMMSAD -2010. LNBIP, vol. 50, pp. 94–107. Springer, Heidelberg (2010). https://doi.org/10.1007/978-3-642-13051-9_9
17. Marjanovic, O., Orlowska, M.: On modeling and verification of temporal constraints in production workflows. Knowl. Inf. Syst. 1(2), 157–192 (1999)
18. Oberrauner, A.: Expressiveness of temporal constraints in process models. Master's thesis, Alpen-Adria-Universitaet Klagenfurt, Austria (2018)
19. van der Aalst, W.M.P., Ter Hofstede, A.H., Kiepuszewski, B., Barros, A.P.: Workflow patterns. Distrib. Parallel Databases 14(1), 5–51 (2003)
20. Vidal, T.: Handling contingency in temporal constraint networks: from consistency to controllabilities. J. Exp. Theor. Artif. Intell. 11(1), 23–45 (1999)
21. Vidal, T., Ghallab, M.: Dealing with uncertain durations in temporal constraint networks dedicated to planning. In: Proceedings of the 12th European Conference on Artificial Intelligence (ECAI 1996), pp. 48–54. PITMAN (1996)

Thoroughly Modern Accounting: Shifting to a De Re Conceptual Pattern for Debits and Credits

Chris Partridge[1,3](✉) [ID], Mesbah Khan[2] [ID], Sergio de Cesare[3] [ID],
Frederik Gailly[4] [ID], Michael Verdonck[4] [ID], and Andrew Mitchell[1] [ID]

[1] BORO Solutions Ltd., London, UK
{partridgec,mitchella}@borogroup.co.uk
[2] OntoLedgy Ltd., London, UK
khanm@ontoledgy.io
[3] University of Westminster, London, UK
s.decesare@westminster.ac.uk
[4] Faculty of Economics and Business Administration,
Ghent University, Ghent, Belgium
frederik.gailly@UGent.be, Michael.Verdonck@ugent.be

Abstract. Double entry bookkeeping lies at the core of modern accounting. It is shaped by a fundamental conceptual pattern; a design decision that was popularised by Pacioli some 500 years ago and subsequently institutionalised into accounting practice and systems. Debits and credits are core components of this conceptual pattern. This paper suggests that a different conceptual pattern, one that does not have debits and credits as its components, may be more suited to some modern accounting information systems. It makes the case by looking at two conceptual design choices that permeate the Pacioli pattern; de se and directional terms - leading to a de se directional conceptual pattern. It suggests alternative design choices - de re and non-directional terms, leading to a de re non-directional conceptual pattern - have some advantages in modern complex, computer-based, business environments.

Keywords: De se · De re · Directional terms · Debits and credits
Accounting information systems

> *Miss Dorothy Brown: You're a modern! Millie Dillmount:*
> *Thoroughly!*
>
> *Thoroughly Modern Millie (1967)*

1 Introduction

Double entry bookkeeping is at the core of modern accounting practice. The system was devised by the merchants of Venice and popularised by Pacioli in a book printed in 1494. Its basic principles have largely remained intact over the last five centuries despite business environments becoming significantly more complex and, more

© Springer Nature Switzerland AG 2018
C. Woo et al. (Eds.): ER 2018 Workshops, LNCS 11158, pp. 134–148, 2018.
https://doi.org/10.1007/978-3-030-01391-2_20

recently, the emergence of computing technology. This is testimony to the good design of Pacioli's conceptual pattern (though cynics may say the accounting community's inherent traditionalism also played a part). However, as noted elsewhere (for example Mattessich [1] and McCarthy [2]) the roots in a manual paper-based system (and a simpler business environment) may also indicate that these structures are ripe for change. The question then arises; what kind of change?

This research paper has two goals. Firstly, to provide an analysis of one of the core conceptual patterns that underlie modern accounting information systems. Put differently, we look at two conceptual design choices that permeate Pacioli's approach - leading to what we call a de se directional conceptual pattern. This pattern emerges from the decision to manage financial information from an owner/proprietor's perspective and it permeates the conceptual model. At its heart is the notion of debits and credits. We reflect upon the pressures that would motivate these choices in Pacioli's time.

Secondly, we then ask whether this de se directional pattern still makes sense in the context of modern computing technology and accounting requirements of transnational corporations. We do this by contrasting it with a different conceptual pattern - what we call a de re non-directional conceptual pattern. This pattern is an alternative 'view from nowhere' from which particular de se perspectives and their debits and credits can be generated. We speculate on why this pattern might be better for some modern accounting requirements.

We start by describing, in Sect. 2, the approach that was followed for identifying the conceptual pattern and the design choices that underlie them. In Sect. 3, we show how this conceptual pattern shapes the conceptual model of accounting information systems. In Sect. 4 we review the issues the pattern gives rise to and then in Sects. 5 and 6 look at the structures that materialise from the alternate choices. Finally, the paper ends with a conclusion and some future research directions.

2 Background: Two Related Conceptual Patterns

As noted above, while it seems likely, if not obvious, that Pacioli's approach is ripe for change, it is far from obvious or easy to work out how it should change. We spotted the opportunity for change described here in our legacy system re-engineering work [3]. This has involved the mining of ontologies from a number of accounting and ERP systems. In every case, one aspect of the resulting ontology – a view from nowhere – has struck us as odd. This was that the mining of the debit and credit transactions revealed a picture with no debits and credits [4]. More recently, when implementing the mined ontology, we have noticed that in the implementation we have needed to mark the owner/proprietor in the system and generate debits and credits (essentially adding back perspectival details that the original ontology mining removed). Reflection upon these and some initial research led us to recognise that a well-researched topic in philosophy – the de se – de re distinction [5] – underlay these two phenomena. We reported this in [6] which investigated the general topic of de se and de re. However, in the case of accounting, and specifically debits and credits, we recognised that de se is only part of the picture, that there was another philosophical topic in play – directional attributes [7, 8]. In this paper we give the more detailed picture that takes account of both topics. We focus on the specific case of debits and credits in accounting and two design choices that have surfaced in our legacy reengineering work.

2.1   Directional Terms

Identity plays a big, often unrecognised, role in formalisation [9]; one particular concern is that different views on identity lead to different conceptual models. The conceptual design choice we are interested in here is the decision to use what we call directional terms and the temptation to simply reify these as objects.

This topic has ancient roots. It appears in Aristotle [7] where he discusses the road from Thebes to Athens and contrasts it with the road from Athens to Thebes; where the first is uphill and the second downhill. It seems inconsistent to say that one road is both uphill and downhill, but equally it seems odd to say there are two roads. More than that, as shown in Fig. 1, there are perfectly reasonable ways of showing the gradient of the road without any commitment to uphill and downhill objects.

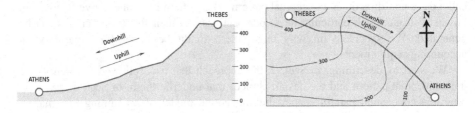

Fig. 1. The road between Athens and Thebes.

Wiggins [8] considers this conundrum. He suggests "that either 'road' means an actual feature of the landscape, in which case 'uphill' collects a term giving the direction and there is a simple relational predicate true of that road, or it means 'journey by road', in which case there is no identity." As Wiggins notes, in the first case, the work that uphill and downhill are doing is not picking out different objects, but building on a bundling together of the road with a direction - making the neutral 'road between Athens and Thebes' into the directional 'road from Athens to Thebes'. This 'directional term' can then be qualified as 'uphill'.

We explain below how much of modern accounting, in particular talk of debits and credits, uses these kinds of directional terms and so provides an opportunity to unbundle the term into the underlying objects.

2.2   De Se (And De Re)

Another important conceptual design choice is when and how to use the de se (Latin 'of oneself') [5]. In ordinary language, whenever we use indexicals like 'me', 'myself' and 'I' we are making a de se formalisation choice - where the self plays the role of the deitic centre (From Greek *deixis*, lit. 'display, demonstration, or reference', meaning point of reference in contemporary linguistics). Technically a de se statement can be translated into a neutral one - called de re (Latin 'of the thing') - by replacing the indexical with a proper noun - what could be called de-re-ifying. So, John saying 'I am left-handed', can be translated into 'John is left-handed'; though it would be odd, but not incorrect, for John to describe himself in the third person. Directional terms and the de se can overlap, in that de se phrases can provide the direction needed for directional term. The traditional example is, 'I am walking uphill' where the deictic centre provides the context for the directional term.

In information systems, the de se appears in guises other than these natural language indexicals, which can make it difficult to identify. To help us with our analysis of the underlying formalisation choices, we can draw on an observation from philosophy and linguistics, (where a good understanding of what differentiates the de se (indexical) and de re (non-indexical) has been developed). A characteristic of pure (de se) indexical utterances is that the reference (and truth) of a sentence can shift from use to use. For instance, if John and Mary both utter the sentence 'I am left-handed', the two utterances refer to different things; namely that (in de re non-indexical terms) 'Mary is left-handed' and 'John is left-handed' respectively. And there is no (logical) inconsistency in one of the utterances being true and the other false. This does not happen with non-indexical de re uses. So, for example, the reference (and truth) of the sentence 'Mary is left-handed' does not change whoever, wherever and whenever it is uttered - each utterance has exactly the same content.

2.3   Accounting Information Systems Implications

Accounting information systems often have de se directional reporting requirements. For example, companies' major financial accounting reports are de se directional; the balance sheet and profit and loss statements report the balances from the 'owner's perspective'. The simplest prima facie design for this reporting is to store the information in de se form and then report it directly (let's call this 'de se storage and reporting'). The alternative design, storing the information in de re form and then querying the de se information (let's call this 'de re storage and de se querying') seems to be unnecessary extra work. Furthermore, the de se storage will appear more parsimonious than the de re as it has no need to make the deitic centre explicit (though one could counter-argue that the implicit deitic centre is not transparent).

However, this assumes that only a single de se perspective is required. If multiple de se reporting over the same de re information is required, then the situation is different. Firstly, it makes more sense to input the information once (whether in de re or de se form) and then calculate the required reporting/presentation forms. Given this, adopting a 'de se storage and reporting' approach here would need multiple processing on input and lead to multiple de se storage silos each with the same information stored in different de se formats. This will open the door to the data anomalies and corruption associated with data redundancy. Here a 'de re storage and de se querying' strategy becomes more attractive. The master de re information is stored only once and the de se queries generated as required. It becomes even more attractive if the de se reporting requirements are volatile, in the sense of new de se perspectives emerging and old de se perspectives retiring (for example, companies joining and leaving a group).

3 De Se and Directional Terms in Accounting

In this section, we look at de se directional conceptual pattern in the double entry accounting conceptual model.

3.1 Pacioli Introduces Modern Accounting

Fra Luca Bartolomeo de Pacioli is known as the father of modern accounting and bookkeeping because he was the first person to publish a detailed description of double-entry bookkeeping that is the foundation for modern accounting. He described this as the method used by Venetian merchants in the final chapter (Particularis de computis et scripturis - About accounts and other writings) of his mathematical text-book Summa de arithmetica, geometria. Proportioni et proportionalita [10] published in Venice in 1494. It was published soon after the introduction of moveable type printing, making it significantly more accessible, and this undoubtedly contributed towards its popularity.

Pacioli's system starts with an inventory and then has a system of internal controls containing three books which has a well-defined process to update the records in a specific order. Whenever there is a transaction, the system starts with a description of the complete transaction in the Day Book or Memorandum (from, as we shall discuss later, the perspective of the owner). From this debit and credit postings for the day are extracted and recorded in the Journal. Finally, these postings are re-recorded under the appropriate account in the Ledger [10].

3.2 Pen and Paper – Designing for Presentation

With pen and paper technology, the storage is external (on paper) but the processing is human. Where the final presentation data is viewed many times, it makes sense to have a system where humans process the data into the final presentation format and then store it (on paper) in that format – then the paper storage can be read directly. This is rather than store the data in its original format and process it into the presentation format each time it is required. Pacioli's design for his process recognizes this. Figure 2 represents the Pacioli process - Pacioli's 'Summa Mathematica' is included in the figure, as it acts as a kind of conceptual model for the manual process.

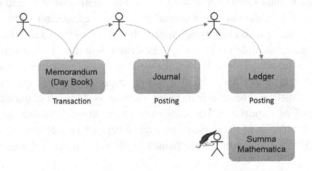

Fig. 2. Pen and paper based accounting process.

This process-once-before-storage approach makes even more sense when one considers the practical concerns with the errors that could arise in the manual recording process. Pacioli takes advantage of the process' algorithmic nature to provide step by step instructions as well as internal controls to mitigate this. From the very beginning, he identifies these internal controls as key to successful bookkeeping [10] and describes a number of them in detail (for example, numbering the pages of the journal to make it easy to identify when a page has been removed and double underlining the journal side entry to mark the two entries in the ledger). There is a significant amount of algorithmic manual processing to keep the entire dataset consistent - with checks to ensure the algorithms were followed correctly.

If the data was not stored on paper in the final presentation format, then this burdensome processing would have to be done multiple times, each time the data was reported. So the final presentation format (which is de se directional) dictates the structure of the paper storage. As we discuss later, these information management concerns were formative for the de se directional conceptual pattern in the design of the system.

3.3   Automation - The Modern Implementation of Pacioli

From our modern perspective, the transcription from Journal to Ledger is just an automatable sort, where the postings are arranged firstly by date (in the Journal) and then by date within account (in the Ledger). And the transcription from Day Book to Journal as an automatable query over the transaction details. So, to our modern eyes the process reveals itself as algorithmic and so automatable. One that we can break down further into a kind of data model for the items being processed - see Fig. 3 - where the leg is the implicit asset (economic resource) that is being posted against.

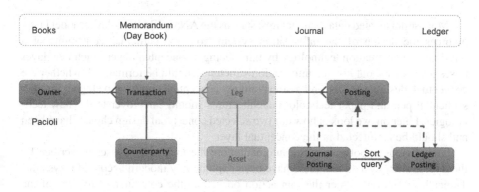

Fig. 3. Breakdown of a typical transaction.

One of the advantages of computing over pen and paper technology is that it enables automated processing inside the computer system rather than manual processing by humans – and the cost of repeating a computer process is insignificant when compared with the cost of repeating the equivalent manual human process. So there is no longer a pressure to store data in the form it will be presented – the computer can reshape the stored data into the format needed for presentation. This enables information architectures where the form of the information that is entered or presented is different from that which is stored - as shown in Fig. 4, compare this with the equivalent Fig. 2 for pen and paper technology.

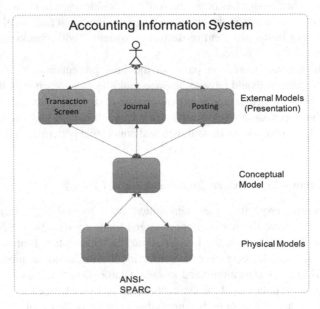

Fig. 4. Accounting information system accounting process

More sophisticated data architectures, such as the ANSI-SPARC 3-layer model [11] - divorces the design of the information system from the particular constraints of a physical implementation technology by introducing a conceptual layer. Each new layer raises questions about how its information structures should be formalized: whether the conceptual design choices and corresponding conceptual model developed for flat, sequential pen and paper technology architectures should carry over to the new technologies. Later on, we look at how our two selected conceptual design choices have been and should be architected in the conceptual layer.

The obvious opportunities for automation are the manual processes prescribed by the double entry system and these have been exploited by modern accounting systems. Typically, users only enter the transaction once, into the computer equivalent of the Day Book. Then the system algorithmically generates the debit and credit postings and posts them into the journal and ledger accordingly. The question we explore here through our investigation of the conceptual patterns in accounting is whether there are less obvious opportunities for a layered implementation of information architecture yet to be exploited.

3.4   Examples of De Se Directional Terms

In this section, we make clear through examples that Pacioli's accounting information system (and so also modern systems) have consistently made a choice for a de se directional conceptual pattern based upon a single owner's perspective. The examples show how this choice extends from the owner, through transactions to debits and credits.

Transaction Party - Counterparty Distinction

In Pacioli's section on day book entries and examples, he writes this description of a transaction: "Purchased from Phillip Ruffon - white silks at 12 ducats each". Presumably, Phillip Ruffon would write in his books: "Sold to Fra Luca Pacioli - white silks at 12 ducats each". This illustrates the de se directional tradition, still used today, of recording the parties to a transaction; where the owner of the books is the deitic centre and is assumed to be one party to the transaction, then only the other party, the counterparty needs to be explicitly recorded. Clearly, this is de se directional: one can only be a counterparty relative to an owner: it would be more accurate to call them owner counterparties. There are no de re counterparties; in the de re 'perspective', transactions have parties, simpliciter: no party has a priority. This situation is shown in Fig. 5.

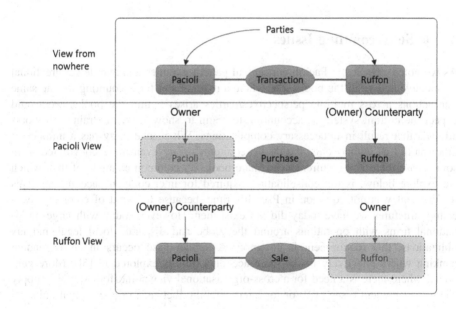

Fig. 5. Parties - viewed in de se as party and counterparty.

The two example descriptions also show how the introduction of a silent owner with a counterparty has a knock-on effect, in that it naturally leads to the use of owner-relative terms such as 'purchase' and 'sale' - which reflect the direction of flow of goods to the owner. These terms make no sense in the de re 'perspective'.

Owner of the Books

The owner is clearly a silent partner in the transaction recorded in the day book. But for this system to work, it needs to be clear who this silent owner is. Pacioli consistently says that his 'internal control system' is for a businessman - this is the silent owner of all the books. This greatly simplifies identifying the owner, it is the same entity for all the transactions in the set of books. In Pacioli's day, the owner's name might be written on the cover of the books. In modern systems, it may be recorded in a system configuration file. However, as we discuss later, this simplification comes at a cost - as the owner has to be a party to all the transactions and they all need to be accounted for from its perspective.

Directional De Se Debits and Credits

The owner's de se directional formalization extends to debits and credits. Typically, a debit from one party's perspective is a credit from the other party's perspective. To return to the original Pacioli example; where he purchased 20 white silks at 12 ducats each - 240 ducats in total. In the Journal he will credit Cash for this payment to Phillip Ruffon. In his counterparty's (Phillip Ruffon) books, this entry would appear in his journal as a debit to Cash for the receipt from Luca Pacioli. This is slightly clearer in Pacioli's original language where he uses the terms 'to' and 'from' (in Italian 'per' and 'a') rather than debit and credit.

4   De Se Accounting Issues

As we noted earlier, the Pacioli conceptual pattern assumes a single de se directional perspective across all the books. So, where a requirement for accounting for the same transactions across multiple perspectives/parties arises within the books, we would expect the current system of accounting to begin to show signs of strain with workarounds that result in unnecessary complications. Where one party has a number of other parties as components, the de se requirements are even more convoluted as in some sense there is a requirement to share books. A common example of this, which we explore below, is the consolidation required for inter-company accounting. This requirement was not common in Pacioli's time, because the kind of company ownership structures we have today did not exist then. However, today with large multinational firms with operations around the globe and disparate local legal and tax obligations, this requirement is common. A similar case occurs in correspondent banking where effectively banks share accounts (this is explored in [6]). More generally, where there is a need for a cross-organisational viewpoint (for example, supply chain management – see [12]) or for interoperability between organisational units the single de se perspective becomes unwieldy.

Today, global value chains account for 80% of global trade[1]. A large proportion of this is intra-firm trade (i.e. international flows of goods and services between parent companies and their affiliates or among these affiliates). Part of this growth in recent times is through consolidation of larger multinationals through acquisitions.

---

[1] http://unctad.org/en/pages/PressRelease.aspx?OriginalVersionID=113.

These consolidations result in heterogeneous accounting systems and charts of accounts being amalgamated using 'band-aid tactics' with the result that the data does not satisfy the requirements of all the primary stakeholders, namely accounting, tax and treasury [13].

The current standards for inter-company accounting, build on the same de se foundations and therefore yield some interesting complications. This is because there is a requirement to take the following perspectives over the transactions:

1. Parent company only de se and directional. This is for accounting for the parent company as a legal entity that transacts with other legal entities (including its subsidiaries).
2. Company group including subsidiaries de se. This is for accounting and financial reporting of the group as a single entity (treating all transactions within the group as internal).
3. Subsidiaries de se. This is for accounting for the subsidiary as an independent legal entity that transacts with other legal entities (including the parent company and other subsidiaries of the parent company).

Figure 6 shows the various directions the transactions can flow in this kind of setting [14].

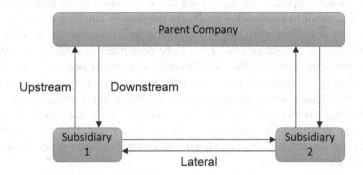

Fig. 6. Directions inter-company accounting.

Therefore, when a downstream transaction is initiated (from parent to subsidiary), each party records the transaction from its perspective as if it is transacting with a third party. Later when the parent company accounts have to be consolidated, the parent company's books have to view the same transactions as if they were internal and therefore have a net-zero effect on the group's assets and liabilities. There is no simple way to reconcile the individual de se transactions into a single parent de se transaction. Calculations have to be carried out outside the books to work out the net changes required to 'hide' the internal transaction. All this overhead is there because there is no single de re, non-directional account of the transaction and each de se account is incompatible with the other.

Worked example: in the case where a parent pays £600 to a subsidiary, the following postings will be made:

- Parent's books credit £600
- Subsidiary's books debit £600
- later parent has to post another debit £600 again into its books to net out the transaction (as if it never happened).

In the above, there is only one movement, but there are three postings, because it is not possible to record the movement in an agent neutral representation. The above example is a very simple transaction leg posting, the actual process of elimination is typically quite intricate, because the transaction's impact has to be externally calculated for each affected account of the parent/subsidiary books. These elimination amounts are then posted into the relevant ledger accounts in order to show either that the 'transaction never happened' (where all legs of the transaction are within the Parent company's owned entity structure) or that the internal changes did not happen but the external component of the transaction did (where there is a third party involved). This process does not stop at the initial transaction but needs to be continued every time any loss or gain needs to be booked. For example, when an asset is sold downstream, depreciation has to be calculated from the subsidiary's perspective using the subsidiary's purchase price. However, from the parent company perspective, because the transaction 'never happened', an adjustment has to be posted in the parent company's depreciation account to account for the Parent company's purchase price.

The current design for managing this intercompany accounting is based on a large number of off-book calculations that have to be performed to establish the values of adjustment postings required. This is additional work that is outside of the standard practices of entity accounting and has to be standardised separately and is implemented differently in different organisations. The manual nature and complexity of the processing introduces risks of mistakes and inconsistencies in the accounting data. It also becomes very difficult to account for all the regulatory and financial requirements imposed by the tax and treasury departments based on their specific perspectives over the financial data.

5 Related Research

This review of the conceptual patterns is raising questions about the foundations of accounting's information architecture. People have been raising these kinds of questions for a while. In the 1960s there was an interest in developing axiomatic foundations for accounting. Mattessich in [15] developed a set theoretical axiomatisation of accounting where full chart of accounts was structured as entries stored in a matrix. More recently the REA approach [2] has aimed at rethinking accounting conceptual models, explicitly suggesting this is driven by the introduction of computing technology. While both these projects challenge some of the elements of the current traditional architecture, neither has made any real attempt to shift away from Pacioli's de se directional formalisation choices apart from changing names; for example, talking of inflows and outflows rather than debits and credits. We look at REA's approach briefly below.

More interestingly, one can see the de se directional approach coming unstuck when faced with the practical challenges of implementing interoperability and reuse. We show this happening in the FIX messaging standard[2] and Universal Enterprise Data Models [16] which both drop the de se structure of owner/counterparty for a simple de re party structure. However, none of these try to unravel the choices inherent in Pacioli's debits and credits. So finally, we briefly outline how this can be done, and show what a de re version of these would look like.

Accounting Theory - The REA Framework

REA [2] has the goal of modelling accounting entities semantically to support data integration at the enterprise level - and sees the Pacioli information structures as irrelevant to this purpose, saying: "The primary contention of this paper is that the semantic modelling of accounting object systems should not include elements of double-entry bookkeeping such as debits, credits and accounts. As noted previously by both Everest and Weber [17] and McCarthy [18], these elements are artefacts associated with journals and ledgers (that is they are simply mechanisms for manually storing and transmitting data). As such, they are not essential aspects of an accounting system." REA, in our view, correctly recognises that the Pacioli structure as driven by the manual demands inherent in pen and paper technology and identifies the opportunity for shifting to a new structure to exploit computing technology. It explicitly discusses how computing technology enables the ANSI-SPARC separation of views. It furthermore recognises that the economic events happen to economic resources (assets). However, despite its stated rejection of the Pacioli information structures, it continues to subscribe to the fundamental de se directional view inherent in its foundations. Prima facie its talk of 'inflow' and 'outflow' as well as 'increments' and 'decrements' are directional. And further inspection shows that it adopts an owner-counterparty view - focusing on the view from a single entity. Hruby et al. [19] makes this point (in Section 1.2.1), where he clarifies that "(t)he terms decrement and increment are relative to the model viewpoint: they depend upon the economic agent which is in the focus of the model" and in the context of his example, "if we modelled the same process from the perspective of the Customer, the transfer of the pizza would be an increment (would be called Purchase) ...". The pattern of the transaction shifting from a sale to a purchase as the perspective changes we described earlier reappears here.

FIX Messages

The Financial Information eXchange protocol is used for real-time exchange of information between organisations in the international securities transactions and markets. It provides us with an example of how the practical requirements of interoperability lead to a de re view of parties. The protocol includes a <Parties> component that is used to identify the transacting parties or any other parties that have a role (broker, clearing firm, exchange, settlement bank) in the transaction. All parties are handled in the same format, and they are allowed to have multiple roles in different transactions. In this complex ecosystem, with a high volume of financial instrument trading, the de re party structure offers a fast, reliable and efficient way for different

---

[2] FIX - www.fixtrading.org.

organisations to exchange transaction data. It is not clear how this could be done in a simple de se way as there is no obvious candidate for the owner - any de se solution would lead to each of the parties having to create data from their individual perspectives, in other words, a multiple de se architecture.

Universal Data Models
Silverston has produced a series of volumes (including [16]) documenting practitioners' experiences in building what he calls a universal data model for all enterprises. This provides a good insight into data modelling. As with FIX, this clearly uses a de re party pattern, there is no evidence of a de se directional owner-counterparty pattern. Indeed, given the overall structure of the model it is difficult to see how a single privileged owner party could be identified. FIX and the universal data model examples show that, at least at the party level, modern computing data architectures are often not suited to a de se architecture.

6 Outline of a Universal (Pure) De Re System

However, the de se aspect of debits and credits is more intransigent; remaining in both accounting theory, data modelling practice and implemented systems. We use ontological analysis to help unravel the historical de se choices and reveal the underlying de re. We do this using the BORO [3] four-dimensional top ontology, which reveals that assets (economic resources) have stages where they are owned by parties. This is shown for the cash element of an exchange in the space-time map in Fig. 7.

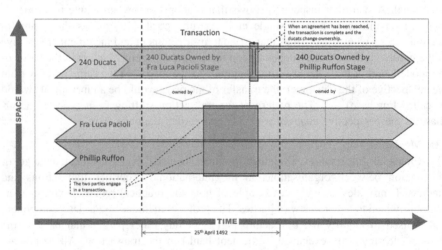

Fig. 7. 240 ducats state change event space-time map.

For a period, a state of the 240 ducats is owned by Fra Luca Pacioli, then the subsequent state is owned by Phillip Ruffon. The boundaries of these states mark changes in ownership. These states and their state boundaries fit into a wider de re system from which a variety of de se reports, with different owners (deitic centres) can be generated - illustrated in Fig. 8.

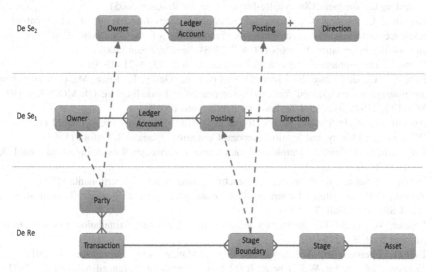

Fig. 8. Transaction viewed in de se/directional terms.

7 Conclusion

We introduced two conceptual design choices and used these to clearly expose how a de se directional choice permeates the whole of Pacioli's accounting conceptual model. We show how this choice stretches from the owner's books, through owner/counterparty transactions to de se directional debits and credits. We recognised that Pacioli's conceptual model was well suited to his contemporary requirements but also marshalled the arguments for change. Suggesting that the evolution of more complex business structures and the emergence of computing technology has changed the landscape to such an extent that there is an opportunity for improvement. We then outlined, using the four-dimensional BORO foundational ontology, a de re non-directional conceptual pattern to replace debits and credits and how this would fit into a complete de re non-directional conceptual model.

References

1. Mattessich, R.: Two Hundred Years of Accounting Research. Taylor & Francis, London (2007)
2. McCarthy, W.E.: The REA accounting model: a generalized framework for accounting systems in a shared data environment. Account. Rev. **57**, 554–578 (1982)
3. Partridge, C.: Business Objects: Re-Engineering for Re-Use (2005)
4. Partridge, C.: A new foundation for accounting: steps towards the development of a reference ontology for accounting. In: Proceedings of the European Conference on Accounting Information Systems (ECAIS 2003), Seville, Spain (2003)
5. Perry, J.: The problem of the essential indexical. Noûs **13**, 3–21 (1979)
6. Partridge, C., de Cesare, S., Mitchell, A., León, A., Gailly, F., Khan, M.: Ontology then agentology: a finer grained framework for enterprise modelling. In: 6th MODELWARE, SCITEPRESS-Science and Technology Publications (2018)
7. Aristotle: Aristotle: Physics Books III and IV. Oxford University Press, Oxford (1993)
8. Wiggins, D.: Identity and Spatio-Temporal Continuity. Blackwell, Oxford (1967)
9. Guizzardi, G., Halpin, T.: Ontological foundations for conceptual modeling. Appl. Ontol. **3**, 91–110 (2008)
10. Pacioli, L.: Summa de arithmetica, geometria. Proportioni et proportionalita (1494)
11. ANSI/X3/SPARC Study Group on Data Base Management Systems: Interim Report. ACM SIGMOD Bull. **7** (1975)
12. Laurier, W., Poels, G.: Invariant conditions in value system simulation models. Decis. Support Syst. **56**, 275–287 (2013)
13. Deloitte: Intercompany Accounting and Process Management - Survey Results (2017)
14. Fischer, P.M., Taylor, W.J., Cheng, R.H.: Fundamentals of Advanced Accounting (2007)
15. Solomons, D., Zeff, S.A.: Accounting Research, 1948-1958: Selected Articles on Accounting History. Garland Pub. (1996)
16. Silverston, L.: The Data Model Resource Book. A Library of Universal Data Models for All Enterprises, vol. 1. Wiley, Chichester (2011)
17. Everest, G.C., Weber, R.: A relational approach to accounting models. Account. Rev. **52**, 340–359 (1977)
18. McCarthy, W.E.: Entity-relationship view of accounting models. Account. Rev. **54**, 667–686 (1979)
19. Hruby, P., Kiehn, J., Scheller, C.V.: Model-Driven Design Using Business Patterns. Springer, New York (2006). https://doi.org/10.1007/3-540-30327-2

Evaluation of the Cognitive Effectiveness of the CORAS Modelling Language

Eloïse Zehnder, Nicolas Mayer[(⊠)], and Guillaume Gronier

Luxembourg Institute of Science and Technology,
5, avenue des Hauts-Fourneaux, 4362 Esch-sur-Alzette, Luxembourg
{eloise.zehnder,nicolas.mayer,
guillaume.gronier}@list.lu

Abstract. Nowadays, Information System (IS) security and Risk Management (RM) are required for every organization that wishes to survive in this networked and open world. Thus, more and more organizations tend to implement a security strategy based on an ISSRM (IS security RM) approach. However, the difficulty of dealing efficiently with ISSRM is currently growing, because of the complexity of current IS coming with the increasing number of risks organizations need to face. To use conceptual models to deal with RM issues, especially in the information security domain, is today an active research topic, and many modelling languages have been proposed in this way. However, a current challenge remains the cognitive effectiveness of the visual syntax of these languages, i.e. the effectiveness to convey information. Security risk managers are indeed not used to use modelling languages in their daily work, making this aspect of cognitive effectiveness a must-have for these modelling languages. Instead of starting defining a new cognitive effective modelling language, our objective is rather to assess and benchmark existing ones from the literature. The aim of this paper is thus to assess the cognitive effectiveness of CORAS, a modelling language focused on ISSRM.

Keywords: Security · Risk management · Visual syntax · Physics of notations

1 Introduction

Risk management is today a steering instrument used in many domains, such as, for example finance, insurance, environment or security. From a process perspective, risk management has been standardized for years, at a generic level like in ISO 31000, as well as in specific domains such as ISO/IEC 27005 for information security. However, at the product level (i.e. the result(s) obtained as output of the different steps of the risk management process), a great variability can be observed, going from tables to detailed and specific conceptual models (e.g., in the security domain, attack-trees [1] or enterprise architecture models [2]). To propose conceptual models to deal with risk management, especially in the information security domain, is today an active research topic, and our aim is to contribute to it by improving existing languages.

In this context, this paper ties in a broader project that aims at integrating conceptual models with Information System Security Risk Management (ISSRM) [3]. This

© Springer Nature Switzerland AG 2018
C. Woo et al. (Eds.): ER 2018 Workshops, LNCS 11158, pp. 149–162, 2018.
https://doi.org/10.1007/978-3-030-01391-2_21

integration seems to us as a promising approach to deal with issues related to the complexity of organizations and associated risks, especially the difficulty to have a clear and manageable documentation for ISSRM activities. One of our main concerns is to take into account the target users' group of our research results that is information security risk managers. This target group is not used to use conceptual models in its daily work, and the associated modelling languages need so to be effective to this target group to convey information, i.e. to be cognitively effective.

In this frame, we have already assessed the "Risk and Security Overlay" (RSO) of the ArchiMate language [2], ArchiMate being a standardized modelling language developed by The Open Group to provide a uniform representation for diagrams that describe Enterprise Architecture (EA). The conclusion established was that the RSO can decently not be considered as an appropriate notation from a cognitive effectiveness point of view and that there is room to propose a notation better on this aspect [4]. CORAS is another well-known modelling language for ISSRM [5]. It is thus a second candidate we want to evaluate. The research question addressed in this paper is then: how cognitive effective is the CORAS language to support the users in their ISSRM activities?

The remainder of the paper is structured as follows. In the next section, the background of our work is described: it introduces cognitive effectiveness and the "Physics of Notation" (PoN), a set of nine principles we use to assess cognitive effectiveness, and then the CORAS approach itself. Section 3 is about related work. Section 4 presents the assessment of the cognitive effectiveness of CORAS: the approach followed and the results obtained. Section 5 is the discussion about the results. Finally, Sect. 6 concludes about our current work and presents our future work.

2 Background

2.1 Cognitive Effectiveness and the "Physics of Notation"

Conceptual models are now widely used to visually communicate a great deal of information (about processes, systems, etc.) to users. Despite their interest in the communication of complex information, conceptual models can, however, give interpretation problems to their users. Indeed, several empirical studies have already shown that they can be misunderstood [6–9]. To improve the understanding of conceptual models, one of the most relevant approach comes from the cognitive sciences and is called 'cognitive effectiveness'. Cognitive sciences refer to the theories of information processing, which specifically include many concepts from cognitive psychology such as perception, memory (short and long term), attention or information processing. Thus, conceptual models are understood as an interaction between a user and a visual representation. In this context, cognitive effectiveness is embodied in the ability of conceptual models to support appropriate translations between cognitive and visual models [10].

Cognitive effectiveness in software engineering has been addressed for years [11] but it became a more active research topic since Moody's work. In order to evaluate and obtain a better quality in the design of modelling languages, Moody has established nine principles, called the "Physics of Notation" (PoN) [12]. These principles have already been applied to assess the cognitive effectiveness of many different modelling languages [13–16] and are used in this paper to assess the one of CORAS. The nine principles, who will be detailed later, are: semiotic clarity, graphic economy, perceptual discriminability, visual expressiveness, dual coding, semantic transparency, cognitive fit, complexity management, and cognitive integration.

2.2   The CORAS Approach

CORAS is an approach for ISSRM consisting basically of a modelling language, a method and a tool [5]. A particular focus is done in this paper on the modelling language of CORAS and associated models. The CORAS language is composed of five types of (so-called) diagrams. The first one is the asset diagram, describing the focus of the analysis and coming with the context establishment. Then, the threat diagram supports the risk identification and the risk estimation steps. Threat diagram describes "scenarios which may cause harm to the assets". An example of threat diagram is proposed in Fig. 1. Next, risk diagram basically summarizes the risks presented in threat diagrams. Finally, treatment overview diagram proposes treatments to risks. The CORAS language also includes three extensions. High-Level CORAS supports abstraction and comprehensible overviews or large risk models. Dependent CORAS supports documentation of assumptions and provide tools to separate the target of study from assumptions. Finally, Legal CORAS supports documentation of legal aspects (legal risks and legal norms) and their impact.

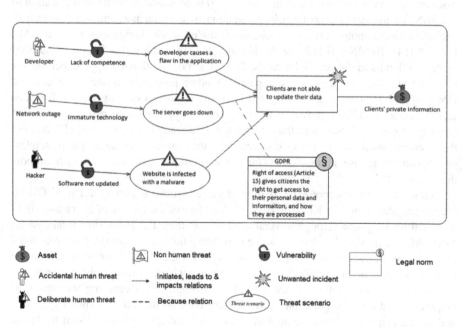

Fig. 1. Example of a threat diagram, including the Legal CORAS extension

For illustration purpose, Fig. 1 is an example of threat diagram. In this example, a network outage (non-human threat) may harm the clients' private information (asset) by making the server going down (threat scenario) caused by the use of an immature technology (vulnerability). The GDPR (legal norm) actually puts a legal risk because citizens shall have access to their data and they shall be able to update them. The fact that the clients are not able to update their data (unwanted incident) can also occur because of deliberate or accidental human threats, respectively because a developer causes a flaw in the application because of lack of competence or because a hacker exploits a non-updated software to infect a website with a malware.

In this paper, our analysis is on the CORAS language as a whole (i.e. all provided diagrams) and is based on the reference book entitled: "Model-Driven Risk Analysis: The CORAS Approach" [5]. It has the advantage of presenting the CORAS modelling language in a self-contained, up-to-date and detailed manner. We are also aware of the existence of the more recent ISMS-CORAS method [17] which adds more detailed steps in the process and more diagrams and symbols related to the implementation of an Information Security Management System (ISMS). However, we decided to stick at the core CORAS approach and not to study this context-specific extension.

3 Related Works

According to a recent study based on a Systematic Literature Review [18], Moody's framework and associated principles are one of the most cited and used approach in the assessment of visual notations. In this review, 70 papers dealing with the application of the PoN for the development and/or evaluation of a visual notation were identified, excluding overlapping versions of already included work. Languages such as UML [13], $i*$ [14], BPMN 2.0 [15], or ArchiMate [16] (to name just a few) were already evaluated thanks to the principles of the PoN. The PoN is globally appreciated for his scientific support and accuracy compared to other proposals in the same domain. However, the PoN approach is targeted by some criticisms coming mainly from the lack of clear guidelines provided for its practical application to assess an existing language. Operationalisation of the principles have already been attempted [19, 20], but they remain unsatisfactory because incomplete (they do not include all the principles) or impractical to use in real cases for some principles according to their authors themselves [18].

Some other frameworks exist to support visual syntax assessment. The SEQUAL framework, developed by Krogstie *et al.* [21], addresses the quality of every aspects of a modelling language through several qualities. Putting the focus on visual syntax, SEQUAL is considered as lacking some concrete guidelines to design effective visual notations, as argued by Genon [22]. Frank addresses the visual notation design in the context of the Domain-Specific Modeling Language which represents the concepts and constraints of a well-defined domain-level knowledge [23]. Regarding the design of graphical notations, the guidelines provided are built on the author's own experience and on the respective literature analysis. However, the approach developed by Frank aim at being applied during the design phase of the graphical notation. The Cognitive Dimensions (CDs) of notations is another framework aiming at improving design practice by focussing on the usability aspects of artefacts [11]. It provides different dimensions (from 9 to 13 depending on the users) that can be used during exploratory design of a modelling language. Guizzardi *et al.* propose an ontology-based assessment and design method of domain-specific visual modelling languages [24]. This method aims at evaluating, on the one hand, the language ability to support the users in communicating and reasoning with the produced models and on the other hand, its truthfulness and appropriateness to the domain which it is supposed to represent. Kleppe introduces and explains the factors that influence an effective domain specific language design and proposes a design strategy for language creation [25]. Once again, these approaches are developed to be used during design time of a modelling language and are less suited than the PoN for the evaluation of an existing visual notation.

4 Assessment of the Cognitive Effectiveness of CORAS

4.1 Methodological Approach

Since the PoN principles are today considered as one of the most advanced framework to evaluate an existing visual modelling language [22], our analysis will be based on these principles. However, as argued in the previous section, their operationalisation remains a complex and open issue. What they lack the most is usability, which, according to ISO 9241-11 can be defined by "the degree to which a product can be used, by means of identified users, to achieve defined goals with efficiency, effectiveness and satisfaction, in a context of specified use". Indeed, Moody's principles are neither proposing quantitative benchmarks allowing us to know the limitations of the syntax to elaborate, nor proper guidelines to use them. Therefore, we will follow recommendations established thanks to lessons learnt from previous usages of these principles: try to operationalise the principles in quantitative metrics to objectify the analysis, whenever it is possible, and apply the PoN with care by providing accountability with design rationale of the evaluated language [18, 20].

To do so, we first operationalised all the principles into tables to get a better view of them, gathering their definition, with the specific characteristics of each of them (for example, cognitive integration regroups conceptual and perceptual integration), and associated metrics established from the definition of the principle. In parallel, we gathered every symbol of CORAS in a document and we designed an illustrative example for every type of diagram included in the language in order to have a concrete example supporting our evaluation (not included in this paper because of space constraints).

Aiming simplicity and usability in its format, our evaluation approach is intended to be used under the guise of the expert or heuristic evaluation. The heuristic evaluation is "an informal method of usability analysis where a number of evaluators are presented with an interface design and asked to comment on it" [26]. With this method, end-users are not involved in the evaluation process, the evaluation being only based on expert analysis. According to their authors, heuristic evaluation can be improved significantly by involving multiple evaluators. Thus, three evaluators (that are the authors of this paper) were involved in the review of the CORAS language: one of them is an expert in security and risk management, the two others are expert in cognitive science. Heuristic evaluation was performed by each evaluator who inspected the CORAS language alone. Then, the individual findings were discussed and aggregated. The next section summarizes the results obtained.

4.2   Results

Hereafter, we analyse the cognitive effectiveness of the RSO through the nine principles elaborated by Moody. We first remind a short definition for each principle, extracted from the PoN reference article [12]. Then, we report on how CORAS meets this principle, based on associated metrics.

Principle of Semiotic Clarity. According to the semiotic clarity principle, there should be a one-to-one correspondence between semantic constructs and graphical symbols. CORAS is a language composed of 17 different graphical symbols and 20 semantic constructs overall. There is a 1:1 correspondence between semantic constructs and graphical symbols for almost all constructs. The only exception is for the following constructs: *initiates*, *leads-to*, *impacts* and *harm* relations are all represented with the same arrow as graphical symbol. According to Moody, there is thus an anomaly with regards to the semiotic clarity principle called symbol overload (when two different constructs are represented by the same graphical symbol) occurring 6 times (one for each couple of two different constructs represented by the same graphical symbol). The other anomalies with regards to semiotic clarity have an occurrence of 0, as reported in Table 1.

Table 1. Anomalies with regards to semiotic clarity.

Anomaly	Definition	Occurrence
Symbol redundancy	Multiple graphical symbols can be used to represent the same semantic construct	0
Symbol excess	Graphical symbols do not correspond to any semantic construct	0
Symbol overload	Two different constructs are represented by the same graphical symbol	6
Symbol deficit	There are semantic constructs that are not represented by any graphical symbol	0

Principle of Perceptual Discriminability. Regarding perceptual discriminability, different symbols should be clearly distinguishable from each other. Discriminability is primarily determined by the visual distance between symbols. Visual distance between symbols occurs when these symbols differ on a sufficient number of visual variables (e.g., shape, size, colour, position, etc.) For this purpose, we compared the visual distance between every symbol two-by-two in a grid (not reported here because of space constraints). CORAS takes profit of three different visual variables in its language (colour, texture and shape; see the visual expressiveness principle for additional information), leading to a visual distance that could be between 0 and 3. As an example, the differences between the *risk* and the *stakeholder* symbols are the shape (warning sign versus human shape in a suit) and the colour (red and white versus white and brown). This gives us a visual distance of 2 for this couple. Among a total of 190 couples analysed, the summary of the visual distance between constructs is reported in Table 2.

We found that symbols are mostly separated by a visual distance of 2 and 6 interactions have a visual distance equal to 0, which are obviously the ones for which symbol overload has been found for the principle of semiotic clarity.

Table 2. Visual distance between the semantic constructs.

Visual distance	Occurrence
Visual distance equal to 0	6
Visual distance equal to 1	26
Visual distance equal to 2	125
Visual distance equal to 3	33

Principle of Semantic Transparency. Semantic transparency corresponds to the use of visual representations whose appearances suggest their real meaning without inducing another and/or false one. In other words, the meaning of a symbol should be understood by looking at its representation.

Determining when a symbol suggests its meaning (or not) is a quite subjective task since it can depend on a lot of individual variables such as culture or education. We thus decided to count the number of symbols (conceptual forms you can link to a concept, a referent) and the number of signs (non-representational symbol, arbitrarily assigned with a wholly learned connection to a referent) according to the work of Zender and Mejía [27]. The limit we try to define regarding semantic transparency resides in "what has to be learned or not in order to be understood".

We found 12 symbols (*unwanted incident, asset, indirect asset, threat scenario, treatment scenario, risk, vulnerability, stakeholder, deliberate human threat, accidental human threat, non-human threat, referring scenario*), and 5 signs (*legal norm, border line, initiates, leads-to, impacts and harm relations – having all four the same symbol, treats relation* and *because relation*). As an example, the legal norm and its '§' sign, according to the authors, does not explicitly refers to a legal symbol. None can guess the legal norm symbol refers to something at least legal unless it is learned. On the other hand, the unwanted incident is represented by an "exploding" symbol, which could be linked to a fire, an explosion, an incident, etc. Therefore it can be classified as a symbol.

Principle of Complexity Management. According to Moody, the notation should include explicit mechanisms for dealing with complexity between the different diagrams, basically modularization and/or hierarchy. The goal of complexity management is to reduce the cognitive overload while reading or creating diagrams. For this principle, we assessed the presence or not of modularization and hierarchy. There is actually no way in CORAS to introduce hierarchy, but modularization is admitted by the High-Level CORAS extension and its referring and referred symbols which allows to expand events happening in any kind of scenario.

Principle of Cognitive Integration. Explicit mechanisms to support integration of information from different diagrams should be included. This principle only applies when multiple diagrams are used to represent a system (this is the case for CORAS) and is closely related to the principle of complexity management, when modularity is used. However, it can still apply if modularity is not used in order to integrate diagrams of different types. We assessed the principle of cognitive integration through the presence or absence of related mechanisms (see Table 3). Concerning perceptual integration (perceptual cues to simplify navigation and transitions between diagrams), CORAS includes no identification (labelling of diagrams), no navigational cues, and no navigational map. The text linked to the symbols is not considered here as a label but more as a description (for example, the text in a threat scenario describes the scenario itself, not the symbol that is a threat scenario). There are also no indication or rules about the level numbering. Regarding conceptual integration (mechanisms to help the reader assemble information from separate diagrams into a coherent mental representation of the system), we can find as 'summary diagram' the treatment overview diagram, but it is only focused on risk treatment aspects. Contextualization is considered absent from the CORAS language since we have not identified any, through symbols or diagrams mechanisms.

Table 3. Mechanisms to simplify navigation and transitions between diagrams.

Perceptual integration	Definition	Presence/Absence
Identification	Labelling of diagrams	Absence
Level numbering	Orientation information	Absence
Navigational cues	Sign-posting	Absence
Navigational map	To show all diagrams and the navigation paths between them	Absence
Conceptual integration	Definition	Presence/Absence
Summary diagram	Provides a view of the system as a whole	Presence
Contextualization	The part of a system of current interest is displayed in the context of the system as a whole	Absence

Principle of Visual Expressiveness. The full range and capacities of visual variables should be used. Visual variables are shape, size, colour, brightness, orientation, and texture for retinal variables, and horizontal and vertical position for planar variables. It is worth to note that perceptual discriminability and visual expressiveness are very close to each other but however different. The first aims at measuring the visual variation between constructs in pairs, and the second aims at measuring the visual variation across the entire visual vocabulary. Among the 17 graphical symbols, 9 different shapes (human shape, flag, warning triangle, oval shape, money bag, explosion, padlock symbol, arrowhead link, and square) were identified. Two different textures (full lines and fine strokes) and seven different colours (white, black, red, brown, yellow, green, light blue) are used. However, no kind of orientation are suggested. Furthermore, we could not find any semantic meaning for the vertical or horizontal position. As a conclusion, across the whole visual vocabulary, we identified 3 visual variables, as depicted in Table 4.

Table 4. Visual variables and associated variations.

Visual variable	Total of visual variations	Variations
Shape	9	Human shape, flag, warning triangle, oval shape, bag, explosion, padlock symbol, arrowhead link, square (high level CORAS gates)
Texture	2	Full lines, fine strokes
Brightness	–	
Size	–	
Colour	7	White, Black, red, brown, yellow, green (treatment), light blue (legal norm)
Orientation	–	
Horizontal position	–	
Vertical position	–	

Principle of Dual Coding. Dual coding relates to the use of text to complement graphics. In the CORAS language, text is present to describe what a symbol may represent, like a precision (describing a law or an article), but it doesn't refer to the associated construct (legal norm). The symbols and the notation of CORAS are purely graphic. Therefore, CORAS is not allowing dual coding.

Principle of Graphic Economy. The number of different graphical symbols should be cognitively manageable. The graphic complexity is defined by the number of different graphical conventions used in a notation (i.e. number of legend entries). CORAS is composed of 17 different graphical symbols, but not all of them are used in every diagram. The reference book of CORAS [5] clearly states what are the elements that are used in the five different diagrams. For example, an asset diagram can contain only stakeholders, (direct or indirect) assets, and harm relations. The maximum number of legend entries is thus 4 for asset diagram. We used this notion of maximum since the number of legend entries can differ from one model to another: some constructs allowed to be used in a given diagram may not be used in a specific instance of this kind of diagram, reducing thus the number of legend entries in this case. For example, in a threat diagram, identified threats can be only human, when no non-human threat have been identified. Moreover, we can include CORAS extensions (High-Level CORAS, Dependent CORAS, Legal CORAS) and their symbols in the different diagrams, increasing the maximum number of legend entries. The graphic economy scores for the five different types of diagrams in CORAS are depicted in Table 5.

Table 5. Graphic economy scores

Types of diagrams	Max number of legend entries
Asset diagram	4
Threat diagram	9
Risk diagram	5
Treatment diagram	12
Treatment overview diagram	7
High-level CORAS	+1
Dependent CORAS	+1
Legal CORAS	+2

Principle of Cognitive Fit. For a satisfying cognitive fit, a visual language is supposed to make a good use of different visual dialects for different tasks and audience. The aim is to come up with a good ability to communicate through peers for the language and to be usable for different situations. CORAS proposes one kind of dialect, with five types of diagrams, for more than one type of audience (are cited: analysis leader, analysis secretary, representatives of the customer with decisions makers, technical expertise and users) and eight different tasks to complete a risk assessment (see Table 6).

Table 6. Cognitive fit characteristics.

Cognitive fit elements	Occurrences
Number of dialects for this language	1
Number of different audiences	1+
Number of different tasks admitted	8
Other	5 types of diagrams

5   Discussion

As already argued earlier in this paper, although the PoN is one of the most elaborated approach to evaluate cognitive effectiveness of a modelling language, it is difficult, if not impossible, to establish clear-cut conclusions with this framework. Indeed, because of the lack of reference scales or explicit outcome to achieve associated to the principles of the PoN, it is not possible to firmly claim a language is considered as satisfactory (or not). However, the conclusions we draw from the results obtained after the evaluation of the cognitive effectiveness of CORAS are the following:

- The notation of CORAS seems not very difficult to apprehend by security risk managers as a larger number of symbols compared to signs are used, making the notation pretty transparent and, therefore, understandable. The semiotic clarity principle is largely fulfilled, as only four constructs are represented with the same graphical symbol, this aspect constituting the sole anomaly with regards to semiotic clarity. Furthermore, perceptual discriminability of the language is globally allowing users to visually differentiate and compare symbols, the visual distance between the different constructs of the language being 2 or more in more than 80% of the cases. The lack of dual coding and the use of additional visual variables (e.g., spatial position or size of constructs) are however ways for improvement;
- The use of the language to support a whole security risk assessment would suffer from the low support proposed by CORAS for cognitive integration, despite the presence of the High-level CORAS extension as complexity management mechanism. Considering the complexity of current IS associated to the large number of risks to manage (usually more than hundred based on our experience), it is clearly difficult to consider that CORAS would be efficient at this level;
- Regarding the graphic economy principle, CORAS seems cognitively manageable for most diagrams, that means individual models produced are usually manageable by our short-term memory. Indeed, Miller came up with the fact that our short-term memory can manage seven, plus or minus, two items [28]. As depicted in Table 5, asset diagram, threat diagram, risk diagram and treatment overview diagram have plus or minus seven legend entries in the general case (this result may however be modified when some allowed legend entries are not introduced or when some extensions are used). Nevertheless, memory span can differ depending on the hierarchical organisation of these items in chunks. Therefore, we cannot really draw a line here on what we consider as cognitively manageable or not. An experience on short-term memory management of visual items and chunk strategy would be necessary in order to put clearer limits in an appropriate context;
- The fact that there is actually only one kind of dialect (cognitive fit) could represent some sort of challenge for the various audiences addressed. Security risk managers are the target group of CORAS model designers, but CORAS model users in general are broader, ranging from security experts to people generally weakly skilled and aware on security aspects such as the top management of a company.

During our evaluation, we identified two main threats to validity of the results obtained. The first one is that the analysis performed remains subjective, because performed (only) by three evaluators who are the authors of this article. To reduce the biases coming from this aspect, we used as much as possible quantitative and verifiable metrics during our analysis. However, it is clear to us that most of the principles would benefit from an approach based on users, who would be involved in the evaluation of the language. This "user centric approach" could be performed by semi-structured interviews, focus groups composed of actual users, and, more specifically, user testing, where users are invited to do tasks, while their behaviours are observed to identify flaws of the language. The problems found with user testing are true problems in the sense that at least one user encountered each identified problem [29]. In contrast, the problems found with heuristic evaluation, as we made in this study, are potential problems: the evaluators suspect that something may be a problem to users.

A second threat to validity is that the discussion and conclusions are not based on some 'reference cognitive effectiveness scale' to which we could compare our quantitative results. Several quantitative analysis of modelling languages could lead us to establish such a calibration scale in order to better compare languages between them and get a benchmark of them for specific applications/contexts.

6 Conclusion

In this paper, we evaluated the cognitive effectiveness of the modelling language of CORAS, an approach for ISSRM. As evaluation framework, we used the PoN, a comprehensive set of principles based on a synthesis of theories from, e.g., the psychology and cognitive science fields, that can be used in order to analyse the cognitive effectiveness of existing visual notations, or aid the design of new ones. The need for such an evaluation comes from some drawbacks we observed in traditional ISSRM methods, especially the difficulty to have a clear and manageable documentation for ISSRM activities. Our insight is to introduce conceptual models as support of ISSRM activities and, in this context, the cognitive effectiveness of the produced models is considered as a must-have.

Based on the conclusion drawn during the evaluation performed, CORAS seems not very difficult to apprehend by security risk managers and individual models should generally be manageable by users. The main weaknesses identified are the difficulty to support a whole security risk assessment due to the weak inclusion of cognitive integration and complexity management mechanisms, and the absence of consideration of the different audiences that can be involved in a risk assessment. Compared to the RSO, the other ISSRM language we evaluated thanks to the PoN [4], CORAS can be considered from a general point of view as more cognitive effective, mainly because of a broader use of iconic shapes, a better semiotic transparency and a better perceptual discriminability.

Regarding future work, to complete the heuristic evaluation, we want to adopt a User Centred Design (UCD) approach, in order to take into account the capabilities of actual users, as well as their skills and cognitive limitations. To do that, we plan to apply some methods, like interviews and personas, which establish user needs and specifications. We also plan to use the experience map method, allowing us to draw and understand the walkthrough of the security risks managers when they assess the risks of an organisation. UCD should allow us to suggest recommendations and improvements aligned with actual needs of users, and to make decisions on the necessary trade-offs about our visual syntax, taking care of a specific context.

Acknowledgments. Supported by the National Research Fund, Luxembourg, and financed by the ENTRI project (C14/IS/8329158).

References

1. Kordy, B., Mauw, S., Radomirović, S., Schweitzer, P.: Foundations of attack–defense trees. In: Degano, P., Etalle, S., Guttman, J. (eds.) FAST 2010. LNCS, vol. 6561, pp. 80–95. Springer, Heidelberg (2011). https://doi.org/10.1007/978-3-642-19751-2_6
2. Band, I., Engelsman, W., Feltus, C., Paredes, S.G., Hietala, J., Jonkers, H., Massart, S.: Modeling Enterprise Risk Management and Security with the ArchiMate® Language. The Open Group (2015)
3. Mayer, N., Grandry, E., Feltus, C., Goettelmann, E.: Towards the ENTRI framework: security risk management enhanced by the use of enterprise architectures. In: Persson, A., Stirna, J. (eds.) CAiSE 2015. LNBIP, vol. 215, pp. 459–469. Springer, Cham (2015). https://doi.org/10.1007/978-3-319-19243-7_42
4. Mayer, N., Feltus, C.: Evaluation of the risk and security overlay of archimate to model information system security risks. In: IEEE 21st International Enterprise Distributed Object Computing Conference Workshops (EDOCW), pp. 106–116. IEEE (2017)
5. Lund, M.S., Solhaug, B., Stolen, K.: Model-Driven Risk Analysis: The CORAS Approach. Springer, Heidelberg (2010). https://doi.org/10.1007/978-3-642-12323-8
6. Hitchman, S.: Practitioner perceptions on the use of some semantic concepts in the entity–relationship model. Eur. J. Inf. Syst. 4, 31–40 (1995)
7. Hitchman, S.: The details of conceptual modelling notations are important - a comparison of relationship normative language. Commun. Assoc. Inf. Syst. 9, 167–179 (2002)
8. Nordbotten, J.C., Crosby, M.E.: The effect of graphic style on data model interpretation. Inf. Syst. J. 9, 139–155 (2001)
9. Shanks, G.: The challenges of strategic data planning in practice: an interpretive case study. J. Strateg. Inf. Syst. 6, 69–90 (1997)
10. Figl, K., Derntl, M., Rodriguez, M.C., Botturi, L.: Cognitive effectiveness of visual instructional design languages. J. Vis. Lang. Comput. 21, 359–373 (2010)
11. Green, T.R.G., Petre, M.: Usability analysis of visual programming environments: a 'Cognitive Dimensions' framework. J. Vis. Lang. Comput. 7, 131–174 (1996)
12. Moody, D.: The "Physics" of notations: toward a scientific basis for constructing visual notations in software engineering. IEEE Trans. Softw. Eng. 35, 756–779 (2009)
13. Moody, D., van Hillegersberg, J.: Evaluating the visual syntax of UML: an analysis of the cognitive effectiveness of the UML family of diagrams. In: Gašević, D., Lämmel, R., Van Wyk, E. (eds.) SLE 2008. LNCS, vol. 5452, pp. 16–34. Springer, Heidelberg (2009). https://doi.org/10.1007/978-3-642-00434-6_3
14. Moody, D.L., Heymans, P., Matulevičius, R.: Visual syntax does matter: improving the cognitive effectiveness of the i* visual notation. Requir. Eng. 15, 141–175 (2010)
15. Genon, N., Heymans, P., Amyot, D.: Analysing the cognitive effectiveness of the BPMN 2.0 visual notation. In: Malloy, B., Staab, S., van den Brand, M. (eds.) SLE 2010. LNCS, vol. 6563, pp. 377–396. Springer, Heidelberg (2011). https://doi.org/10.1007/978-3-642-19440-5_25
16. Moody, D.L.: Review of ArchiMate: The Road to International Standardisation. ArchiMate Foundation and BiZZDesign B.V. (2007)
17. Beckers, K., Heisel, M., Solhaug, B., Stølen, K.: ISMS-CORAS: a structured method for establishing an ISO 27001 compliant information security management system. In: Heisel, M., Joosen, W., Lopez, J., Martinelli, F. (eds.) Engineering Secure Future Internet Services and Systems. LNCS, vol. 8431, pp. 315–344. Springer, Cham (2014). https://doi.org/10.1007/978-3-319-07452-8_13

18. van der Linden, D., Hadar, I.: A systematic literature review of applications of the physics of notation. IEEE Trans. Softw. Eng. **PP**, 1 (2018)

19. Störrle, H., Fish, A.: Towards an operationalization of the "Physics of Notations" for the analysis of visual languages. In: Moreira, A., Schätz, B., Gray, J., Vallecillo, A., Clarke, P. (eds.) MODELS 2013. LNCS, vol. 8107, pp. 104–120. Springer, Heidelberg (2013). https://doi.org/10.1007/978-3-642-41533-3_7

20. van der Linden, D., Zamansky, A., Hadar, I.: How cognitively effective is a visual notation? On the inherent difficulty of operationalizing the physics of notations. In: Schmidt, R., Guédria, W., Bider, I., Guerreiro, S. (eds.) BPMDS/EMMSAD -2016. LNBIP, vol. 248, pp. 448–462. Springer, Cham (2016). https://doi.org/10.1007/978-3-319-39429-9_28

21. Krogstie, J.: Using a semiotic framework to evaluate UML for the development of models of high quality. In: Unified Modeling Language: Systems Analysis, Design and Development Issues, pp. 89–106. IGI Global (2001)

22. Genon, N.: Unlocking Diagram Understanding: Empowering End-Users for Semantically Transparent Visual Symbols (2016)

23. Frank, U.: Domain-specific modeling languages: requirements analysis and design guidelines. In: Reinhartz-Berger, I., Sturm, A., Clark, T., Cohen, S., Bettin, J. (eds.) Domain Engineering, pp. 133–157. Springer, Heidelberg (2013). https://doi.org/10.1007/978-3-642-36654-3_6

24. Guizzardi, G., Pires, L.F., van Sinderen, M.: Ontology-based evaluation and design of domain-specific visual modeling languages. In: Nilsson, A.G., Gustas, R., Wojtkowski, W., Wojtkowski, W.G., Wrycza, S., Zupančič, J. (eds.) Advances in Information Systems Development, pp. 217–228. Springer, Boston (2006). https://doi.org/10.1007/978-0-387-36402-5_19

25. Kleppe, A.: Software Language Engineering: Creating Domain-Specific Languages Using Metamodels. Addison-Wesley Professional (2008)

26. Nielsen, J., Molich, R.: Heuristic evaluation of user interfaces. In: Proceedings of the SIGCHI Conference on Human Factors in Computing Systems, pp. 249–256. ACM, New York (1990)

27. Zender, M., Mejía, G.M.: Improving icon design: through focus on the role of individual symbols in the construction of meaning. Vis. Lang. **47**, 66–89 (2013)

28. Miller, G.A.: The magical number seven, plus or minus 2: some limits on our capacity for processing information. Psychol. Rev. **63**, 81–97 (1956)

29. Lauesen, S., Pave Musgrove, M.: Heuristic evaluation of user interfaces versus usability testing. In: User Interface Design - A Software Engineering Perspective, pp. 443–463 (2005)

Symposium on Conceptual Modeling Education (SCME) 2018

Preface to SCME 2018

6th Symposium on Conceptual Modeling Education (SCME 2018) at the 37th International Conference on Conceptual Modeling (ER 2018)

Xi'an, China
October 22–25, 2018

Organized by
Isabelle Comyn-Wattiau and Hui Ma

The 6th Symposium on Conceptual Modeling Education (SCME 2018) is held in conjunction with the 37th International Conference on Conceptual Modeling (ER 2018), in Xi'an (China). SCME aims at providing a forum for discussing education and teaching concepts conceptual modeling. It is an opportunity to exchange on the reasons why conceptual modeling must be taught, the context in which it may take place, and the "what" and the "how" of teaching conceptual modeling.

We would like to thank all of the authors who submitted papers to SCME 2018 for their efforts in promoting concepts and methods for teaching conceptual modeling and sharing their experience. Each paper was carefully reviewed by at least three members of the program committee. We thank all the program committee members for their constructive and timely reviews of the papers, meeting tight deadlines. Finally, we are grateful to the ER 2018 Conference, Program and Workshop Chairs, for their help in organizing this Symposium.

We hope that the interesting presentations of SCME 2018 will contribute to the efficiency and efficacy of the reader's teaching of conceptual modeling and will motivate future editions of the SCME. The first paper entitled *Teaching physical database design* describes an active learning approach experimenting how students can better acquire the main skills of database design, allowing them to choose between different physical design alternatives in the context of relational databases. The second paper entitled *The Simple Enterprise Architecture Framework: Giving Alignment to IT Decisions* presents a methodological framework facilitating the definition of enterprise architectures, aligned with IT decisions and organizational needs. Its experimentation for teaching enterprise architecture is described.

We hope that much fruitful conversation and exchange will occur during the session, and so, we express special appreciation to everyone who will participate in the workshop.

October 2018

Isabelle Comyn-Wattiau
Hui Ma
Program Co-chairs

Teaching Physical Database Design

Karen C. Davis[✉]

Computer Science and Software Engineering Department, Miami University,
Oxford, OH 45056, USA
karen.davis@miamioh.edu

Abstract. Database design is traditionally taught as a process where requirements are captured in a conceptual model, then forward engineered into a logical
model, followed by implementation in a physical model. Conceptual models are
intended to be abstract and platform-independent, thus expressing aspects of the
application being modeled without narrowing the design choices prematurely.
The early stages of the process are more established in pedagogy, while the last
stage, physical design, seems largely unexplored in the computing education
literature. Moreover, if the choice for the logical model is relational, there are
several possible implementation models; the design space widens again after the
phase of transforming a conceptual model into a logical model. This paper
explores teaching students about the design space for physical modeling. The
contents of learning modules on physical design are presented, including scaf-
folding of technical content in an abstract (conceptual) manner, followed by
connection to real-world analogues, culminating in a project that requires
application of conceptual knowledge to explore the impact of physical design
alternatives. The achievement of learning outcomes with and without the final
project is assessed for courses taught in two consecutive academic terms. Future
areas of research are discussed, such as expanding and refining the physical
design space, the student preparation, and the types of impact investigated.

Keywords: Teaching database design · Physical design · Index selection
Cost models

1 Introduction

The ACM/IEEE Joint Task Force on Computing Curricula (2013) defines knowledge
areas for computer science undergraduate education. The area that encompasses conceptual modeling and database design is Information Management. The Information
Management area includes topics such as data modeling and abstraction, physical data
models, and selecting appropriate techniques that address design concerns. The
learning outcomes associated with the topics include:

- identifying appropriate indexes for a given relational schema and query set,
- estimating the time to retrieve information with and without indexes,
- working with primary, secondary, and clustering indexes, including B+trees, and
- determining how physical database design affects transaction efficiency.

© Springer Nature Switzerland AG 2018
C. Woo et al. (Eds.): ER 2018 Workshops, LNCS 11158, pp. 165–175, 2018.
https://doi.org/10.1007/978-3-030-01391-2_22

While the ACM/IEEE curriculum guidelines consider all of these topics to be elective subjects for computer science students, other recent task forces have considered data management as a critical component in the study of data science.

Widespread interest in developing and enhancing undergraduate data science education is evidenced by the interim report *Envisioning the Data Science Discipline: The Undergraduate Perspective* (2017) by the US National Academies of Science, Engineering, and Medicine. The National Academies report states that "the need to manage, analyze, and extract knowledge from data is pervasive across industry, government, and academia." The report acknowledges that data science is a hybrid area of study, drawing on diverse fields including statistics and computer science. Their list of critical curricular topics includes principles of effective data management, techniques for data description and curation, and data modeling approaches, among many other skills and areas necessary for students to achieve competency in data science.

The US National Science Foundation funded a study entitled *Realizing the Potential of Data Science* (2016) that defines data science as "processes and systems that enable the extraction of knowledge and insights from data in various forms, either structured or unstructured." In their lifecycle view of data science, data management permeates all phases: acquisition, cleaning, using/reusing, publishing, and preserving/destroying data.

I believe the interest in studying database design will continue to increase as a result of the popularity of data science; particularly relevant is how to store and retrieve data effectively. I also contend that what we term *physical modeling* in database education is sufficiently abstract to be considered a kind of *conceptual modeling* (but not in the traditional use of the term as applied to a database design process). To keep the phases of database design distinct, in this paper I will refer to the conceptual part of physical modeling as *abstract modeling.*

The options for physical storage and retrieval of data have proliferated recently with the advent of NoSQL systems. The question about which logical/physical model to use for data management is a pressing issue for many projects and organizations, and sometimes the answer is still the relational model; however, even for the relational model there are numerous possibilities for physical designs that impact performance. In a recent conversation with a colleague who was using MySQL for a large research data set, he said that it took 2.5 h to load his data using the MySQL backend InnoDB; he switched to an older backend system, MyISAM, and it took 5 min. I was intrigued and I wanted to investigate the reason why with my database class. They were using MySQL for their database implementation project. We were concurrently studying physical storage design, including abstract models of the structure and performance of different kinds of indexes. The students had some experience with the Google Cloud Platform and I was also intending to use Google's Big Query in an assignment. I devised a new assignment to apply what they were learning in class to a scenario based on real-world systems. I hoped relating the abstract physical modeling to actual systems would help motivate them to learn the topic and that there would be a corresponding increase in their achievement of course learning outcomes.

This paper focuses on an assignment addressing the impact of physical design choices that reflect real-world relational systems. The abstract models for physical design that are studied are: a multi-level primary index, a B+tree index, and a column store with data distribution and parallel query execution. These correspond to the real-world implementations of MySQL's MyISAM, MySQL's InnoDB, and Google's Big Query.

Section 2 reviews physical design topics covered in selected database textbooks and research about teaching physical design from the computer science educational literature. Section 3 outlines an aspirational educational pedagogy that is partly implemented in the learning activities described in Sect. 4. Section 5 examines results of student learning in the same database course offered in two academic terms. Section 6 discusses lessons learned and directions for future enhancements and investigation.

2 Related Work on Teaching Physical Design

Table 1 gives an overview of typical coverage of physical design topics in some well-known general database textbooks. For convenience, the numbered citations are [1] Connolly and Begg (2009), [2] Elmasri and Navathe (2016), [3] Garcia-Molina et al. (2000), [4] Ramakrishnan and Gehrke (2000), and [5] Silberschatz et al. (1997). Connolly and Begg (2009) focus on database design and offer the only textbook to include a discussion of forward engineering a logical schema design to a physical schema design in SQL DDL; they do not include details of index implementation. Both Ramakrishnan and Gehrke (2000) and Garcia-Molina et al. (2000) include extensive coverage of the physical database design space, query optimization, and performance tuning. For database textbooks that cover physical design, it is standard practice to include the indexes that are utilized in the assignment described in Sect. 4. Only one of the textbooks surveyed includes a discussion of columnar storage on query performance (note that this may be due to older editions of the other books used because of their ready availability to the author.) Table 1 is included here to illustrate that the focus of the lesson discussed in Sect. 4 is typical for introductory database courses; it is not intended as an evaluation of textbooks.

There are research papers discussing teaching of database design, but digital searches of the ACM, IEEE, and Google Scholar databases did not identify any relevant papers about teaching physical design. A manual search through the table of contents of the *ACM Transactions on Computing Education* journal and the *Journal on Information Systems Education* for the last 10 years did not identify any related literature. A manual search of ten years of the ACM SIGCSE Technical Symposium on Computer Science Education proceedings contents yielded papers describing:

- an undergraduate database course where students develop a database management system, including implementing B-trees (Sotomayor and Shaw 2016), and
- an undergraduate database course where students investigate database internals that mostly focuses on transactions processing (Sciore 2007).

These two papers are the closest match to teaching of physical design concepts and neither addresses physical database design; they are more targeted toward the development of systems rather than design. To the best of my knowledge, there is no other educational research investigating the teaching of physical database design.

Table 1. Physical design topics in database textbooks

Topic	[1]	[2]	[3]	[4]	[5]
File structures	•	•	•	•	•
External hashing	–	•	•	•	•
Indexes: primary	–	•	•	•	•
Clustered	–	•	•	•	•
Secondary unique	–	•	•	•	•
Secondary non-unique	–	•	•	•	•
B+tree (or variant)	–	•	•	•	•
Multi-level indexes	–	•	•	•	•
Column storage model	–	•	–	–	–
Index selection (cost-model based)	–	•	•	•	–

• indicates coverage; - indicates no coverage

3 Active Learning Pedagogy

The merits of active learning (as opposed to passive listening to a lecture) are well-documented. A meta-study by Freeman et al. (2014) examined 225 STEM education research efforts and came to the conclusion that the evidence for impact on student learning is overwhelmingly positive for active learning techniques. One of the best models I have encountered, supported by research and practice, is given in the recent book *Flipped Learning* by Talbert (2017). Instructors do not have to commit to a fully flipped model to use active learning, but the flipped model can provide good student outcomes and it frees up time to do more active learning.

Two important ideas for the flipped model are the individual space and the group space. The individual space is where a student learns on their own, which ideally prepares a student to engage more fully in the group space activities by first encountering and becoming familiar with simpler learning outcomes. The instructor should provide guided practice resources and exercises for the individual space.

In the group space, usually in the classroom, the instructor can focus on more complex learning outcomes. The time and activities in the classroom can be broken down into several phases: (1) opening minutes: connecting with pre-class preparation, keeping students accountable for the preparation, and clarifying misconceptions, (2) middle of class: conducting the group activity, and (3) the closing minutes: debriefing the activity and reflection on issues and learning.

The post-group activity can be a continuation or extension of the in-class activity. The reader is referred to Talbert's book (2017) for an extensive discussion of the history and pedagogy of flipped learning as well as detailed instructions about how to design lessons and troubleshoot common problems. The next section describes a partial application of the model, with particular emphasis on a new post-group space exercise.

4 Active Learning Applied to Physical Database Design

The introductory database course described and studied here is a first course taken by undergraduate computer science and software engineering students, students pursuing a minor in either of those two disciplines, or students pursuing an analytics co-major offered jointly by Statistics and Information Systems and Analytics departments. The only prerequisite for the course is a data structures course. The course was offered as a blended course rather than a fully flipped course (a combination of mini-lectures and in-class active learning group exercises.)

4.1 Individual Space: Student Preparation

Ideally in the flipped model, first exposure to the material comes through individual preparation using a guided practice established by the instructor. In the blended model, the first exposure may be through a mini-lecture in class, although resources are provided in advance. If students are not accountable in any way for preparation, it is likely that they will not prepare. In the case of the course under discussion here, the resources posted in advance on the course learning management system are lecture slides and textbook readings. Students were occasionally required to complete online quizzes prior to class, but this was not consistently assigned. An example of the learning outcomes and resources posted in advance in the course learning management system is given in Fig. 1. This learning module is one of two related to physical design.

Additional learning outcomes for the other physical design learning module are summarized here:

1. compute performance characteristics (size of indexes and number of accesses) for a secondary index with unique values.
2. sketch a secondary index with a unique index key for sample data files.
3. compute performance characteristics (size of index and number of accesses) for a B +tree.
4. search and insert into a B+tree.
5. discuss advantages and disadvantages of all indexes discussed in the course.

4.2 Group Space: Active Learning Classroom Exercises

The course was offered in blended mode, so the first exposure to the new material was generally through mini-lectures delivered in class, followed by working through a typical problem the students would be asked to do in class. Students worked in small groups, generally pairs or trios, while the instructor and TA would walk around and provide guidance. In the case of widespread misconceptions or other difficulties, the instructor would generally work a problem on a document camera or whiteboard to clarify the issues. Students were provided the opportunity to engage in peer instruction while working on low-stakes exploration of the topic, with expert help on hand to assist. Students were expected to work through to a correct solution, although the grading was effort-based rather than correctness-based. Correct solutions were generally provided on the board by the end of class, although we sometimes ran out of time and had to continue in the next class. Ideas for improving the lessons are discussed in Sect. 6.

Physical Storage Characteristics and Indexing Sorted Data

Learning Outcomes

Upon successful completion of this module, you will be able to:

1. given a logical schema and physical organization, compute performance characteristics including
    1. record size
    2. blocking factor
    3. number of records in a file
    4. number of blocks to store a file
2. compute the number of blocks retrieved to answer an exact match query for
    1. unsorted data
    2. sorted data
    3. a primary index
3. sketch a primary index and illustrate sample data and index entries.

Readings

- 6th edition: Ch. 17-17.2, 17.4, 17.6-17.7, Ch. 18-18.1.2
- 7th edition: Ch. 16-16.2, 16.4, 16.6-16.7, Ch. 17-17.1.2

Lectures

- physical storage design: disks.pptx
- organizing data on disk: data-org.pptx
- answering a query: retrieval.pptx
- indexing sorted data: indexing-sorted.pptx
    - primary-index-worksheet.docx

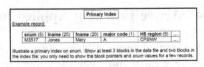

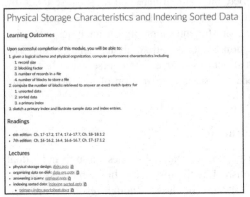

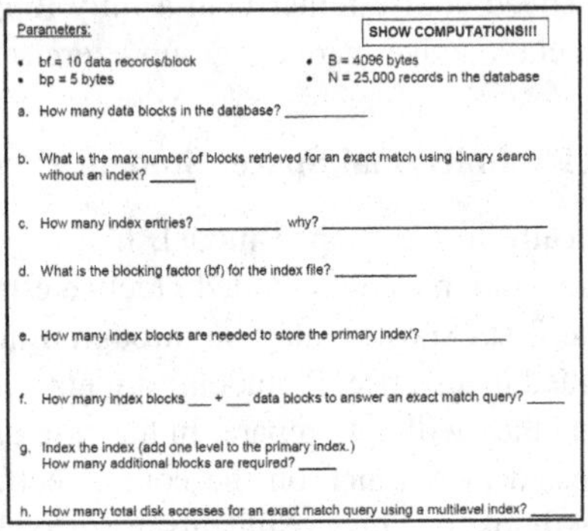

Fig. 1. Introducing physical design in the course learning management system

Primary Index

Example record:

snum (5)	lname (25)	fname (20)	major code (1)	HS region (5)	...
M3517	Jones	Mary	A	CPSNW	...

Illustrate a primary index on *snum*. Show at least 3 blocks in the data file and two blocks in the index file; you only need to show the block pointers and *snum* values for a few records.

Fig. 2. Example problem for sketching a primary index

Parameters:

SHOW COMPUTATIONS!!!

- bf = 10 data records/block
- bp = 5 bytes
- B = 4096 bytes
- N = 25,000 records in the database

a. How many data blocks in the database? ________

b. What is the max number of blocks retrieved for an exact match using binary search without an index? _____

c. How many index entries? ________ why? ________________

d. What is the blocking factor (bf) for the index file? ________

e. How many index blocks are needed to store the primary index? ________

f. How many index blocks ___ + ___ data blocks to answer an exact match query? _____

g. Index the index (add one level to the primary index.) How many additional blocks are required? _____

h. How many total disk accesses for an exact match query using a multilevel index? _____

Fig. 3. Cost modeling for primary and multilevel indexes

An excerpt from an in-class problem set is given in Fig. 2. Figure 3 shows a cost model computational exercise. These are the problems the instructor would work through, followed by the students working similar problems together and with guidance readily available.

4.3 Post-class Assignment

A new assignment was created in a recent course offering to reinforce the abstract physical design concepts and connect them to real-world systems. In the assignment, students were allowed to work in pairs to apply their understanding of data in sorted files and unsorted files, multi-level primary indexes and B+trees, and centralized and distributed/parallel data access.

A mini-lecture introduced the real-world analogues for abstract models as follows:

- multi-level primary index: MySQL MyISAM[1]
- B+tree file and index: MySQL InnoDB[2]
- column store (centralized and distributed): Google Big Query[3]

[1] https://dev.mysql.com/doc/refman/8.0/en/myisam-storage-engine.html.

[2] https://dev.mysql.com/doc/refman/8.0/en/innodb-storage-engine.html.

[3] https://cloud.google.com/blog/big-data/2016/01/bigquery-under-the-hood.

Students were already familiar with the necessary background to perform computations with these three abstract models, but they could now map their understanding to existing systems. The assignment still remained in the abstract realm, however; the main point of the assignment was for students to see that their course learning outcomes were reflections of contemporary systems. The students were asked to think of themselves as professional consultants and justify the answers they computed for each of the techniques. The introduction to the problem, parameters to use, and the first computation are shown in Fig. 4. Questions 2 and 3 (not shown) increase the number of records to 10,000,000 and 1,000,000,000, respectively.

Figure 5 shows questions 4 and 5 of the assignment. Question 4 asks students to consider adding more nodes (processors) to distribute the data in a columnar storage format. Big Query does not utilize any indexes, it scales out horizontally to improve performance. Since the data is unsorted, the expected time to answer a single exact match query on average is $n/2$, where n is the number of blocks at the node. The students were asked to determine how many nodes were needed for the Big Query storage (a vertical fragment of the data with just one column) to have better performance than either of the centralized, indexed data files. The students were allowed to solve the problem manually, in a spreadsheet, or write a program. The final question, Question 5, asks the students to reflect on the style of their solution choice and how easy it would be to alter the parameters of the problem and respond to their client in a timely fashion.

5   Evaluation of Learning Outcomes

There are two sets of student assessment data relevant to physical database design that were collected as a normal part of the course offerings. Neither was established specifically to evaluate the impact of the new assignment. A closer alignment of the assessment with the learning outcomes remains as future work. One data set consists of pre- and post-assessment of a student's perception of their competency regarding the learning outcomes for the course. The second is the instructor's assessment of learning outcomes based on the final exam for the course.

A client of Gizmonic Consultants, Inc. wants to know which database system they should choose. They want to know which configuration gives the best performance for an exact match query with the parameters specified below. The choices they have are MySQL with the ISAM engine, MySQL with the InnoDB engine, and Google's Big Query (with one node). For ISAM, assume that the data is sorted. For the others, assume it is not sorted.

- a block is 16K bytes (use 16,000 for your computations)
- the indexed field is a BIGINT field (8 bytes)
- the record size is 200 bytes
- a block pointer is an INT field (4 bytes)

Create a report for your client to explain (justify) your answers for Questions 1-4. Be sure to bill the client for your hours (create a timelog). Summarize your results on a cover page using the provided tables.

1. For n = 100,000 records, how many blocks are retrieved for an exact match query using

a. ISAM	
b. InnoDB	
c. Big Query	

Fig. 4. Assignment introduction

4. Increase the number of nodes in Big Query to 2, then 3, then 4, and so on, until the performance exceeds ISAM and InnoDB. Use the number of blocks for one Big Query node as the performance indicator since the processing would be done in parallel for multiple nodes.

   a. How many nodes are necessary to exceed the performance of ISAM for each number of records? (fill in the table below)

   b. How many nodes are necessary to exceed the performance of InnoDB for each number of records? (fill in the table below)

	100,000	10,000,000	1,000,000,000
a. ISAM			
b. InnoDB			

5. How well would your solution technique scale for different record sizes, key sizes, and block sizes?

Fig. 5. Comparison of centralized solutions to distributed/parallel solutions

The pre- and post-assessment surveys completed by students have a number of general course outcomes established by a departmental curriculum committee. The question most closely related to physical design asks the students to rate their agreement with the following statement: *I can explain the file structures that are used by Relational Database Management Systems.*

The student scores range from 1 = *strongly disagree* and 5 = *strongly agree*. Data was collected in two consecutive academic terms where the course was taught by the same instructor. Results of a paired t-test for each term indicate that statistically significant improvement is shown from the pre-assessment at the beginning of the course to the post-assessment at the end of the course. Only observations for students who gave consent to use their data and who completed both pre- and post-assessment results are reported here (about 1/3 to 1/2 of the students enrolled in each course). The summary statistics are shown in Table 2.

Table 2. Self-reported competency for RDBMS file structures

	n	avg (pre)	stdev (pre)	avg (post)	stdev (post)	p-value
Course A	18	1.78	0.41830	4.00	1.05882	0.000000084072
Course B	24	1.96	1.08514	4.25	0.36957	0.000000007174

It is interesting to note that the variation decreased in the post-responses for course B while the average increased. This might possibly indicate that the additional practice with a project based on real-world database storage models had a positive impact on learning, or at least a positive impact on increasing students' confidence that they understood the concepts, compared to course A students. Further study will investigate this idea and continue to improve the assignment as well as the student's preparation and instructor's feedback.

Course A students did not have the project introduced in Sect. 4.3, while course B students did. In order to examine the impact on abstract learning, the scores for an exam question are compared for course A and course B using a one-tailed t-test for unequal variances. The points awarded for the question are normalized to a 100 point scale. The exam question involved deciding which of the three indexes could be built over a sample data set and then sketching the selected indexes. For the exam, students were asked to determine which of the indexes they studied apply to the data set, and then they are asked to sketch an index, if applicable, for the given data set. The statistical summary is given in Table 3. Only scores for students providing consent to use their data in this study are utilized.

Table 3. Exam question scores

	n	Average	stdev	p-value
Course A	36	61.30	34.72594	0.033017
Course B	42	74.29	24.83862	

The results indicate that students in course B performed significantly better than students in course A on the abstract physical design exam question. It is encouraging that the average score is higher and the standard deviation is lower, however, there is still a great deal of room for improvement in the scores. It is anticipated that additional refinement of the assignment and increased preparation guidelines, additional in-class project time, and instructor follow-up after completion of the assignment and before the exam will have increased positive impact on achievement of student learning outcomes, although this remains to be demonstrated.

6 Reflections

Teaching physical database design embodies the usage of conceptual models for communication, discussion, and most especially for analysis. A new assignment was created in an introductory database class to teach physical design concepts. The assignment was motivated by a desire to create a post-group space learning experience that would relate abstract physical storage models to actual systems. Some positive impact on learning was measured in the course that featured the assignment compared to a prior offering of the course without the new assignment. However, the assignment and the method of teaching the concepts can be improved, as can the assessment methods used to study student learning. In terms of organizing the lesson, suggested revisions include

- create individual space guided practice (Talbert 2017; Garcia 2018) for basic outcomes emphasizing familiarity with the terminology and having some accountability on the part of the student,
- allocate more time for group space activities (in-class work on problem sets and the project itself), and
- debrief after the assignment, including time for questions, verifying solutions, and student reflection.

The assignment itself could be better supported by creating a small assignment to write queries using Big Query. This would not require any additional SQL knowledge beyond the course outcomes and activities for SQL already in place, but it would make the discussion about columnar storage for distributed query processing more pertinent. Another option to allow students more practice with the physical design assignment and to give them better feedback before their exam would be to offer the assignment in phases. The conceptual modeling assignment the students do earlier in the semester leverages two phases: an individual phase (where they receive copious feedback) and a paired phase, where they revise and merge their individual designs into a consensus design. The scores on the second phase are several letter grades higher than on the first phase, so the second one contributes more to their grade. A marked improvement on students' exam performance was also observed (Davis 2014).

The study could be strengthened by assessing the impact on learning with more precise questions, for example, revising the pre- and post-statements toward the activities emphasized in the course. Additional exam questions about the computational aspects of performance that relate directly to the assignments could be examined as well.

References

ACM/IEEE Joint Task Force on Computing Curricula: Computer science curricula 2013: curriculum guidelines for undergraduate degree programs in computer science. Technical report, Association for Computing Machinery (ACM). IEEE Computer Society (2013)

Envisioning the Data Science Discipline: The Undergraduate Perspective: Interim Report, The National Academies of Science/Engineering/Medicine (2017)

Realizing the Potential of Data Science, Final Report from the National Science Foundation Computer and Information Science and Engineering Advisor Committee Data Science Working Group, co-chairs: Francine Berman and Rob Rutenbar, December 2016

Connolly, T.M., Begg, C.E.: Database Systems: A Practical Approach to Design, Implementation, and Management, 5th edn. Pearson Education, New York (2009)

Davis, K.C.: Teaching conceptual design capture. In: Parsons, J., Chiu, D. (eds.) ER 2013. LNCS, vol. 8697, pp. 247–256. Springer, Cham (2014). https://doi.org/10.1007/978-3-319-14139-8_26

Elmasri, R., Navathe, S.B.: Fundamentals of Database Systems, 7th edn. Pearson Education, New York (2016)

Freeman, S., et al.: Active learning increases student performance in science, engineering, and mathematics. Proc. Natl. Acad. Sci. 111(23), 8410–8415 (2014)

Garcia, S.: Improving classroom preparedness using guided practice. In: Proceedings of the 49th ACM Technical Symposium on Computer Science Education, pp. 326–331. ACM, February 2018

Garcia-Molina, H., Ullman, J.D., Widom, J.: Database System Implementation. Prentice Hall, Upper Saddle River (2000)

Ramakrishnan, R., Gehrke, J.: Database Management Systems, 2nd edn. McGraw Hill, Boston (2000)

Sciore, E.: SimpleDB: a simple Java-based multiuser system for teaching database internals. In: ACM SIGCSE Bulletin, vol. 39, no. 1, pp. 561–565. ACM, March 2007

Silberschatz, A., Korth, H.F., Sudarshan, S.: Database System Concepts, 4th edn. McGraw-Hill, New York (1997)

Sotomayor, B., Shaw, A.: chidb: building a simple relational database system from scratch. In: Proceedings of the 47th ACM Technical Symposium on Computing Science Education, pp. 407–412. ACM (2016)

Talbert, R.: Flipped Learning: A Guide for Higher Education Faculty. Stylus Publishing, Sterling (2017)

The Simple Enterprise Architecture Framework: Giving Alignment to IT Decisions

Giovanni Giachetti[1(✉)], Beatriz Marín[2], and Estefanía Serral[3]

[1] Universidad Tecnológica de Chile INACAP, Santiago, Chile
ggiachetti@inacap.cl
[2] Facultad de Ingeniería y Ciencias, Universidad Diego Portales, Santiago, Chile
beatriz.marin@mail.udp.cl
[3] Faculty of Economics and Business, KU Leuven, Leuven, Belgium
estefania.serralasensio@kuleuven.be

Abstract. *Context*: Enterprise Architecture seeks to align organizational objectives with decisions associated to people, processes, information and technology. Different frameworks have been proposed for designing an enterprise architecture, such as TOGAF or Zachman. However, defining an enterprise architecture following these frameworks is not an easy task since it requires the alignment of organizational needs with technological decisions. *Goal*: This paper presents a methodological framework, called Simple Enterprise Architecture (SEA), that eases the definition of an enterprise architecture and provides concrete proof of the alignment between IT decisions and organizational needs. *Method*: The SEA framework has been developed by integrating components from existing ones in order to generate a concrete process that guides analysts in the correct definition of an enterprise architecture. *Results*: This framework has been used in teaching system architecture master courses for 5 years with positive results.

Keywords: Enterprise architecture · Information technology · Framework
Teaching · Lessons learned · IT decisions

1 Introduction

Enterprise Architecture (EA) is a field intended to improve the management and functioning of complex enterprises and their information systems by aligning organizational objectives with decisions associated to people, processes, information and technology. EA helps an organization to transform its business vision and strategy into effective enterprise change through a clear understanding of its current state (as-is) and its expected future state (to-be) [12].

Numerous frameworks have been created for defining EA [19], such as the Open Group Architecture Framework (TOGAF) [6] or the Zachman framework [24]. They provide a set of practices or activities that guide the EA definition. However, correctly defining an EA using one of these frameworks (or combination of them) is still a complex task, since it requires understand each framework specification, its use to represent the set of key concepts related to the organizational context, and finally, understanding how to align the organizational needs with technological decisions that is represented by a conceptual specification. This complexity makes the quality of an EA modeling design directly dependent on the experience of the work team or external consultants that creates it.

In this paper, we present the Simple Enterprise Architecture (SEA) Framework that is oriented to guide analysts in the correct definition and validation of an enterprise architecture for IT projects. The SEA framework builds on and integrates different practices from the well-known EA frameworks TOGAF and the Zachman framework, with system architecture and measure specification methods, the Attribute-Driven Design [23] (ADD), the Architecture Tradeoff Analysis Method (ATAM) [11], and the Goal Question Metric (GQM) strategy [1]. Moreover, some lessons learned from the use of SEA framework in system architecture master courses for novel and senior engineers during five years are presented, which can be a reference for practitioners related to use and/or teaching of enterprise architectures.

The rest of the paper is organized as follows. Section 2 presents the state of the art and related work on EA architectures, Sect. 3 introduces the SEA framework. Section 4 presents the analysis and lessons learned from applying the SEA framework. Finally, Sect. 5 presents our main conclusions and future work.

2   Related Work

The initial idea about considering different dimensions of an enterprise was developed simultaneously within different disciplines in the early nineties. This inevitably led to the emergence of several EA frameworks (EAFs) [2]. An EAF tries to map the enterprise organizational goals to the work processes and IT infrastructure that are needed to reach those goals. EA frameworks support a broad range of objectives and enable decision making on different levels. According to [7, 13], more than 50 EAFs are available to date. Some examples are: the Zachman framework, ARIS, TOGAF, the Federal Enterprise Architecture (FEA) framework, the Department of Defense Architecture Framework (DoDAF), the Treasury Enterprise Architecture Framework (TEAF), Enterprise Architecture Planning (EAP), or more recently, ArchiMate, and the integrated electronic Requirements Information Management framework (eRIM). For EAFs' comparisons and analyses, the authors refer to [2, 13, 16, 17, 25].

The Zachman Framework [24] is organized as a matrix with two dimensions. The first dimension represents six perspectives or views: Planner, Owner, Designer, Builder, Subcontractor, and User. The second dimension deals with the following six basic questions: what, how, where, who, when and why. These questions allow to gain insights into different aspects of an enterprise. The framework does not provide guidance on sequence, process, or implementation, but rather focuses on ensuring that all views are well established, regardless the order in which they are created.

ARIS [18] is a framework for modeling business information systems from a process-based perspective. The framework is close to a classic software development process consisting of the sequence of requirements elicitation, design specification, and implementation description, applied to the different views of a business information systems: organization, data, function, output, and control.

TOGAF [6] provides a structure and classification schema used as a reference for architecture development. It gives a comprehensive description of the relevant elements of EA and provides a clear distinction of the business oriented description and the derived technological implementation. A key element of TOGAF is the Architecture Development Method (ADM), which is an iterative method for developing EA. TOGAF explains rules for developing three levels of principles: (1) support decision making across the entire enterprise, (2) provide guidance of IT resources, and (3) support architecture principles for development and implementation.

The FEA framework [10] consists of a set of interrelated reference models that provide standardized categorization for strategic, business, and technology models. The main goal of FEAF is to organize and promote sharing of Federal information for the entire Federal Government. FEA allows for flexibility in the use of methods, work products, and tools to be used by the individual federal agencies.

DoDAF [4] is a holistic framework and conceptual model for EA development particularly in Department of Defense agencies. The framework builds on three sets of views: Operational, System, and Technical Standards. These views are linked in a fourth view by means of a dictionary to define terms and by providing context, summary, or overview-level information. This framework provides descriptions of final products as well as guidance and rules for consistency to facilitate comparison and integration of systems and architectures.

TEAF [3] aims at facilitating the integration, information sharing, and exploitation of requirements across the Department of the Treasury [3]. Similar to DoDAF, TEAF includes descriptions of work products for documenting and modeling enterprise architectures. The framework was created to map the interrelationships among the organizations of the department in order to manage IT resources.

EAP [20] proposed by Capgemini contains activities and processes in order to achieve a To-Be architecture by considering four perspectives: Business, Data, Application, and Infrastructure.

The ArchiMate Standard [5] is a synthesis of ideas coming from previous EAFs. It introduces an integrated language for describing EAs. ArchiMate fits into the TOGAF framework as it provides concepts for creating a model that correlates to its three architectures (layers).

eRIM [9] was defined for managing information about client requirements across all phases of a construction project and through-life of a built facility. The framework defines an information-centric EA approach that is process and service-oriented.

Some works have also been proposed to integrate different EA frameworks in order to better support the design of an EA. The Integrated Architecture Framework (IAF) [22] is based on ideas from Zachman and EAP and it provides practical guidelines to apply it. In [14], a framework is presented for the definition of EA of government organizations. The framework integrates TOGAF ADM with the Service Architecture Network Architecture (SONA) for connecting network infrastructure, network services, and application. In [25], a framework that integrates Zachman, Four-domain (processes, procedures, business tools, and dependencies required to support business functions), TOGAF, and the Reference Model of Open Distributed Processing (RM-ODP) is introduced to be used as a common framework to communicate about EAFs of different businesses and relate them to each other.

While some of these frameworks (e.g., DoDAF, FEAF, and TEAF) have been designed to address specific needs, others (e.g., Zachman and TOGAF) can be more widely applied. In any case, these frameworks are still too abstract, they lack of straightforward guidelines to apply them, which makes difficult their understanding and accurate application [17, 21]. Moreover, no framework provides a concrete process that guides analysts in the definition of an EA, which gives evidence on how IT decisions are aligned with the organizational goals. All these issues provoke that students cannot properly use these frameworks, resulting in a frustrating experience. Next section presents the SEA framework, which overcome the issues identified.

3   The Simple Enterprise Architecture (SEA) Framework

The SEA framework builds on the three levels shared by TOGAF and Zachman (see Fig. 1). The Business Architecture, shown on top of the pyramid, defines the business goals and processes, and is used as the driving artefact to align the two levels below: the Information Systems Architecture, which defines the design of the information systems that must support the defined business processes, and the Technical Architecture, which defines the technical infrastructure (e.g., hardware, networks, etc.) that will support those information systems.

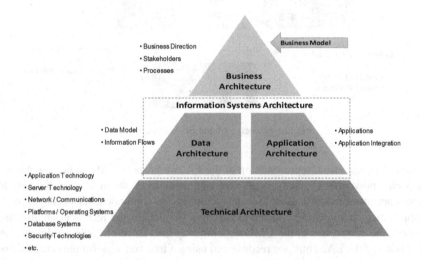

Fig. 1. Architectural levels of the EAF framework from TOGAF

To guide the alignment of these levels, quality attributes must be considered to evaluate to what extent the organizational requirements are met. These attributes are used to drive the information systems and technical decisions (at lower levels of the pyramid) that support the identified business processes. SEA applies the Attribute-Driven Design [23] (ADD) method and the utility tree form Architecture Tradeoff Analysis Method (ATAM) to use these attributes as architectural drivers that guide analysts to refine the first version of the information system and technical architectures. Thus, the EA will be depicted in a way that all decisions of the three levels are aligned to these architectural drivers.

In order to validate and refine this version, we use tradeoffs analysis template from ATAM [11], which complements the previous step. This step allows obtaining a set of indicators that can be used to evaluate and support the architectural decisions.

Finally, the indicators obtained are used to define specific measures that demonstrate the alignment between the three levels, mainly the alignment of the technical decisions (technical architecture) with the business objectives. To obtain these measures, we use the GQM (goal, question, metric) approach.

Figure 2 shows the complete SEA process. It is iterative and incremental, starting with a first definition of the EA guided by the business strategy. This definition is iteratively evaluated and refined by the application of ADD, ATAM and GQM.

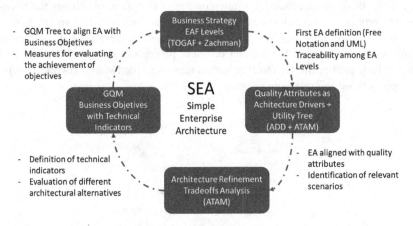

Fig. 2. Elements of the SEA framework

Regarding the visual representation of the EA, we propose UML, a language that is very well known by IT analysts. We recommend using the use case diagram for business architecture, class model for data architecture, component (or package) model for application architecture, and the deployment model for the technical architecture. However, the correct use of the UML notation may also be an obstacle for defining the first version of the EA. Thus, we recommend using a free and very simple notation first, e.g., a notation based on boxes and arrows, and use the UML to give a detailed and refined EA definition once a first EA version is obtained.

3.1   Business Strategy: EAF Levels

The starting point of SEA is to stablish a first proposal for the EA taking as guidance the business strategy, by following the next steps: definition of the business architecture, definition of the data architecture, definition of the application architecture, and definition of the technical architecture.

The business architecture determines the data and application architecture, but also the implementation of new technology can modify the data and the business architectures: the business strategy and the technology of an organization are tightly related and greatly influence each other. Therefore, once the business, the data, and the application architectures are defined, they are refined starting the process in the opposite order (see bottom of Fig. 3): first considering the application architecture, then the data architecture, and finally the business architecture. In this second iteration, the first EA proposal is analyzed to evaluate whether additional elements are necessary. A new component in the application architecture can create the need to support additional data, and consequently the need of additional processes that support the management of those data. Normally, new elements for adding functionality or services not previously supported and needed to perform the organizational change are identified in this second iteration, such as applications to integrate legacy systems, to deal with security issues, or to support distributed services or user interfaces (Web, mobile, etc.).

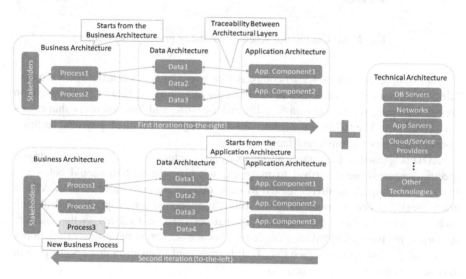

Fig. 3. Right-to-left definition schema for EA levels

The right-to-left process finishes when no new elements are identified. Once these three architectures are defined, a first definition of the technical architecture is created where the technical components to support the correct execution of the applications and data management are identified.

3.2   Quality Attributes as Architectural Drivers

In order to organize the different architectural drivers, we extend the utility tree proposed by ATAM [11] (see Fig. 4). This tree identifies a set of quality attributes in the first level. These attributes are: Performance, Modifiability, Availability, and Security. In the second level of the tree, the different architectural drivers are detailed, and finally the intended scenarios for each driver are presented. In the SEA framework, we propose two changes to this utility tree. The first is to consider the six ISO 9126 [8] quality attributes and their sub-characteristics for internal and external quality instead of the four attributes originally proposed by ATAM.

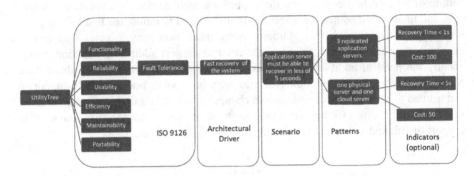

Fig. 4. SEA utility tree schema (Reliability branch is detailed)

As example, consider a Retail company that wants to develop a website for online sales. Important quality attributes in this example following the ISO 9126 classification are: *security* that comes from *functionality* and *usability;* and *fault tolerance* that comes from *reliability*. The following architectural drivers can be identified: payment security, ease of use, response time, and fault recovery.

ADD is applied to use quality attributes as architectural drivers. (a) for each driver, specific scenarios are described to make explicit how the system should behave to successfully support the driver related to some quality attribute (third level of ATAM utility tree); (b) one or more technical solutions (patterns) are proposed to deal with each scenario. To show the technical alternatives (patterns) related to the different scenarios, we have included an extra level after the scenarios definition in the utility tree of the ATAM approach (see Fig. 4). It is important to mention that a same pattern may support more than one scenario.

3.3   Tradeoff Analysis and Technical Indicators

In order to choose the appropriate patterns for the different scenarios, the SEA framework uses the tradeoffs analysis template proposed by ATAM [11]. For instance, from the information of Table 1, if the cost of having the server down for 1 min on Black Fridays is 500 then, the most appropriate pattern will be the 3 replicated servers. However, the architects can also choose to combine patterns, in case it is possible. For instance, another alternative is to consider the three replicated servers together with the cloud server since the cost having the server down is much higher and the cloud server can mitigate the risk R1 for pattern P1.

Table 1. Example of ATAM template for defining scenarios.

Architectural Driver: Fast recovery of the system				
Scenario: Application server must be able to recover in less of 5 min				
Environment: Execution with more than 10.000 concurrent customers (Black Friday)				
Stimulus: The main server is not responding				
Response: The system resumes in less of 5 min				
Architectural Patterns	**Sensitivity**	**TradeOff**	**Risk**	**No Risk**
P1: 3 replicated servers		T1	R1	
P2: local and cloud server	S1	T2		
Reasoning	T1: Lower recovery time (1 min), but higher cost (100) T2: Lower cost (50), but higher recovery time (5 min) S1: Dependency of the cloud provider R1: The three servers may fail			

3.4 Measures for Aligning Technical Architecture and Strategy

In order to make explicit the relation between the technical architecture indicators and the organizational goals, we use the GQM approach. GQM allows us to generate measures that use these indicators as input GQM proposes the following three levels to align objectives with specific measures: Conceptual Level – Goals, Operational Level– Questions, and Quantitative Level – Measures.

At the conceptual level, the goals that have driven the EA are identified. In the Retail example, to increase the number of sales is a goal. At the operation level, we define the questions to evaluate to what extent the goal is reached. A key aspect to increase sales is to offer a good customer service. Thus, a question associated to this goal can be, *e.g.*, is the company offering a good customer service? Questions can be linked to more than one goal, *e.g.*, knowing if a company offers a good customer service is also important for the goal of increasing customer satisfaction.

Finally, the quantitative level set the measures that can evaluate the questions in a quantitative manner. Figure 5 shows that the Recovery time indicator enables the definition of the measures cost of system down and number of unattended requests. This indicator is obtained from the tradeoffs analysis in the previous (ATAM) step.

We recommend applying GQM by following a hybrid approach (between top-down and bottom-up). Figure 5 shows the indicator I1 (recovery time), which allows the definition of the measure number of unattended request and costs of system down. In this case, the lower is the recovery time, the lower is the cost to deal with the system failure as well as the not processed requests.

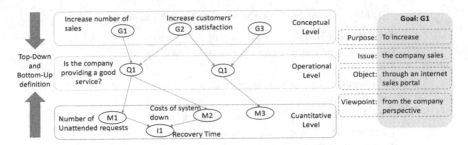

Fig. 5. GQM for generation of measures aligned with business objectives

4 Analysis and Lessons Learned

Previous to the use of SEA framework, former students of the system architecture master course fail in the proper alignment of the IT architecture with the organizational goals, in the representation of the architectural models, and finally, they were not able to explain how the defined architecture fits to the organizational needs. Moreover, we have observed that students put the main focus on the intended system functionality, omitting how the system interacts with other organizational processes and stakeholders.

During the last five years, SEA has been used by more than 400 students of the system architecture course of a master program. The SEA framework breaks the lack of organizational vision, providing a systematic process and tools that guide the students in the definition of technical solutions driven by the business strategy.

The use of a well-known notation as UML facilitates the communication among the technical team. However, it can be difficult to start with the architecture representation in a specific notation since students pay more attention to properly represent the model more than to determine if all the relevant elements are included. To solve this issue, we propose to start with a free notation (boxes and arrows) for representing the initial EA model, mainly centered on identifying the relevant concepts, and, in a further step, to translate the initial model with a concrete notation.

We have observed another interesting effect of translating the EA from a free notation model to a concrete notation model: inconsistencies are identified during the translation process. Normally, these inconsistencies are non-identified actors (stakeholders) or hidden uses cases (process). Thus, this translation step also works as a verification activity for the EA specification using SEA.

5 Conclusions and Future Work

This paper has presented a simple framework called SEA that integrates best practices from well-known EAFs and system architecture and measure specification methods. The SEA framework has been designed to systematically guide analysts in the correct definition of EAs. The SEA framework facilitate the proper design of technical solutions, and also provides evidence that allows these solutions to be evaluated in terms of how they contribute to the achievement of organizational goals.

The aim of the SEA framework is to facilitate the teaching of EA courses at computer engineering careers, due to existing approaches, such as [15], where too complex according to the students perception. The evaluations surveys of the courses where SEA was applied has scores of 90% or above regarding comprehension, support and guidance material, ability of apply the knowledge acquired, and the feasibility of use SEA in professional practice. In contrast, similar curses without SEA have an average between 50–70%. This provides an initial evidence of the perception of the students about the use of SEA framework to correctly define an enterprise architecture. Nevertheless, the validation of the SEA framework and industrial empirical evaluation of SEA is planned as future work.

Finally, we are preparing an online course to teach the SEA framework and facilitate its free use by practitioners and students, this is the Open SEA initiative. Some online lectures have already been recorded (such as https://vimeo.com/album/2472336/video/72053189).

References

1. Basili, V., Caldeira, G., Rombach, H.D.: The Goal Question Metric Approach. Wiley, Hoboken (1994)
2. Buckl, S., Schweda, C.M.: On the State-of-the-Art in Enterprise Architecture Management Literature. Technische Universität München, Munich (2011)
3. Goethals, F.: An overview of enterprise architecture framework deliverables (2003)
4. Group, D.o.D.A.F.W.: Department of Defense Architecture Framework, January 2003
5. Group, T.O.: The Open Group, ArchiMate 2.0 Specification, Berkshire, UK (2012)
6. Haren, V.: TOGAF Version 9.1. Van Haren Publishing (2011). ISBN: 9087536798
7. Hinkelmann, K., Gerber, A., Karagiannis, D., Thoenssen, B., Van der Merwe, A., Woitsch, R.: A new paradigm for the continuous alignment of business and IT: combining enterprise architecture modelling and enterprise ontology. Comput. Ind. **79**, 77–86 (2016)
8. ISO: ISO/IEC 9126-1. In: Software engineering – Product quality – Part 1: Quality Model. ISO (2001)
9. Jallow, A.K., Demian, P., Anumba, C.J., Baldwin, A.N.: An enterprise architecture framework for electronic requirements information management. Int. J. Inf. Manag. **37**(5), 455–472 (2017)
10. Ji, W.-L., Xia, A.-B.: Federal enterprise architecture framework. Comput. Integr. Manuf. Syst. Beijing **13**(1), 57 (2007)
11. Kazman, R., Klein, M., Clements, P.: ATAM: Method for Architecture Evaluation. Software Engineering Institute, Carnegie-Mellon University, Pittsburgh, PA (2000)
12. Lapkin, A., et al.: Gartner clarifies the definition of the term 'enterprise architecture'. Gartner Res. (2008)
13. Matthes, D.: Enterprise Architecture Frameworks Kompendium: Über 50 Rahmenwerke für das IT-Management. Springer, Heidelberg (2011). https://doi.org/10.1007/978-3-642-12955-1
14. Muhamad-Firmansyah, C., Bandung, Y.: Designing an enterprise architecture government organization based on TOGAF ADM and SONA. In: 2016 International Conference on Information Technology Systems and Innovation (ICITSI), pp. 1–6. IEEE (2016)
15. Pereira, C.M., Sousa, P.: A method to define an Enterprise Architecture using the Zachman Framework. In: ACM Symposium on Applied Computing, pp. 1366–1371. ACM (2004)

16. Romero, D., Vernadat, F.: Enterprise information systems state of the art: past, present and future trends. Comput. Ind. **79**, 3–13 (2016)
17. Rouhani, B.D., Mahrin, M.N.R., Nikpay, F., Najafabadi, M.K., Nikfard, P.: A framework for evaluation of enterprise architecture implementation methodologies. Int. J. Soc. Behav. Educ. Econ. Bus. Ind. Eng. **9**(1) (2015)
18. Scheer, A.-W., Nüttgens, M.: ARIS architecture and reference models for business process management. In: van der Aalst, W., Desel, J., Oberweis, A. (eds.) Business Process Management. LNCS, vol. 1806, pp. 376–389. Springer, Heidelberg (2000). https://doi.org/10.1007/3-540-45594-9_24
19. Schekkerman, J.: How to Survive in the Jungle of Enterprise Architecture Frameworks: Creating or Choosing an Enterprise Architecture Framework. Trafford Publishing, Bloomington (2004)
20. Spewak, S.H., Tiemann, M.: Updating the enterprise architecture planning model. J. Enterp. Architect. **2**(2), 11–19 (2006)
21. Urbaczewski, L., Mrdalj, S.: A comparison of enterprise architecture frameworks. Issues Inf. Syst. **7**(2), 18–23 (2006)
22. Van't Wout, J., Waage, M., Hartman, H., Stahlecker, M., Hofman, A.: The Integrated Architecture Framework Explained: Why, What, How. Springer, Heidelberg (2010). https://doi.org/10.1007/978-3-642-11518-9
23. Wojcik, R., et al.: Attribute-driven design (ADD), version 2.0. Software Engineering Institute, Carnegie-Mellon University, Pittsburgh, PA (2006)
24. Zachman, J.A.: Concepts of the framework for enterprise architecture (1996)
25. Zarvić, N., Wieringa, R.: An integrated enterprise architecture framework for business-IT alignment. Des. Enterp. Architect. Framew. Integr. Bus. Process. IT Infrastruct. **63**, 9 (2014)

Empirical Methods in Conceptual Modeling (Emp-ER) 2018

Preface to EmpER'18

1st Workshop on Empirical Methods in Conceptual Modeling (EmpER'18) at the 37th International Conference on Conceptual Modeling (ER 2018)

Xi'an, China
October 22–25, 2018

Organized by
Jennifer Horkoff and Sotirios Liaskos

Conceptual modeling has enjoyed substantial growth over the past decades in diverse fields such as Information Systems Analysis, Software Engineering, Enterprise Architecture, Business Analysis and Business Process Engineering. A plethora of conceptual modeling languages, frameworks and systems have been proposed, promising to facilitate activities such as communication, design, documentation or decision-making.

Success in designing a conceptual modeling system is, however, predicated on demonstrably attaining such goals through observing their use in practical scenarios. At the same time, the way individuals and groups produce and consume models gives raise to cognitive, behavioral, organizational or other phenomena, whose systematic observation may help us better understand how models are used in practice and how we can make them more effective.

The inaugural International Workshop on Empirical Methods in Conceptual Modeling (EmpER'18), co-located with the 37th International Conference on Conceptual Modeling (ER 2018), aimed at bringing together researchers with an interest in the empirical investigation of conceptual modeling systems and practices. The workshop invited three kinds of papers: finished empirical studies, proposed empirical studies and theoretical, review or experience papers on the topic of empirical research in conceptual modeling. The workshop particularly welcomed proposed empirical studies that are in their design stage so that authors can benefit from early feedback and adjust their designs prior to a potentially effort- and resource-intensive administration. A review checklist based on standard validity concerns, geared primarily towards experimental studies, was constructed and distributed to the reviewers to be used – at their discretion – for reviewing the papers and organize their feedback to the authors. A total of twenty (20) reviewers were invited to serve in this first offering of the workshop based on their record of past contributions in the area of empirical conceptual modeling.

Overall, a total of three (3) papers were accepted out of the six (6) that were reviewed. Two of the accepted papers describe proposed experimental studies and one proposes a methodological tool. The workshop involves presentations of the papers followed by extensive discussion and audience feedback to the authors.

We would like to thank all authors, reviewers and workshop chairs of ER'2018 for their valuable work in making this workshop possible. We are hoping that this only the first small step of a long and exciting journey of community building and mutual learning in the area of Empirical Methods in Conceptual Modeling.

July 2018

Jennifer Horkoff
Sotirios Liaskos
Program Co-chairs

Towards an Empirical Evaluation of Imperative and Declarative Process Mining

Christoffer Olling Back[1]([⊠])(iD), Søren Debois[2](iD), and Tijs Slaats[1](iD)

[1] Department of Computer Science, University of Copenhagen,
Emil Holms Kanal 6, 2300 Copenhagen S, Denmark
{back,slaats}@di.ku.dk
[2] Department of Computer Science, IT University of Copenhagen,
Rued Langgaards Vej 7, 2300 Copenhagen S, Denmark
debois@itu.dk

Abstract. Process modelling notations fall in two broad categories: declarative notations, which specify the *rules* governing a process; and imperative notations, which specify the *flows* admitted by a process. We outline an empirical approach to addressing the question of whether certain process logs are better suited for mining to imperative than declarative notations. We plan to attack this question by applying a flagship imperative and declarative miner to a standard collection of process logs, then evaluate the quality of the output models w.r.t. the standard model metrics of precision and generalisation. This approach requires perfect fitness of the output model, which substantially narrows the field of available miners; possible candidates include Inductive Miner and MINERful. With the metrics in hand, we propose to statistically evaluate the hypotheses that (1) one miner consistently outperforms the other on one of the metrics, and (2) there exist subsets of logs more suitable for imperative respectively declarative mining.

Keywords: Process mining · Modelling paradigms
Statistical evaluation · Declarative models · Imperative models
Hybrid models · Evaluation metrics

1 Introduction

Workflow notations are commonly categorised as falling within either the *imperative* or *declarative* paradigm [1]. Imperative notations use flow-based constructs to explicitly model the *paths* through a process [2]. Declarative notations use

T. Slaats—This work is supported by the Hybrid Business Process Management Technologies project (DFF-6111-00337) funded by the Danish Council for Independent Research, and the EcoKnow project (7050-00034A) funded by the Innovation Foundation.

C. Woo et al. (Eds.): ER 2018 Workshops, LNCS 11158, pp. 191–198, 2018.
https://doi.org/10.1007/978-3-030-01391-2_24

constraint-based constructs to model the *rules* of a process. A declarative model allows all paths not forbidden by the constraints, and therefore the behaviour of the model is implicit in the rules and needs to be deduced by the system or users [3–5]. While the imperative paradigm is more mature, both paradigms have seen industrial adoption [6–8].

A recent trend in both academia and industry is to extract models from real-life data via *process discovery* [9], where an *output model* is automatically constructed from an *event log* of observed process executions. Research into this approach has focused primarily on the discovery of imperative models, but substantial energy has been directed towards algorithms that discover declarative models as well [10–12].

Thus, when constructing process models by process discovery, we have a choice regarding which paradigm to use. Does one approach return better models than the other? Would such a difference be universal or depend on the particular input log?

This paper outlines an approach to empirically evaluating the effect of miner paradigm (*independent variable*) on output model quality (*dependent variable*) [13,14]. We propose to measure model quality using notation-agnostic metrics for *precision* and *generalisation* from [15], which apply equally to imperative and declarative models. These metrics are analogous to the standard data mining metrics of the same name and are intended to capture the degree to which a model is underfitting or overfitting the data, respectively. Other quality metrics exist, such as fitness, simplicity/understandability, and soundness; but we are forced to keep these as *controlled variables* due to the fact that this formulations of precision and generalisation restricts our choice of miners, which in turn restricts our ability to include other metrics as dependent variables. Namely, since precision and generalisation require output models either be perfectly fitting or that data be aligned to the model to account for "noise", and since we have chosen to exclude the alignment procedure as a *confounding variable* in the first iteration of this evaluation approach (see Sect. 2.2), we are restricted to perfectly fitting output models.

We propose Inductive Miner [16] and MINERful [11] as representatives of the imperative and declarative paradigms, respectively, as they fulfill our requirements and are widely considered to be at the cutting edge of their respective fields. Evaluating the miners on publicly available, real-life logs, we test the following hypotheses:

Hypothesis 1: One miner consistently outperforms the other on one of the metrics:

(a) outperformance on precision   (b) outperformance on generalisation

Hypothesis 2: There exist subsets of logs:

(a) more suited for imperative mining   (b) more suited for declarative mining

That is, there exists a subset of logs which when mined either declaratively or imperatively represents a Pareto improvement over the other; *and* this deviation from the zero mean lies outside of the bounds of what can be accounted for by random chance.

A Pareto improvement simply denotes an improvement on at least one metric without sacrificing performance on the remaining metric. The zero mean is the mean of the probability distribution associated with the null hypothesis, and represents no performance difference between models produced by different miners from the same log.

Note that in the most extreme case, a subset may consist of a single log which is best suited to one paradigm, with the remaining logs showing only an insignificant difference in precision and generalisation, or requiring a tradeoff between the metrics, thus not a Pareto improvement. We are, in fact, testing multiple sub-hypotheses for hypothesis 2: one for each log. To compensate for the increased likelihood of making a type I error (false positive), we perform the appropriate adjustment to statistical significance testing: a *Bonferroni correction*. We leave as future work the task of identifying the characteristics of event logs which distinguish them as best suited for a given paradigm, in the interest of first rigorously establishing a clear framework for evaluation.

2   Methods

2.1   Log Selection

In the interest of reproducibility, we base log selection on the criteria of public availability, drawing upon the IEEE Task Force on Process Mining Real-life Event Log Collection[1], with the addition of one additional real-life log originating from our own industrial contact, the Dreyer Foundation in Denmark [17]. The logs stem from diverse sectors, including healthcare related processes, fine management, permit, loan and grant applications, as well as production, software engineering, and robotic vehicle related processes. The logs vary in degree of structure, number of activities, and trace length.

2.2   Process Discovery

We will mine the selected logs both imperatively and declaratively, selecting miners according to the following criteria:

1. Miners must be configurable to always produce perfectly fitting models.
2. Miners must be configurable to produce models of a given simplicity, save one which can serve as a benchmark.

The first criterion follows from both precision and generalisation requiring perfect fitness of the output model. It would have been an option to allow non-fitting output, and then use model-log alignment [15, 18], but without domain

---

[1] http://data.4tu.nl/repository/collection:event_logs_real.

knowledge or access to an expert, we cannot know which exact alignment is more appropriate for the real-world log. This means that we would be evaluating not just the mining algorithm, but the combination of mining algorithm and alignment function. In particular, we would not know whether to attribute a result in favour of one miner over the other to the miner itself, or to a fortunate choice of alignment for that particular miner.

The second criterion follows partly from a tendency of declarative miners to produce output models containing excessive numbers of constraints: for large logs, on the order of *hundreds of thousands*. More importantly, we require model simplicity to be held constant, so that the choice of mining algorithm remains the only independent variable.

The two criteria left us only two miners: The Inductive Miner and MINERful.

Inductive Miner is an imperative miner developed by Leemans et al. [16] which uses a divide-and-conquer approach to generate sound, block-structured process models output as process trees or Petri nets. With only one parameter, noise threshold, which for our purposes must be held at 1.0 to ensure perfect fitness, the model generated by Inductive Miner provides a baseline model from which to set a threshold on model simplicity.

MINERful is a *declarative* miner developed by Di Ciccio et al. [11]. It uses a two-phase approach: in the first phase, a knowledge base of statistical information on the log is built; in the second, this knowledge is queried in order to infer the constraints of the process. The output is a Declare model, possibly including negative constraints.

MINERful has configurable thresholds for *support, interest factor*, and *confidence*. By iteratively adjusting these settings until a model is found which has the highest possible number of constraints without exceeding the complexity of the imperative model, we ensure that the imperative and declarative models are of comparable simplicity. We note that while many measures of simplicity have been proposed for imperative models, there exists no widely accepted method for comparing the simplicity of imperative and declarative models. For this reason, we begin by simply comparing the number of edge elements: edges between transitions vs. constraints between activities.

2.3 Computing Metrics

Defining standard measures for precision and generalisation remains an open research challenge for two main reasons. First, in process mining, data is generally not assumed to be labelled, i.e. event logs contain examples of what *did* happen, not what should *not* happen. This means that the standard definitions used in data mining and statistics cannot be applied to process discovery, since they rely on defining true and false positives, and true and false negatives. Second, the prevalence of unbounded loops in process models means that they often describe an infinite set of allowed behaviour. Therefore, definitions of precision and generalisation which take into account all of the of behaviour allowed by the model are not applicable in practice. Instead most metrics aim to reduce the measured behaviour of the model to a finite set of traces.

Metric Selection. To compare imperative and declarative models, we require metrics that can be applied to both equally. This means that they need to be defined on either the level of languages or transition systems. Accordingly, we have chosen to employ the metrics introduced in [15], in particular:

Precision [15, p. 10] measures the degree to which a model is "underfitting" or "allowing too much behaviour" relative to the input log. This particular metric is based on the notion of *escaping edges*, which represent a point at which the model allows behaviour not seen in the log. The measured amount of additional behaviour is kept finite by only considering the first divergent activity. I.e. an escaping edge may lead to a loop representing an infinite set of traces that did not occur in the log, but only the trace ending with the first divergent activity will be counted.

Generalisation [15, p. 11], on the other hand, measures the degree to which a model is "overfitting": is there behaviour not allowed by the model and not exhibited in the log, but that can be reasonably expected to occur in the future? This particular metric approximates generalisation by estimating for each state in the model the likelihood that a new, hitherto unseen, activity will occur. This estimation is based on the number of activities that have been observed, and how often the state was visited. Two alternatives are offered: *event-based generalisation* takes into account the number of visits to a state, *state-based generalisation* does not.

Implementation. The widely used process mining framework, ProM, contains a plugin for computing the metrics of [15] on Petri nets, but does not offer support for declarative models. Also, we seek the ability to run tests in batches and easily pipeline several operations (mining, metrics computation, analysis) on multiple logs. Therefore we developed our own evaluation framework[2]. The code, methods and results are straightforward to inspect and reproduce by following instructions provided on the associated wiki. The framework was tested against the examples and results reported in [15].

Challenges arise when computing precision and generalisation: mainly regarding time and space efficiency, but also handling nondeterminism arising from silent transitions present in models produced by Inductive Miner. When identifying enabled activities in a given marking, a greedy algorithm will naively follow silent transitions until encountering a non-silent transition, potentially firing silent transitions unnecessarily and associating incorrect markings with an event: subsequently excluding activities which should be enabled. Using the shortest path to a non-silent transition prevents this.

---

[2] Available at: https://bitbucket.org/coback/qmpm.

To minimize redundancy, a prefix tree is built from the event log, replaying each trace on the given model as it is added to the trie. In each node (corresponding to an event in the log), the state of the model is saved, unless the node has been visited previously, in which case a counter associated with the node is incremented, recording the number of occurrences of that prefix. Finally, a map containing model states (markings) as keys, and sets of nodes (events) as values, is maintained in order to facilitate the calculation of state-based generalisation. Given an event, the enabled activities in the log simply correspond to that node's children, while the enabled activities in the model are obtained by querying the model using the model state associated with that node. This approach minimizes redundancies, keeping state-space enumeration to a minimum.

3   Conclusion

We outline an approach to systematically compare the performance of imperative and declarative process mining algorithms based on notation-agnostic quality metrics for precision and generalisation defined in [15]. We will investigate two hypotheses: first, one miner performs better on precision and/or generalisation; second, there exist logs on which either miner provides a statistically significant Pareto improvement.

To the best of our knowledge, this will be the first study comparing imperative and declarative process discovery techniques using this approach. Future evaluations incorporating other aspects of the process mining life-cycle, e.g. alignment, will be able to use this approach as a point of reference. Not least, we contribute a comprehensive software framework and tackle a number of methodological and implementation challenges, providing a foundation upon which further work can build.

Finally, we believe that the proposed study will be extremely valuable to the field of hybrid process mining [19–21], which aims to combine the strengths of the two paradigms. Research into which characteristics identify a portion of a log as more suitable to one paradigm have been hampered by the lack of an objective procedure on which to compare models across paradigms [22]. Our approach lays the groundwork for addressing this shortcoming.

References

1. Reijers, H.A., Slaats, T., Stahl, C.: Declarative modeling–an academic dream or the future for BPM? In: Daniel, F., Wang, J., Weber, B. (eds.) BPM 2013. LNCS, vol. 8094, pp. 307–322. Springer, Heidelberg (2013). https://doi.org/10.1007/978-3-642-40176-3_26
2. Van der Aalst, W.M.P.: Verification of workflow nets. In: Azéma, P., Balbo, G. (eds.) ICATPN 1997. LNCS, vol. 1248, pp. 407–426. Springer, Heidelberg (1997). https://doi.org/10.1007/3-540-63139-9_48
3. van der Aalst, W.M.P., Pesic, M., Schonenberg, H., Westergaard, M., Maggi, F.M.: Declare. Webpage (2010). http://www.win.tue.nl/declare/

4. Debois, S., Hildebrandt, T.T., Slaats, T.: Replication, refinement & reachability: complexity in dynamic condition-response graphs. Acta Informatica **55**, 489–520 (2017)
5. Hull, R., et al.: Business artifacts with guard-stage-milestone lifecycles. In: DEBS 2011, pp. 51–62 (2011)
6. Object Management Group: Business Process Modeling Notation Version 2.0. Technical report, Object Management Group Final Adopted Specification (2011)
7. Marquard, M., Shahzad, M., Slaats, T.: Web-based modelling and collaborative simulation of declarative processes. In: Motahari-Nezhad, H.R., Recker, J., Weidlich, M. (eds.) BPM 2015. LNCS, vol. 9253, pp. 209–225. Springer, Cham (2015). https://doi.org/10.1007/978-3-319-23063-4_15
8. Object Management Group: Case Management Model and Notation, version 1.0. Webpage, May 2014. http://www.omg.org/spec/CMMN/1.0/PDF
9. Van der Aalst, W.M.P.: Process Mining: Data Science in Action. Springer, Heidelberg (2016)
10. Maggi, F.M., Bose, R.P.J.C., van der Aalst, W.M.P.: Efficient discovery of understandable declarative process models from event logs. In: Ralyté, J., Franch, X., Brinkkemper, S., Wrycza, S. (eds.) CAiSE 2012. LNCS, vol. 7328, pp. 270–285. Springer, Heidelberg (2012). https://doi.org/10.1007/978-3-642-31095-9_18
11. Di Ciccio, C., Mecella, M.: On the discovery of declarative control flows for artful processes. ACM Trans. Manag. Inf. Syst. **5**(4), 24 (2015)
12. Debois, S., Hildebrandt, T.T., Laursen, P.H., Ulrik, K.R.: Declarative process mining for DCR graphs. In: Proceeding of the Symposium on Applied Computing, SAC 2017, pp. 759–764 (2017)
13. Buijs, J.C.A.M., van Dongen, B.F., van der Aalst, W.M.P.: On the role of fitness, precision, generalization and simplicity in process discovery. In: Meersman, R., et al. (eds.) OTM 2012. LNCS, vol. 7565, pp. 305–322. Springer, Heidelberg (2012). https://doi.org/10.1007/978-3-642-33606-5_19
14. Buijs, J.C.A.M., van Dongen, B.F., van der Aalst, W.M.P.: Quality dimensions in process discovery: the importance of fitness, precision, generalization and simplicity. Int. J. Coop. Inf. Syst. **23**(1), 1440001 (2014)
15. van der Aalst, W.M.P., Adriansyah, A., van Dongen, B.F.: Replaying history on process models for conformance checking and performance analysis. Wiley Interdisc. Rew. Data Min. Knowl. Disc. **2**(2), 182–192 (2012)
16. Leemans, S.J.J., Fahland, D., van der Aalst, W.M.P.: Discovering block-structured process models from event logs - a constructive approach. In: Colom, J.-M., Desel, J. (eds.) PETRI NETS 2013. LNCS, vol. 7927, pp. 311–329. Springer, Heidelberg (2013). https://doi.org/10.1007/978-3-642-38697-8_17
17. Debois, S., Slaats, T.: The analysis of a real life declarative process. In: CIDM 2015, pp. 1374–1382 (2015)
18. Adriansyah, A., Munoz-Gama, J., Carmona, J., van Dongen, B.F., van der Aalst, W.M.P.: Alignment based precision checking. In: La Rosa, M., Soffer, P. (eds.) BPM 2012. LNBIP, vol. 132, pp. 137–149. Springer, Heidelberg (2013). https://doi.org/10.1007/978-3-642-36285-9_15
19. Slaats, T., Schunselaar, D.M.M., Maggi, F.M., Reijers, H.A.: The semantics of hybrid process models. In: Debruyne, C. (ed.) OTM 2016. LNCS, vol. 10033, pp. 531–551. Springer, Cham (2016). https://doi.org/10.1007/978-3-319-48472-3_32
20. Maggi, F.M., Slaats, T., Reijers, H.A.: The automated discovery of hybrid processes. In: Sadiq, S., Soffer, P., Völzer, H. (eds.) BPM 2014. LNCS, vol. 8659, pp. 392–399. Springer, Cham (2014). https://doi.org/10.1007/978-3-319-10172-9_27

21. Schunselaar, D.M.M., Slaats, T., Maggi, F.M., Reijers, H.A., van der Aalst, W.M.P.: Mining hybrid business process models: a quest for better precision. In: Abramowicz, W., Paschke, A. (eds.) BIS 2018. LNBIP, vol. 320, pp. 190–205. Springer, Cham (2018). https://doi.org/10.1007/978-3-319-93931-5_14

22. Back, C.O., Debois, S., Slaats, T.: Towards an entropy-based analysis of log variability. In: Teniente, E., Weidlich, M. (eds.) BPM 2017. LNBIP, vol. 308, pp. 53–70. Springer, Cham (2018). https://doi.org/10.1007/978-3-319-74030-0_4

Artifact Sampling in Experimental Conceptual Modeling Research

Roman Lukyananko[1], Jeffrey Parsons[2(✉)], and Binny M. Samuel[3]

[1] HEC Montreal, Montreal, QC, Canada
roman.lukyanenko@hec.ca
[2] Memorial University of Newfoundland, St. John's, NL, Canada
jeffreyp@mun.ca
[3] University of Cincinnati, Cincinnati, OH, USA
samuelby@uc.edu

Abstract. Experimental research in conceptual modeling typically involves comparing grammars or variations within a grammar, where differences between experimental groups are based on a focal construct of interest. However, a conceptual modeling grammar is a collection of many constructs and there is a danger that grammatical features other than those under consideration in an experiment can influence or confound the results obtained. To address this issue, we propose the use of *artifact sampling* as a way to systematically vary non-focal grammatical features in experimental conceptual modeling research to control for potential confounds or interactions between constructs of interest and other grammatical features. In this paper, we describe the approach and illustrate its application to the design of a large-scale study to compare alternative notations within the Entity-Relationship family of grammars.

Keywords: Artifact sampling · Conceptual modeling · Experimental research

1 Introduction

Conceptual modeling grammars serve as important boundary objects in the development and use of information systems. Many grammars have been proposed for such diverse applications as modeling data, business processes, and objects.

Historically, the design of grammars has paid scant attention to justifying the selection and visual design of modeling constructs – an issue brought into focus by Moody's concept of the "physics of notations" [1]. Design decisions made in creating a modeling grammar (e.g., why diamonds for relationship types in the original Entity-Relationship (ER) model) are often not justified during the development and articulation of a grammar.

Empirical research in conceptual modeling, particularly that focusing on experimental evaluation of the effects of grammars (or constructs within a grammar) on dependent variables of interest (such as comprehension), generally compares grammars based on variations in some focal construct(s) of interest. Examples include whether or not optional attributes are permitted [e.g., 2], or whether domain familiarity affects understanding [e.g., 3]. However, in such comparisons, most components of the

© Springer Nature Switzerland AG 2018
C. Woo et al. (Eds.): ER 2018 Workshops, LNCS 11158, pp. 199–205, 2018.
https://doi.org/10.1007/978-3-030-01391-2_25

grammar are taken as given, and their potential impact on results is not explicitly considered. For example, does it matter if the effect of optional versus mandatory attributes is tested using the ER grammar or using the UML class diagram grammar? As a consequence, it may be difficult to disentangle meaningful results in such studies from the capricious effects of incidental features of the grammar that might confound or interact with the independent variables of interest. There is evidence from other fields that: (1) "*subtle differences* in interface can cause *major differences*" [4] in a dependent variable of interest [for discussion, see 5]; (2) the "[r]eal-world impact of Information Systems" may be hindered by the potentially unpredictable large effects of "seemingly small design choices" [6]; and (3) "[u]nless sound criteria for evaluating instantiation validity of IS design research is applied, doubts remain whether results are due to extraneous factors or attributable to idiosyncratic" design choices [7].

In this research, we propose a large-scale study using *artifact sampling* [8] to illustrate the value of systematically varying features of a conceptual modeling grammar as a way of studying their effect on a focal construct of interest in an experimental study. We begin by describing artifact sampling. We then tailor the artifact sampling approach to a specific empirical study of conceptual modeling grammars. Next, we outline our empirical study. We conclude by summarizing the expected contributions of our study.

2   Artifact Sampling

2.1   The Need for Artifact Sampling

Any conceptual modeling script generated for use in an experimental study is an artifact instantiated from some design principles – specifically, the rules of a conceptual modeling grammar. However, typically only a few variations of the artifact are used in a particular experimental study. For example, in a study to examine whether allowing or prohibiting optional properties on entity types affects ER diagram comprehension, there might be as few as two scripts compared in a study – one depicting optional attributes and the other replacing optional attributes with subclasses having only mandatory attributes [e.g., 2]. Thus, it is impossible to determine whether any observed effects are due solely to the treatment, or are influenced by interactions between the treatment and the way attributes are depicted in the grammar (e.g., as bubbles attached to an entity type or as a list inside the entity type box). Likewise, such incidental choices might mitigate potential effects of a treatment. We contend that examining such potential effects is needed to ensure confidence in the results of conceptual modeling experiments that examine modeling grammars or scripts, and to understand the boundary conditions for any observed effects.

2.2   Artifact Sampling Foundations

Sampling theory underlies much of scientific experimental work [9]. Fundamental to the theory is the principle that one may generalize the results of observations only to those subjects or objects that have been sampled [10]. However, while sampling theory has been readily applied to research subjects, it has rarely been used for experimental stimuli [11].

While the benefits of involving multiple subjects in experiments and surveys have been widely recognized, applying this to research objects has been largely overlooked, casting doubts on the validity of conclusions drawn from such studies. To increase the validity of experimental studies, some researchers call for *stimuli sampling* – selecting objects at random from the theoretical feature space [12, 13].

We extend this suggestion of sampling object stimuli (experimental or questionnaire items) to the sampling of artifacts such as scripts (or diagrams) used in experimental conceptual modeling research. As the scripts generated using conceptual modeling grammars are relatively complex artifacts (note, psychologists suggest stimuli sampling even for simple line drawings, see [14]), the interpretation of diagrams by model readers might be influenced by incidental decisions about how to visualize a modeling construct or how to arrange symbols in a script. Therefore, many ways of depicting these constructs should be tested to enable conclusions to be drawn about the generalizability of findings or the extent to which they are limited to particular forms of representation.

3 Experimental Design

To demonstrate the utility of artifact sampling for experimental conceptual modeling research, we plan to conduct an exemplar artifact sampling study to compare an exhaustive set of variants of grammar features. By testing whether these features lead to different results on dependent variables of interest, we expect to determine the utility and practical limitations of artifact sampling.

3.1 Experimental Context

We choose as an experimental setting a replication of Khatri et al. [15], who studied the effects of IS domain knowledge and application domain knowledge of conceptual modeling scripts generated using the ER and EER grammars on comprehension of the semantics conveyed in scripts (comparisons were only made within one grammar and not between the grammars). In that context, specific diagrams were used to represent the semantics of familiar and unfamiliar domains. However, no attempt was made to vary the stimuli (e.g., depiction of elements on a diagram). Figure 1 contains four variants of a portion of one of the original diagrams used in the study. Starting in the top left corner of the Fig. 1 and moving clockwise:

(1) Figure 1a directly replicates their work
(2) Figure 1b alters Fig. 1a by using a UML cardinality notation
(3) Figure 1c alters Fig. 1b and removes the relationship diamond
(4) Figure 1d alters Fig. 1c by moving the attributes inside the entity type

To determine whether the findings are robust across a range of diagram variants, we propose to use artifact sampling to replicate the study under a variety of conditions.

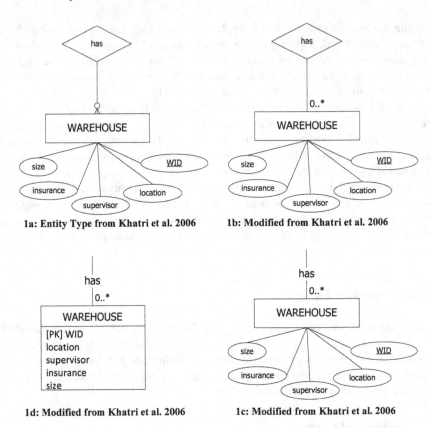

1a: Entity Type from Khatri et al. 2006   1b: Modified from Khatri et al. 2006

1d: Modified from Khatri et al. 2006   1c: Modified from Khatri et al. 2006

Fig. 1. Sample diagram variations based on Khatri et al. [15]

3.2   Artifact Sampling Method

Following Lukyanenko et al. [8, 16], we adopt an artifact sampling method based on four steps. First, we identify the *artifact population space*; in our study it is the space of all possibilities for conducting data modeling in practice. As it is impossible to implement every variation within this space (there could be thousands of ways to draw a data modeling diagram for a domain), reducing this space is necessary to reduce the number of diagrams to a manageable level.

Second, we *theoretically* sample from the artifact population space by considering the constructs within an ER-based conceptual model which we conjecture (based on theory, past research and intuition) to potentially interact with the theoretical constructs in Khatri et al. [15]. This created an *artifact instantiation space* – a theoretical enumeration of artifact properties within population space we are interest in implementing and varying. Table 1 identifies the constructs we considered and the variations within each. Note that other constructs could be considered, such as the size of diagrams or whether they are real diagrams or ones artificially constructed for an experiment – this choice depends on the intent of the study and researcher's knowledge of potential confounds.

Table 1. ER constructs and variants to be considered in the experiment

Construct	Variations	Number of conditions
Cardinalities	Martin, UML[a]	2
Attributes	Outside entity type, Inside Entity Type	2
Relationships	Diamond, Line	2
Domain	Familiar, Unfamiliar, Void	3
Type hierarchies	ER-style, EER-style, UML-style	3

[a]Labels refer to different notations, which depict cardinalities in visually different ways

Third, as there could be multiple ways to instantiate the instantiation space [for discussion, see 5], we perform the second round of sampling. As indicated in Table 1, we attempt to sample systematically across several dimensions for which visual representations are handled differently in various data-oriented conceptual modeling grammars. In doing so, note that the diagrams generated in some conditions will not correspond to modeling notations as published by the developers of a particular grammar. This does not mean the sampling is invalid. On the contrary, by systematically varying features to capture the range of visual notations used for constructs, we will be able to discern whether, how, and under what conditions such variants influence results.

Fourth, we created variants of diagrams to instantiate each condition determined by combinations of variants shown in Table 1 (72 conditions in total). Based on general recommendations in experimental research and uncertainty about the effect size in each condition, we plan to use 40 human subjects for each of the 72 conditions [17].

We are currently designing the study and intend to collect data in the upcoming academic year for analysis. Thus, we expect to complete the remaining steps of the artifact sampling method [8] in time to present preliminary results at the EmpER workshop.

4 Conclusions

Experimental research in conceptual modeling is vital to obtain a clear understanding of how modeling grammars affect important outcomes related to the ability of model readers to understand the semantics of the domain as represented in a script. However, prior experimental research in conceptual modeling has largely ignored issues related to possible unanticipated effects of design choices unrelated to the focal construct of interest on outcomes of interest such as comprehension. In this work, we propose using an artifact sampling approach to systematically vary elements of visual notation in a conceptual modeling grammar to understand how elements might interact or interfere with hypothesized relationships about the effect of design decisions on dependent variables in a study. We have designed a study in the context of ER-style conceptual modeling, with a view to replicating and better understanding a published study in which artifact sampling was not undertaken.

We expect this research to both lead to a better understanding of how to design modeling grammars to best represent constructs of interest in a visual manner, and to provide evidence of the importance of artifact sampling in the methodological toolkit of conceptual modeling researchers and scholars in other disciplines that rely on artifacts for theory testing and other analysis.

References

1. Moody, D.L.: The "Physics" of notations: toward a scientific basis for constructing visual notations in software engineering. IEEE Trans. Softw. Eng. **35**, 756–779 (2009)
2. Bodart, F., Patel, A., Sim, M., Weber, R.: Should optional properties be used in conceptual modelling? A theory and three empirical tests. Inf. Syst. Res. **12**, 384–405 (2001)
3. Parsons, J.: An experimental study of the effects of representing property precedence on the comprehension of conceptual schemas. J. Assoc. Inf. Syst. **12**, 441–462 (2011)
4. Briggs, R.O., Nunamaker, J.J., Sprague, R.: 1001 unanswered research questions in GSS. J. Manag. Inf. Syst. **14**, 3–21 (1997)
5. Lukyanenko, R., Parsons, J.: Reconciling theories with design choices in design science research. In: vom Brocke, J., Hekkala, R., Ram, S., Rossi, M. (eds.) DESRIST 2013. LNCS, vol. 7939, pp. 165–180. Springer, Heidelberg (2013). https://doi.org/10.1007/978-3-642-38827-9_12
6. Ableitner, L., Tiefenbeck, V., Hosseini, S., Schöb, S., Fridgen, G., Staake, T.: Real-world impact of information systems: the effect of seemingly small design choices. In: Workshop on Information Technologies and Systems (WITS 2017) (2017)
7. Lukyanenko, R., Evermann, J., Parsons, J.: Instantiation validity in IS design research. In: Tremblay, M.C., VanderMeer, D., Rothenberger, M., Gupta, A., Yoon, V. (eds.) DESRIST 2014. LNCS, vol. 8463, pp. 321–328. Springer, Cham (2014). https://doi.org/10.1007/978-3-319-06701-8_22
8. Lukyanenko, R., Parsons, J., Samuel, B.M.: Artifact sampling: using multiple information technology artifacts to increase research rigor. In: Proceedings of the 51st Hawaii International Conference on System Sciences (HICSS 2018), Big Island, Hawaii, pp. 1–12 (2018)
9. Lohr, S.L.: Sampling: Design and Analysis. Cengage Learning, Boston (2009)
10. Hammond, K.R., Stewart, T.R.: The Essential Brunswik: Beginnings, Explications Applications. Oxford University Press, Oxford (2001)
11. Brunswik, E.: Organismic achievement and environmental probability. Psychol. Rev. **50**, 255 (1943)
12. Fontenelle, G.A., Phillips, A.P., Lane, D.M.: Generalizing across stimuli as well as subjects: a neglected aspect of external validity. J. Appl. Psychol. **70**, 101 (1985)
13. Wells, G.L., Windschitl, P.D.: Stimulus sampling and social psychological experimentation. Pers. Soc. Psychol. Bull. **25**, 1115–1125 (1999)
14. Snodgrass, J.G., Vanderwart, M.: A standardized set of 260 pictures: norms for name agreement, image agreement, familiarity, and visual complexity. J. Exp. Psychol. [Hum. Learn.] **6**, 174–215 (1980)

15. Khatri, V., Vessey, I., Ramesh, V., Clay, P., Park, S.-J.: Understanding conceptual schemas: exploring the role of application and IS domain knowledge. Inf. Syst. Res. **17**, 81–99 (2006)
16. Lukyanenko, R., Evermann, J., Parsons, J.: Guidelines for establishing instantiation validity in IT artifacts: a survey of is research. In: Donnellan, B., Helfert, M., Kenneally, J., VanderMeer, D., Rothenberger, M., Winter, R. (eds.) DESRIST 2015. LNCS, vol. 9073, pp. 430–438. Springer, Cham (2015). https://doi.org/10.1007/978-3-319-18714-3_35
17. VanVoorhis, C.R.W., Morgan, B.L.: Understanding power and rules of thumb for determining sample sizes. Tutor. Quant. Methods Psychol. **3**, 43–50 (2007)

Design of an Empirical Study for Evaluating an Automatic Layout Tool

Haoyuan Zhang, Tong Li$^{(\boxtimes)}$, and Yunduo Wang

Beijing University of Technology, Beijing, China
zhyuan97@163.com, litong@bjut.edu.cn, wangyunduochn@gmail.com

Abstract. Generating meaningful layout of iStar models is a challenging task, which currently requires significant manual efforts. However, it is time-consuming when dealing with large-scale iStar modeling, rising the need of having an automatic iStar layout tool. Previously, we have proposed an algorithm for laying out iStar SD models and have implemented a corresponding prototype tool. In this paper, we report our ongoing empirical work which aims to evaluate the effectiveness and usability of the prototype tool. In particular, we present a research design which is applied to compare manual layout and automatic layout in terms of efficiency and model comprehensibility. Based on such a design, we are planning to carry out empirical studies accordingly in the near future.

Keywords: iStar models · Automatic layout · Controlled experiment
Prototype tool

1 Introduction

iStar modeling language is a wildly used social goal modeling language, based on which hundreds of requirements analysis approaches have been proposed [1]. As a modeling language, one of the key performance indicators is enabling efficient communication among requirements engineers and stakeholders. To this end, it is essential to layout iStar models in a meaningful manner, as a bad layout could bring tremendous troubles to model comprehension [2].

Manually laying out iStar models is time-consuming, especially for large-scale models[1] that consists of hundreds of nodes. Unfortunately, there are no effective algorithms for automatically laying out iStar models, which motivates us to propose a corresponding solution [3]. Moreover, our proposed layout algorithm has been implemented by a prototype tool, which is supposed to significantly facilitate iStar modeler's work. In particular, the tool can automatically lay out an iStar model that is in the format of iStarML [4], an example of which is shown in Fig. 1(a). In addition, our prototype tool also support manual adjustment on top of its automatic layout, enabling a semi-automatic layout strategy. Figure 1(b) presents an corresponding example.

---

[1] Large iStar models can be found at https://zenodo.org/record/581653.

© Springer Nature Switzerland AG 2018
C. Woo et al. (Eds.): ER 2018 Workshops, LNCS 11158, pp. 206–211, 2018.
https://doi.org/10.1007/978-3-030-01391-2_26

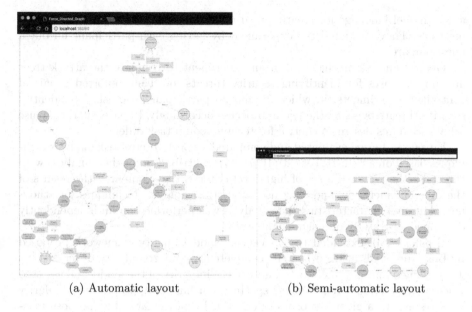

(a) Automatic layout   (b) Semi-automatic layout

Fig. 1. Prototype tool demonstration

In this paper, we report our ongoing empirical work which aims to evaluate the effectiveness and usability of the prototype tool. In particular, we present the research design of a controlled experiment in detail, which will be carried out in the near future to compare manual layout and automatic layout in terms of efficiency and model comprehensibility. In addition, our empirical study also includes case studies and performance testing for evaluating usability and scalability of the tool, respectively.

2   Related Work

Santos et al. have proposed and designed a comprehensive experiment to evaluate the layout of the iStar model in terms of the understandability of social goal models [2]. They performed a quasi-experiment where participants were given two understanding and two reviewing tasks. Both tasks involved a model with a bad layout and another model following the iStar layout guidelines. They evaluated the impact of layouts by combining the success level in those tasks and the required effort to accomplish them. Effort was assessed using time, perceived complexity with regard to NASA TLX (Task Load Index), and eye-tracking data.

Matulevičius and Heymans have presented an experiment to compare the quality of two goal languages (i.e., iStar and KAOS) by the means of the semiotic quality framework [5]. The experiment consisted of three steps: interviewing, creating goal models, and evaluating models and languages. They quasi-equally divided 19 students into four groups, and randomly assigned each group with a goal model language for creating goal models by using corresponding modeling tools. They evaluated the experiment by letting each participant fill out a questionnaire.

Opdahl and Sindre have taken an experiment to compare the attack trees and misuse cases for identifying security threats [6]. They reported a pair of controlled experiments, in which 28 and 35 participants are asked to identify security threats by using the two approaches accordingly. In particular, they use a Latin-Squares design to control for technique and task order.

Buyens & Joosen evaluated and compared a selected set of risk analysis techniques based on a realistic case study [7]. The contribution for their analysis were threefold: they defined a set of high-level criteria, they compared the results of the different methods, and they used statistical analysis techniques for studying additional characteristics. The analysis was performed on an independently developed case study of a significant scale.

España et al. extended the conceptual model quality framework proposed by Lindland et al. in regard to the completeness and granularity of models [8]. They defined four new metrics to improve the integrity of the model and assessed whether a functional requirement specification has an appropriate granularity with respect to a given set of unity criteria. In particular, They propose measuring the degree of functional encapsulations completeness and the degree of linked communications completeness with respect to a reference model (which is agreed by an expert modelling committee).

3   Research Design

3.1   Controlled Experiments

The major research goal of our study is to investigate whether the models that are automatically laid out by our tool can be well comprehended like models that are manually laid out.

Selection of Participants. Participants for the controlled experiments are ideally recruited among iStar modeling experts. Pragmatically, in order to ensure the number of participants to draw comparatively sound conclusions, we plan to involve 60 third-year bachelor students who majored in computer science in the experiment. In particular, the students will be taught about iStar modeling language during their software engineering course for one hour or so. Their performance of the students during the experiment will be associated with their final marks in order to better motivate their participation.

Experimental Materials and Tasks. We plan to use publicly available iStar models[2] as experimental materials. The models have been divided into two sets, i.e., smaller models and larger models, and we will randomly pick up 3 models

---

[2] https://zenodo.org/record/581653 and https://zenodo.org/record/581654.

from each set. Note that such models all have been manually laid out in a reasonable manner, and our tool will import those model files and automatically produce an alternative layout according to our algorithm. Participants will be assigned with particular models for read, after which they will be asked to answer a series of questions regarding the semantics of iStar models.

Variables and Hypotheses. There are three independent variables that will be controlled in our experiment: model scales, layout approaches and time. In particular, we have two scales of models, large-scale models and small-scale models. It is worth noting that such scales are determined according to the two model sets we use. Generally speaking, small models contain less than 100 elements, while the large models involve hundreds of elements. The layout approaches compared in the experiment are manual layout created by experienced iStar modelers and automatic layout created by our tool. Moreover, we plan to have two time settings of the experiment, fixed time and unlimited time.

After participants perform the model comprehension tasks, there are two dependent variables will be measured. First, we will measure the accuracy of questions (AOQ) to evaluate to which extent the participants can comprehend the models. Second, we will measure the time used by the participants when the experiment setting is *unlimited time* in order to evaluate to which extent the model is difficult to read.

Our studies aim at evaluating the following hypotheses:

- H1 Given a fixed time span and small models, participants read automatically laid out models can have similar (+/−5%) AOQ with participants read manually laid out models.
- H2 Given a fixed time span and large models, participants read automatically laid out models can have similar (+/−5%) AOQ with participants read manually laid out models.
- H3 Given unlimited time and small models, both participants read automatically laid out models and participants read manually laid out models can achieve 100% AOQ.
- H4 Given unlimited time and small models, participants read automatically laid out models will spend similar (+/−5%) time with participants read manually laid out models.
- H5 Given unlimited time and large models, both participants read automatically laid out models and participants read manually laid out models can achieve 100% AOQ.
- H6 Given unlimited time and large models, participants read automatically laid out models will spend similar (+/−5%) time with participants read manually laid out models.

Experimental Design. We design a series of experiments according to the independent variables we want to control, the details of which is shown in Table 1 In addition, we will divide the participants into 4 groups of 15 participants each.

Each group is supposed to read small and large models, respectively. Taken into account the learning effect, all groups will start with small model experiments. In particular, group 1 and group 3 are always under a fixed time setting, while the other two groups are always under an unlimited time setting.

Table 1. Experiment design

	Auto	Manual
Small-scale models & Fixed time	Group1	Group3
Small-scale models & Unlimited time	Group2	Group4
Large-scale models & Fixed time	Group1	Group3
Large-scale models & Unlimited time	Group2	Group4

Threats to Validity. As there are different ways of classifying validity in the literature, we here adopt the classification used by Runeson et al. [9], which has an focus on case study research in software engineering.

- External Validity. As the developers of the tool, we are able to train participants in a lot of more details than others, which may hinder the generalization of our conclusion. Thus, we plan to develop a comprehensive guideline and enable a standardized training.
- Construct Validity. The accuracy of questionnaire may or may not be able to measure the level of model comprehension, depending on the quality of the questions. Thus, we plan to have experience iStar expert to design such questions and iteratively refine them via pilot study.
- Conclusion Validity. The motivation of participants to join the experiment will dramatically affect the conclusion validity. Thus, we associate the performance of the participants with their final marks of the software engineering course to ensure they will carefully read the models.

3.2   Case Study

We would conduct in-depth case studies to further evaluate the usability of our tools. In particular, participants will be asked to use our tool to model a particular scenario, after which they need to describe the steps they took and the difficulties encountered during the practices. Moreover, we plan to use eye-tracking tools to further understand the participants' mind while they dealing with a specific layout. As for abnormal phenomena obtained from qualitative analysis, we will interview the participants to deeply understand his rationale.

3.3   Performance Test

To further evaluate the performance of our prototype tool, we will produce a set of iStar models in different scales with particular parameters (such as number of actors and dependencies). Specifically, we plan to test models consist of 100 elements, 1000 elements, and 10000 elements, respectively. For one thing, we record the time used for laying out the model. For another, we also plan to check whether large-scale models can be arranged in a meaningful manner.

4   Conclusions and Future Work

In this paper, we present the design of empirical studies that we plan to carry out to verify the effectiveness of our proposed layout algorithm, as well as the usability and performance of our prototype tool. Specifically, We plan to include controlled experiments, case studies, and performance testing in our study. As for future work, we plan to first implement our experiment, based on which iteratively optimize our layout algorithm. In addition, we also plan to deploy and publish our tool in order to be practically used by domain experts.

Acknowledgements. This work is supported by National Key R&D Program of China (No. 2017YFC08033007), the National Natural Science of Foundation of China (No. 91546111, 91646201) and Basic Research Funding of Beijing University of Technology (No. 040000546318516).

References

1. Horkoff, J., et al.: Goal-oriented requirements engineering: an extended systematic mapping study. Requir. Eng. 1–28 (2017)
2. Santos, M., Gralha, C., Goulao, M., Araújo, J., Moreira, A., Cambeiro, J.: What is the impact of bad layout in the understandability of social goal models? In: IEEE 24th International Requirements Engineering Conference (RE) 2016, pp. 206–215. IEEE (2016)
3. Du, X., Li, T., Wang, D.: An automatic layout approach for istar models (2017)
4. Cares, C., Franch, X., Perini, A., Susi, A.: istarml: an xml-based interchange format for i* models. In: Proceedings of the 3rd International i* Workshop, Recife, Brazil, February, pp. 11–12. Citeseer (2008)
5. Matulevičius, R., Heymans, P.: Comparing goal modelling languages: an experiment. In: Sawyer, P., Paech, B., Heymans, P. (eds.) REFSQ 2007. LNCS, vol. 4542, pp. 18–32. Springer, Heidelberg (2007). https://doi.org/10.1007/978-3-540-73031-6_2
6. Opdahl, A.L., Sindre, G.: Experimental comparison of attack trees and misuse cases for security threat identification. Inf. Softw. Technol. 51(5), 916–932 (2009)
7. Buyens, K., De Win, B., Joosen, W.: Empirical and statistical analysis of risk analysis-driven techniques for threat management. In: The Second International Conference on Availability, Reliability and Security, ARES 2007, PP. 1034–1041. IEEE (2007)
8. España, S., Condori-Fernandez, N., González, A., Pastor, Ó.: An empirical comparative evaluation of requirements engineering methods. J. Braz. Comput. Soc. 16(1), 3–19 (2010)
9. Runeson, P., Höst, M.: Guidelines for conducting and reporting case study research in software engineering. Empir. Softw. Eng. 14(2), 131 (2009)

Conceptual Modeling in Requirements and Business Analysis (MREBA) 2018

Preface to MREBA 2018

5th Workshop on Conceptual Modeling in Requirements Engineering and Business Analysis (MREBA 2018) at the 37th International Conference on Conceptual Modeling (ER 2018)

Xi'an, China
October 22–25, 2018

Organized by
Jennifer Horkoff, Renata Guizzardi and Jelena Zdravkovic

It is our pleasure to welcome you to the fifth edition of the International Workshop on Conceptual Modeling in Requirements Engineering and Business Analysis (MREBA) as a co-located event of the 37th International Conference on Conceptual Modeling (ER 2018), in the city of Xi'an, China.

The MREBA workshop aims to provide a forum for discussing the interplay between Requirements Engineering and Business Analysis topics and Conceptual Modeling. Requirements Engineering (RE) and Business Analysis (BA) are common practices within organizations, often applied in tandem. In particular, the workshop focuses on how requirements modeling can be effectively used as part of Business Analysis and Systems Engineering.

MREBA builds on the success of the previous four instances, as well as an evolution of the previous RIGiM (Requirements Intentions and Goals in Conceptual Modeling) Workshop (2007–9, 12–13). While RIGiM was specifically dedicated to goal modelling and the use of intentional concepts in RE, MREBA handles any kind of modelling notation or activity in the contexts of RE and BA.

This year, MREBA started with a keynote by Prof. João Araujo from New University of Lisbon, entitled "On the Quality of Requirements: the Case of Goal Models". The workshop proceeded with the presentation and discussion of five high-quality full-accepted papers.

Each of the submitted papers went through a thorough review process with at least three reviews from our program committee. We deeply thank the authors of the submitted papers for their high-quality papers. We also thank the Program Committee members and additional reviewers for their effort and dedication in the review of the submitted works. And we finally thank the ER workshop chairs, PC chairs and the remaining of the organizing committee for their support.

October 2018

Jennifer Horkoff
Renata Guizzardi
Jelena Zdravkovic
Program Co-chairs

On the Quality of Requirements: The Case of Goal Models

João Araujo

NOVA LINCS, Departamento de Informática, Faculdade de Ciências e Tecnologia,
Universidade Nova de Lisboa, Lisbon, Portugal
joao.araujo@fct.unl.pt

Abstract. Requirements models have been developed for the requirements engineers and stakeholders work, providing abstraction mechanisms to, for example, facilitate the communication among them by providing better structuring of requirements, thus helping with their analysis. Nevertheless, the extent to which requirements modelling languages are adequate for communication purposes has been somewhat limited. Several quality aspects have contributed to that, ranging from lack of abstraction mechanisms to address model's complexity, to the impact of layout of models or the actual notation adopted. For example, in large-scale systems, building requirements models may end in complex and/or incomplete models, which are harder to understand and modify, leading to an increase in costs of product development and evolution. Consequently, for large-scale systems, the effective management of complexity and completeness of requirements models is vital. Moreover, it is undeniable that the communication potential of requirements modeling languages is not entirely explored, as their cognitive effectiveness is often not boosted. For example, choosing an adequate layout for requirements models may be a relevant issue, as a bad layout may compromise the adequacy of the models. Also, although visual notations are often adopted (as they are perceived as more effective for conveying information to nontechnical stakeholders than text), their careful design is often not considered. Not taking all this into account, in the long run, may result in poorly understood requirements, leading to problems in artifacts produced in later stages of software development. So, in this talk, I will discuss in detail these issues based on the results of experiments where metrics were collected to evaluate and discuss some quality aspects of requirements models, in particular requirements goal models (increasingly popular in the requirements community), such as complexity, completeness, understandability and semantic transparency.

Representing and Analyzing Enterprise Capabilities as Specialized Actors - A BPM Example

Mohammad Hossein Danesh[1(✉)] and Eric Yu[1,2]

[1] Department of Computer Science, University of Toronto, Toronto, Canada
danesh@cs.toronto.edu, eric.yu@utoronto.ca
[2] Faculty of Information, University of Toronto, Toronto, Canada

Abstract. The notion of capability is used by practitioners and researchers alike to enable better understanding of business trajectories and the role of IT in achieving them. Building on the origins of the concept from strategic management, this paper lays out the requirements for capturing enterprise-specific and social characteristics of capabilities. The paper proposes adoption of a goal-driven agent-oriented modeling approach to satisfy the requirements. The ability of such an approach to explicate social and technical design alternatives and enable decision making on their tradeoffs is illustrated on a BPM capability.

Keywords: Enterprise capability · Dynamic capability
Agent-oriented modeling · i* framework · Enterprise modeling

1 Introduction

As organizations face ongoing challenges to take advantage of advancing technologies, they are increasingly using notions of capability to conceptualize how emerging technologies can be deployed and integrated into their existing organizational and technological fabric [1–3]. For example, capability heatmaps are widely used in industry practices to identify gaps in enterprise capabilities and to prioritize investments [2, 3].

Research and academic communities are actively contributing to advance the state of the practice for modeling capabilities. For example, in the ArchiMate modeling language, the capability extension facilitates analysis of how resources and capabilities are aligned with strategic goals [4]. The Capability Driven Development (CDD) approach [1], provides methods and techniques to enable better adaptation of information systems and software services to changes in the contextual situation of enterprise.

The concept of capabilities has long been studied in the field of strategic management with the aim to attain and sustain competitive advantage [5–8]. In this paper, an approach to representing and analyzing capabilities drawing on insights from the strategic management literature about how capabilities are developed and evolve within the context of an enterprise is presented. In an earlier work, an integrative meta-model representing enterprise capability and its related concepts was presented [9]. In this paper, the focus is on the treatment of Enterprise Capability (EC) as a specialized kind

© Springer Nature Switzerland AG 2018
C. Woo et al. (Eds.): ER 2018 Workshops, LNCS 11158, pp. 217–227, 2018.
https://doi.org/10.1007/978-3-030-01391-2_27

of strategic actor modeled in the i* language [10] that enables reasoning about: (a) couplings of resources and processes to form ECs, (b) how the couplings account for unique characteristics that can result in competitive advantage, (c) the organizational and social setting that co-evolve with ECs, and (d) the network of complementary ECs that are orchestrated to offer services/products.

In Sect. 2, the socio-technical characteristics and requirements of modeling ECs are discussed. Section 3 offers arguments on the suitability of an agent-oriented modeling approach that can represent and reason on EC development and evolution meeting the requirements set out in Sect. 2. In Sect. 4, the related work and alternative representations of EC are reviewed. The paper is concluded in Sect. 5 with discussion on benefits of the agent-oriented paradigm and future research direction.

2   Characterizing Enterprise Capabilities

Drawing on the strategic management literature [7, 8, 11, 12], Enterprise Capability (EC)[1] is defined as *an intentional combination of firm-specific assets, organizational routines (business processes), and human knowledge (skillset/know-how) that take advantage of complementary relations and are created and evolved overtime through social collaboration and learning.* Five characteristics of ECs are evident in the proposed definition. First, ECs are **intentionally** built and evolved to pursue continuously changing enterprise objectives that serve strategic trajectories and create/sustain competitive standing [6, 8, 11, 13]. Second, an EC emerges as a result of **intelligent coupling** of enterprise-specific resources and processes, implying (a) enterprise resources and processes are heterogenous across firms, and (b) the form and format of their coupling has significant impact on the output [6, 8, 12–14].

Third, ECs are most impactful when developed and used in a harmonizing and synergetic way as a series of **complementary and cospecialized** capabilities [7, 8, 13]. Fourth, ECs have a **social** characteristic because the social capital, reputation and relationships of the managers and teams responsible for them have an immense impact on their development and evolution options [12, 15]. Fifth, ECs are continuously applying meta-level **learning** processes to codify and extend their knowledge base serving as the foundation of their operation and evolution [12, 14, 16]. As such, modeling ECs requires a socio-technical approach that can express these characteristics and support reasoning at the right abstraction level to satisfy the following requirements:

Req-1: Enabling representation and reasoning on technical, business and organizational goals (aspects of intentionality) in a fashion to enable co-design of IT and business is essential. The need is heightened as in today's environment most capabilities, particularly the competitive differentiators, are IT-enabled capabilities [17–19]. Capturing intentions from multiple perspectives (a) enhances stakeholder onboarding

---

[1] The term enterprise capability is purposefully used to distinguish this research from some other more general treatment of the notion of capability. This enterprise focus is supported by the management literature and its definition and usage of the concept.

and communication, (b) enables requirement specification and co-design of business and IT, and (c) enhances the understanding of the context in which decisions are made.

Req-2: Decision making about resource and process couplings that form and shape an EC is not trivial, yet entails (often) long-lasting commitments referred to as path dependency [5, 12]. To enable reasoning on such decisions one needs to (a) specify and elaborate the means of coupling firm-specific resources and processes to satisfy objectives, and (b) identify the conditions under which an alternative is preferable to another while considering multiple perspectives (Req-1).

Req-3: ECs often create value in combination, forming complementary and inter-locking relationships [7, 11, 13]. As enterprises today are heavily dealing with dynamic and continuously evolving ecosystems with loose boundaries, enterprises and conse-quently their ECs are ever more required to participate in co-creating networks even beyond organizational boundaries [11, 18]. On the other hand, to keep up with the fast-paced changing environment and foster innovation, capabilities and their decision making process should be decentralized to the extent of near autonomy [7, 18].

As a result, one needs to represent and reason about ECs that are nearly autono-mous in their decision making, yet highly interdependent in creating value. The combination of multi-perspective intentions, autonomous nature, fast-paced evolution, and highly interdependent network(s) results in emergent behaviors and desires, cre-ating a sense of identity for an EC. Therefore, any conceptual representation of the capability should be able to account for localized intentions and their trajectory, interdependent goals and needs, and alignment between the two.

Req-4: Expression of the social setting [12, 14], cultural aspirations [12, 14, 15], and organizational and team structure [15] is an essential step for decision making as they can inhibit or enable certain alternatives for the development and evolution of capa-bilities. Therefore, the ability to (a) represent and analyze options for organizing teams, (b) provide understanding of social relationships among organizational actors, and (c) identify and steer the impact of social norms and cultural values are important requirements for making decisions about the development and evolution of ECs.

3   Agent-Oriented Modeling of Enterprise Capabilities

Agent-oriented modeling approaches have been well-studied for capturing requirement of social, technical and interdependent networks of software systems [20]. Particularly the internal makeup of agents in terms of goals, means-ends, quality attributes, con-tributions, and tradeoffs are of significant benefit when considering requirements of modeling EC. Representation of actors' external relationships through dependencies and associations will empower understanding of the social and complementary requirements of modeling ECs. As such, this section demonstrates and assesses the benefits of an agent-oriented representation and illustrates the extent to which the set-out requirements of Sect. 2 are satisfied. This is achieved by building on the proposed extension to the i* framework [21] that represents EC as specialized i* actors.

For illustration, we refer to a set of best practices for managing an enterprise wide Business Process Management (BPM) capability proposed by Dumas et al. [22]. The proposal motivates investments in a "mature holistic BPM practice" as shown in the model of Fig. 1. The analysis in this section demonstrates how a specification of a capability meant to prescribe best practices might in fact contain design choices that can lead to vastly different outcomes. In the following, we highlight several distinctive features of a goal-based agent-oriented approach to modeling ECs that can analyze and identify the strengths and weaknesses of capability design strategies.

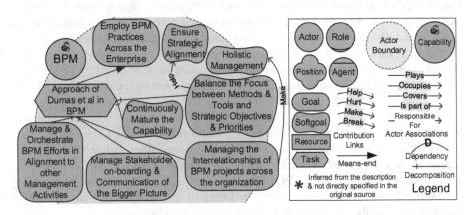

Fig. 1. A high-level view of a BPM capability (For high-quality figures see "http://www.cs. toronto.edu/~danesh/AgentCapBPM.pdf")

3.1 Explicating Alternatives and Quality Attributes to Capture EC Intentions and Perspectives

In Fig. 1, we model the BPM capability as an intentional actor, thus revealing intent in terms of functional and quality objectives. There can be different ways of achieving these objectives. These are shown as tasks pointing to the goal via the means-end link. Here we show only the one approach proposed by Dumas et al. [22]. This approach is modeled as consisting of sub-elements corresponding to the three main components of BPM capability identified by Dumas et al., accompanied by two quality objectives (softgoals in i*). By making goals and means-ends relationships explicit in the model, the EC designer is provoked to think carefully about what the EC is aiming to achieve, and that there can be different ways of achieving those objectives. Such representation of intentions and their qualitative aspects enables satisfaction of Req-1 and Req-2.

Using this model of the intentional structure of the BPM capability, an architect can readily adapt the best practice recommendations, by adding, for example, quality goals *"Design business processes with reusability in mind"* and *"Incentivize development of horizontal business processes"* (not shown in the model of Fig. 1) in the context of a highly vertically optimized enterprise. Such enterprise-specific requirements are considerations that are often not captured in best practices such as Dumas et al. [22].

3.2   Using Actor Boundaries and Associations

Building on the i* agent-oriented approach with ECs represented as specialized actors, the actor boundaries play an important role in the design of EC by enabling reasoning on what to include or exclude from a capability. This treatment enables:

(1) Leveraging Actor Associations & Dependencies to Represent Intentions of Capability Decompositions: In the BPM case, six underlying ECs are discussed as presented in Fig. 2 that are coordinated with one another to ensure a holistic BPM practice. One could choose to have *"Align Strategy"* and *"BPM Governance"* as two distinct capabilities, so that each can develop on its own with autonomy, which requires coordination from the parent capability (BPM). Alternatively, one could make a case of integrating the two into a single capability. The two ways of setting boundaries for these capabilities result in different topologies and have different contributions to softgoals.

In the agent-oriented paradigm, ECs cannot be simply decomposed to sub-capabilities, instead capability relationships are described through dependencies and associations. Representing decompositions with *"is-part-of"* associations will force the modeler to explore and explicate motivations behind the decomposition. The actor boundaries and dependencies will allow analysis on the implications of the decomposition. This treatment of capabilities contributes to the satisfaction of Req-3.

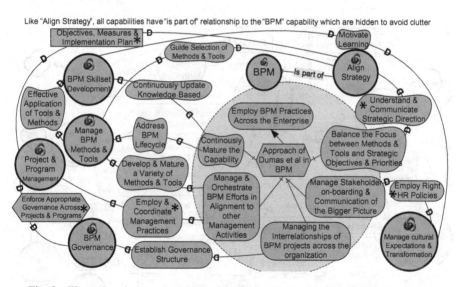

Fig. 2. Illustrating the decomposition of BPM capability – legend provided in Fig. 1

(2) Use of Actor Associations to Represent Organizational Structure and Needs: Dumas et al. [22] recommends separating the development of methods and tools for *"Project & Program Management"* from *"BPM Methods & Tools"* as captured in Fig. 2. The separation allows for more flexibility when evolving each discipline and

enables application of localized expertise. In Fig. 3, we use the means-end links to indicate that there can be (at least) four ways to separate and coordinate the two capabilities. Each of these alternatives will require their own organizational structure (s). For example, alternatives 1 and 2 require one member on the *"BPM Team"* albeit with very different expertise represented as P1 covering two distinct roles in Fig. 3. Alternatives 3 and 4 each requires three team members represented by P2, P3 and P4.

Expressing the organizational structure and setting contributes to the satisfaction of Req-1 and Req-4 by firstly, enabling the identification of enterprise-specific objectives that impact each alternative and their contribution to quality goals, e.g., *"Balanced Cost of Human Resources"* (not shown in the model due to space limitation) as the objective of the *"BPM Team"* and *"Effective Coordination Among Projects and Program(s)"* as the objective of the *"Project & Program Management"* will impact the specified alternatives of Fig. 3. Secondly, allowing explication of the relevant stakeholders and their interests that will ease onboarding and communication, e.g., the three dependencies to the *"Procurement Management"* capability in Fig. 3, represent different stakeholder needs with respect to project management. Lastly, the expression distinguishes the interests of the teams from capabilities while explicating their dependencies. Thus, it enables reasoning on the alignment and co-evolution of the two sets of objectives as decoupling the teams' interests from capability interests is neither desirable nor possible.

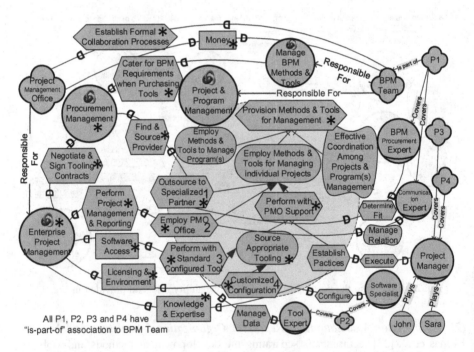

Fig. 3. Illustrating alternatives and organizational structure – legend in Fig. 1

3.3   Using Means-Ends and Dependencies to Illustrate Historical Decisions

Depicting boundaries, explicating intentions of decompositions, and enforcing means-ends relationships will clarify the logic behind (historical) decisions. Furthermore, it will empower reasoning on limitations in the paths ahead when evolving capabilities both from an organizational standpoint, i.e., explaining responsibilities and loss/gain of control through positioning i* elements in actor boundaries, and a technical standpoint, i.e., through comparing alternatives regarding their satisfaction of goals, softgoals and dependencies. Understanding the limitations in the paths ahead plays a primary role in managing resistance when evolving ECs [5, 12, 16].

For example, the decision to rely on the *"Enterprise Project Management"* capability by the *"BPM Team"* as described in Fig. 3, will limit the team's ability to propose new methods and tools for managing BPM projects in the future. The limitation is due to commitments and investments made by the *"Project Management Office"* on establishing processes, preparing and disseminating training material, and acquiring software.

3.4   Building on Strategic Dependencies to Orchestrate Capability Choices

Decision makers need to orchestrate choices in a network of interdependent ECs. As an example, the choice among the four alternatives presented in Fig. 3 should be orchestrated with decisions made on the alternative methods and tools employed by the *"Project Management Office"* to shape the *"Enterprise Project Management"* capability.

Orchestrating alternatives among capabilities will not be possible if (a) alternatives with clear contribution to goals are not explicitly represented, and (b) the intention behind capability relationships are not expressed. The agent-oriented representation can do both as discussed in a prior publication [21] and will enable satisfaction of Req-3.

3.5   Using Means-Ends, Quality Tradeoffs, and Dependencies to Allow Expression of Meta-level Capabilities and Choose Among Design Patterns

Meta capabilities refer to the set of capabilities that initiate and redesign other capabilities [7]. The BPM capability as proposed by Dumas et al. [22] is a meta-level capability that enterprises need in order to trigger and manage changes to business processes. Figure 4, illustrates how an insurance company will employ the BPM capability to evolve its operational capabilities through the dependencies 1 and 2.

The very nature of the changes required by these dependencies are different and therefore require diverse design patterns to be developed and employed. An effective selection of design patterns cannot be achieved without modeling the full context and requirements of the capability. For example, the capability of *"Manage Customer Policies"* in Fig. 4 that relies on the *"BPM"* capability is co-developed by two departments. Therefore, the dependency to *"Improve the Quality of the Process for*

Validating Customer Policies" needs to be refined with respect to the needs of each participating stakeholder. The co-development of *"Manage Customer Policies"* is illustrated by the set of *"is part of"* and *"Responsible For"* relationships.

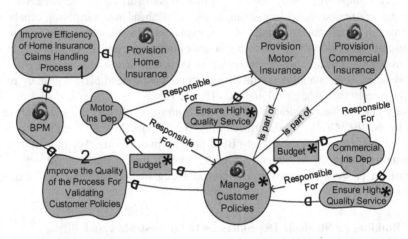

Fig. 4. Illustrating dependencies to the meta-level capability – legend provided in Fig. 1

4 Related Work

In this section, three kinds of representational constructs for the notion of capability are reviewed and their strengths and weaknesses are discussed.

Capability Maps: Capability maps and heatmaps depict the set of prioritized capabilities and their decomposition in a simple view used to communicate investment allocation and technology maturity. The analysis supporting such prioritization is often done behind the scenes and then summarized in a capability heatmap. Many industry practices such as those promoted by the Business Architecture Guild [3] and Gartner [2] employ such maps. The maps are often used to make decisions regarding which capabilities and organizational teams to fund, and how to align business and IT.

The kind of representation lacks the ability to (a) investigate resource and process couplings that form alternatives, (b) reason on the quality goals and their tradeoffs, (c) identify required social capital and organizational choices, and (d) express and reason on complementarities among capabilities.

ArchiMate: A capability according to the ArchiMate extension [4] can be a capability of a person or of an enterprise, and hence viewed as a more general concept than the EC concept discussed in this paper. By introducing the capability concept, the extension focuses on expressing and tracking linkages among capabilities and resources that generate value to layers of an enterprise architecture.

The extension also introduces the concept of a "capability enabling bundle" which can be used to describe couplings of processes and resources and their relation to strategic goals comparable to what is done in Sect. 3.1 in this paper. However, it cannot reason on the quality attributes of the coupling, the social capital needed, and the organizational structure necessary to support a capability. Furthermore, ArchiMate would need to be supplemented with guidelines and methods to prompt modelers to state why relationships among two capabilities are desirable or not, perhaps by extending the "Relationships" between capabilities in the extended ArchiMate metamodel.

The EKD Based Capability Driven Development (CDD) Approach: The CDD approach inherits goal-based features from the Enterprise Knowledge Development (EKD) methodology and representation [1]. It focuses on varieties in the processes that can deliver a capability and enables adaptation of the process in response to changing contextual parameters at run-time. The adaptation relies on the variations in the activities of a business process as oppose to couplings of resources and processes that form enterprise-specific capabilities. Therefore, the approach cannot reason on the qualitative aspects of couplings, unless they are captured under the broad category of contextual parameters. The social capital and its organizational setting supporting the capability are not explored and modeled in the CDD approach.

5 Conclusion and Future Work

To enable co-design of business and IT, researchers and practitioners have raised the abstraction level of design artifacts so they can capture the intertwined relationship of software requirements and organizational/business context [1, 18, 23]. To this end, the notion of capability is used to represent enterprise strengths and weaknesses and its strategic trajectories. Capability modeling practices are developed and used to enable better communication of strategic priorities [2, 3, 23], expression of architecture layers and their business alignment [4], and adaptation to changing contexts [1].

Building on the i* framework, this papers argues for a goal-driven and agent-oriented modeling approach that results in: (1) explication of choices for coupling enterprise-specific resources and processes to enable understanding of capability formation and emerging quality attributes, (2) expression of the social and organizational setting that capabilities reside in to empower analyzing the influences and interests of multiple stakeholders, and (3) representation of interdependent networks of capabilities and actors to enable orchestration of design choices among capabilities, information systems and organizational structure(s).

By enabling identification and expression of the socio-technical situation in which capabilities are designed for, the proposed representation will complement existing practices by serving as an early requirements elicitation approach. The approach empowers enterprise architects to choose and tailor reference models and design patterns to fit enterprise context. Analysis and reasoning techniques presented in earlier publications [21, 24] enable orchestration of design choices for a network of interdependent enterprise capabilities, information systems and organizational actors.

Prior publications [21, 24] have explored the suitability of the actor-based representation on a few IT intensive case studies but further validation is necessary to determine the conditions and types of situations that require a socio-technical analysis.

Methods and guidelines that will steer the modeler in gathering the required information for reasoning on capability formation and evolution with more detailed specification of how i* elements are used and extended is required. The i* language has the capacity to represent organizational and social settings but lacks guidelines and techniques to guide the modeler in doing so which is a critical step in modeling capabilities and should be addressed in the future.

References

1. Stirna, J., Grabis, J., Henkel, M., Zdravkovic, J.: Capability driven development – an approach to support evolving organizations. In: Sandkuhl, K., Seigerroth, U., Stirna, J. (eds.) PoEM 2012. LNBIP, vol. 134, pp. 117–131. Springer, Heidelberg (2012). https://doi.org/10.1007/978-3-642-34549-4_9
2. Burton, B.: Eight Business Capability Modeling Best Practices. Gartner (2012)
3. Business Architecture Guild: A Guide to the Business Architecture Body of Knowledge (2014)
4. Azevedo, C.L.B., Iacob, M.-E., Almeida, J.P.A., van Sinderen, M., Pires, L.F., Guizzardi, G.: Modeling resources and capabilities in enterprise architecture: a well-founded ontology-based proposal for ArchiMate. Inf. Syst. 54, 235–262 (2015)
5. Teece, D.J., Pisano, G., Shuen, A.: Dynamic capability and strategic management. Strat. Manag. J. 18, 509–533 (1997)
6. Helfat, C.E., Peteraf, M.A.: The dynamic resource-based view: capability lifecycles. Strat. Manag. J. 24, 997–1010 (2003)
7. Teece, D.J.: Explicating dynamic capabilities: the nature and microfoundations of (sustainable) enterprise performance. Strat. Manag. J. 28, 1319–1350 (2007)
8. McKelvie, A., Davidsson, P.: From resource base to dynamic capabilities: an investigation of new firms. Br. J. Manag. 20, S63–S80 (2009)
9. Danesh, M.H., Loucopoulos, P., Yu, E.: Dynamic capabilities for sustainable enterprise IT – a modeling framework. In: Johannesson, P., Lee, M.L., Liddle, S.W., Opdahl, A.L., López, Ó.P. (eds.) ER 2015. LNCS, vol. 9381, pp. 358–366. Springer, Cham (2015). https://doi.org/10.1007/978-3-319-25264-3_26
10. Yu, E.: Modeling Strategic Relationships for Process Reengineering (1995). ftp://ftp.cs.toronto.edu/pub/eric/DKBS-TR-94-6.pdf
11. Sirmon, D.G., Hitt, M.A., Ireland, R.D., Gilbert, B.A.: Resource orchestration to create competitive advantage breadth, depth, and life cycle effects. J. Manag. 37, 1390–1412 (2011)
12. Leonard-Barton, D.: Core capabilities and core rigidities: a paradox in managing new product development. Strat. Manag. J. 13, 111–125 (1992)
13. Helfat, C.E., et al.: Dynamic Capabilities: Understanding Strategic Change in Organizations. John Wiley & Sons, New York (2009)
14. Collis, D.J.: Research note: how valuable are organizational capabilities? Strat. Manag. J. 15, 143–152 (1994)
15. Adner, R., Helfat, C.E.: Corporate effects and dynamic managerial capabilities. Strat. Manag. J. 24, 1011–1025 (2003)
16. Zollo, M., Winter, S.G.: Deliberate learning and the evolution of dynamic capabilities. Organ. Sci. 13, 339–351 (2002)

17. Ross, J.W., Sebastian, I.M., Beath, C.M., Jha, L., Technology Advantage Practice of the Boston Consulting Group: Designing Digital Organizations - Summary of Survey Findings. MIT CISR, Boston, MA (2017)

18. Bosch, J.: Speed, data, and ecosystems: the future of software engineering. IEEE Softw. **33**, 82–88 (2016)

19. Nevo, S., Wade, M.: The formation and value of IT-enabled resources: antecedents and consequences. Manag. Inf. Syst. Q. **34**, 163–183 (2010)

20. Horkoff, J., et al.: Goal-oriented requirements engineering: an extended systematic mapping study. Requir. Eng. 1–28 (2017). https://rdcu.be/7sH5

21. Danesh, M.H., Yu, E.: Modeling enterprise capabilities with i*: reasoning on alternatives. In: Iliadis, L., Papazoglou, M., Pohl, K. (eds.) CAiSE 2014. LNBIP, vol. 178, pp. 112–123. Springer, Cham (2014). https://doi.org/10.1007/978-3-319-07869-4_10

22. Dumas, M., Rosa, M.L., Mendling, J., Reijers, H.: Fundamentals of Business Process Management. Springer, Heidelberg (2018). https://doi.org/10.1007/978-3-662-56509-4

23. Homann, U.: A business-oriented foundation for service orientation. MSDN, Microsoft Corporation (2006)

24. Danesh, M.H., Yu, E.: Analyzing IT flexibility to enable dynamic capabilities. In: Persson, A., Stirna, J. (eds.) CAiSE 2015. LNBIP, vol. 215, pp. 53–65. Springer, Cham (2015). https://doi.org/10.1007/978-3-319-19243-7_5

An Approach Toward the Economic Assessment of Business Process Compliance

Stephan Kuehnel[1(✉)] and Andrea Zasada[2]

[1] Martin Luther University Halle-Wittenberg, Halle (Saale), Germany
stephan.kuehnel@wiwi.uni-halle.de
[2] University of Rostock, Rostock, Germany
andrea.zasada@uni-rostock.de

Abstract. Business process compliance (BPC) denotes business processes that adhere to requirements originating from different sources, e.g., laws or regulations. Compliance measures are used in business processes to prevent compliance violations and their consequences, such as fines or monetary sanctions. Compliance measures also incur costs, e.g., for tools, hardware, or personnel. To ensure that companies can work economically even in intensively regulated environments, the economic viability of BPC has to be taken into account. A body of literature is already devoted to the economic assessment of processes and focuses on the business perspective, whereas corresponding approaches for BPC appear to be lacking. Consequently, we introduce a novel approach that allows for an economic assessment of process-based compliance measures. The approach takes monetary consequences of compliance violations into account and is based on the well-known basic workflow patterns for control flows. We demonstrate its applicability by means of an exemplary ordering process affected by Article 32 (1) of the EU General Data Protection Regulation.

Keywords: Business process compliance · Economic assessment
Compliance cost · Workflow patterns

1 Introduction

Business Process Compliance (BPC) denotes business processes that adhere to applicable requirements that originate from various sources, e.g., laws, regulations or internal guidelines [1]. BPC checking approaches are aimed at checking business processes against regulatory requirements, focusing on technical methods to provide (semi)automated support for managing BPC [2]. However, complying with a constantly increasing number of requirements is not only a technical but also an economic challenge. For example, Article 32 (1) of the EU General Data Protection Regulation 2016/679 (GDPR) requires data processors to implement technical and organizational measures for protecting personal data. The implementation of such measures incurs

---

The original version of this chapter was revised: There was an error in the first paragraph of page 236. The correction to this chapter is available at https://doi.org/10.1007/978-3-030-01391-2_38

noticeable costs [3, 4], for example, for software (e.g., encryption tools), hardware (e.g., backup databases), and personnel (e.g., working hours). In turn, refraining from implementing such measures and thus provoking a compliance violation entails economic risks, e.g., fines or penalty fees [5]. To ensure that companies can work economically even in intensively regulated environments, the economic viability of BPC must be taken into account. Several studies [6–10] have demonstrated that business processes offer a sound basis for cost analyses and economic assessment. To the best of our knowledge, the existing approaches are focused on the business perspective [6–10] but not on BPC. Hence, the research goal of this paper is the development of an approach that allows for an economic assessment of process-based compliance measures taking into account the monetary consequences of compliance violations.

The remainder of this paper is structured as follows. Section 2 defines basic terms and integrates the concept of economic assessment into the conceptual framework of BPC. Section 3 presents the formulas used in our approach that are required for the economic assessment of process-based compliance measures. Section 4 exemplifies its applicability to an ordering process affected by Article 32 (1) GDPR. Subsequently, related work is discussed in Sect. 5. Section 6 concludes the paper with a summary and a discussion of limitations.

2 Basic Terms and Conceptual Structure

The conceptual structure of BPC checking and the extension of this structure for economic assessment have already been analyzed in greater detail in a previous study [3]. The conceptual model in Fig. 1 provides a brief overview of the concepts relevant to our approach and illustrates their interrelations. For a more in-depth understanding and for reasons of traceability, a brief description of the concepts is given below.

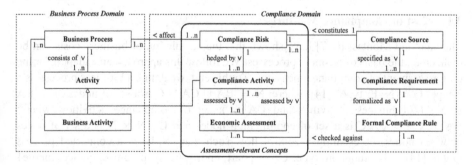

Fig. 1. Basic terms and conceptual structure (based on [3])

A *Compliance Requirement* is a condition or restriction that describes desirable results or binding obligations. Compliance requirements derive from the interpretation of *Compliance Sources*, such as subsections or paragraphs of laws, regulations, contracts, etc. Using logical languages, compliance requirements can be mapped as machine-readable *Formal Compliance Rules* and checked against business processes by means of process-verification tools. A *Business Process* is a sequence of work items (or so-called *Activities*) that takes one or more kinds of input and transforms them into valuable output [11]. Failure to observe a relevant compliance source results in a compliance violation, so that each compliance source constitutes a *Compliance Risk*. A compliance risk is commonly defined quantitatively as the product of an uncertain consequence and its probability of occurrence [3]. The probability of occurrence corresponds to the relative frequency with which a compliance violation occurs in a business process instance. Uncertain consequences result from compliance violations and can take various forms. While monetary consequences such as fines or monetary sanctions [5] can easily be used for calculation purposes, non-monetary consequences such as prison sentences or trade bans can limit risk quantification [3]. Therefore, the approach in this paper is initially limited to the assessment of monetary consequences and leaves the consideration of non-monetary factors as a subject for future research. To protect against a compliance violation and, thereby, the uncertain consequence of a compliance risk, process-integrated compliance measures are used (hereafter referred to simply as compliance measures). Since compliance measures are part of a business process, they can be mapped as compliance activities or activity combinations using business process modeling languages such as the expressive Business Process Model and Notation (BPMN) [12]. A *Compliance Activity* serves the fulfillment of a compliance task that results from a requirement. The *Economic Assessment* of BPC is based on the economic principle and thus on the relationship between an economic result (monetary output) and the required means (monetary input) [3, 13]. The monetary input of BPC is generated in establishing and executing compliance activities and is considered as its total costs. The monetary output of BPC is represented by the proportion of the monetary compliance risk that is prevented by the use of compliance activities.

3   Economic Assessment of Compliance Measures

3.1   Cost of Compliance Measures

Magnani and Montesi [6, 7] have shown that the calculation of business costs can be realized on the basis of business processes. We follow this approach and adapt it for the cost assessment of compliance measures. To this end, we define a business process as a 3-tupel $G = (N, E, type)$ [14], where: $N = BA \cup CA \cup C$ is a set of nodes in G. BA is a set of business activities and CA is a set of compliance activities, where: $BA \cap CA = \varnothing$. C is a set of coordinating nodes and $E \subseteq N \times N$ is a set of edges between nodes representing a control flow such that (N, E) is a connected process graph [14]. The function type: $C \rightarrow \{start, end, split, synchronize, choice, merge\}$ assigns a coordinator type to each coordinating node of G.

Existing approaches dealing with the economic assessment of business processes capture a selection of real processes by means of process classes [6, 7] or selected process patterns [8, 9] to ensure both a simple and broad applicability. We follow the idea of using patterns for assessment, but make use of the well-known workflow patterns of van der Aalst et al. [15], as they have done a comprehensive study on the most frequently recurring process patterns. Since our approach does not serve to assess an entire business process but rather its compliance measures, we only consider compliance activities ca $\in$ CA for assessment.

For our first approach and due to space limitations, we focus on the basic control flow patterns as represented in Table 1. We use these patterns to represent different combinations of $i = 1 \ldots m$ compliance activities $ca_{i,j}$ that are needed to hedge a compliance risk CR_j that arises from a compliance source CS_j ($\forall j = 1 \ldots q$). Van der Aalst et al. [15] describe the *Sequence* pattern as follows: "An activity in a workflow process is enabled after the completion of a preceding activity in the same process." We assume that each $ca_{i,j}$ can be specified by variable costs and allocable average fixed costs. Given that w is the number of workflow instances of a period s and that $cf_{i,j}$ represents the fixed costs of $ca_{i,j}$ incurred in s, then $cf_{i,j} * w^{-1}$ represents the average fixed costs per execution of $ca_{i,j}$. Together with the variable costs $cv_{i,j}$, the total costs of a $ca_{i,j}$ per instance are calculated by $tc_{i,j} = cv_{i,j} + cf_{i,j} * w^{-1}$. Since all $ca_{i,j}$ of a sequence pattern are executed one by one, the total pattern costs $tc_{p,j}$ of a sequence consisting of $i = 1 \ldots m$ activities $ca_{i,j}$ are simply calculated by summing the $tc_{i,j}$'s:

$$tc_{p,j} = \sum_{i=1}^{m} tc_{i,j} \tag{1}$$

In Table 1, the sequence pattern is represented in BPMN, using an example with three $ca_{i,j}$ labeled A, B, and C. For this example, the total pattern costs are calculated by $tc_{p,j} = tc_{A,j} + tc_{B,j} + tc_{C,j}$.

Table 1. Exemplary representation of basic control flow patterns in BPMN

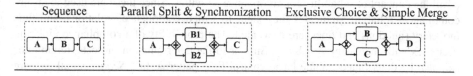

Sequence	Parallel Split & Synchronization	Exclusive Choice & Simple Merge

The *Parallel Split* pattern depicts the divergence of a single thread of control into multiple threads of control that are executed concurrently [15]. For a clearer illustration and due to space restrictions, we mapped this pattern in Table 1 in conjunction with the corresponding converging counterpart, the *Synchronization* pattern, which depicts the convergence of several parallel activities or subprocesses into a single thread of control [15]. Since all branches and, therefore, all activities are executed in a combined pattern of split and subsequent synchronization, the total costs are calculated in the same way as shown in formula (1). Table 1 shows a combination of the parallel split and synchronization pattern, using an example with four $ca_{i,j}$ labeled A, B1, B2 and C, with B1 and B2 being parallel activities. For this example, the total pattern costs are calculated by $tc_{p,j} = tc_{A,j} + tc_{B1,j} + tc_{B2,j} + tc_{C,j}$.

The *Exclusive Choice* pattern depicts a point at which one of several branches is chosen [15]. As with the parallel split, we mapped this pattern in Table 1 together with its corresponding counterpart, the *Simple Merge* pattern, which depicts two or more alternative branches coming together without synchronization [15]. Mathematically, an exclusive choice can be resolved by identifying all $k = 1 \ldots v$ alternative paths P_k, determining the associated path probabilities b_k as well as path costs $tc_{k,j}$, and calculating an expected cost value. As with formula (1), the $tc_{k,j}$ are calculated as the sum of the costs of the activities ($tc_{k,j} = \sum_{i=1}^{m} tc_{k,i,j}$). The b_k can be estimated at design time, derived from empirical data or, if completely unknown, initially assumed to be equally distributed [7]. Thus, the expected pattern costs are calculated by

$$tc_{p,j} = \sum_{k=1}^{v} b_k * tc_{k,j} = \sum_{k=1}^{v} \sum_{i=1}^{m} b_k * tc_{k,i,j} \qquad (2)$$

Table 1 shows the combination of the exclusive choice and simple merge pattern, using an example with four $ca_{i,j}$ labeled A, B, C and D. Since B and C are exclusive activities, there are two alternative paths $P_1 = \{A, B, D\}$ and $P_2 = \{A, C, D\}$. For this example, the pattern costs are calculated by $tc_{p,j} = b_1 * (tc_{A,j} + tc_{B,j} + tc_{D,j}) + b_2 * (tc_{A,j} + tc_{C,j} + tc_{D,j})$.

3.2   Reliability of Compliance Measures and Compliance Risks

Both business and compliance activities are characterized not only by costs but also by a certain reliability of execution $r_{i,j}$. In the context of the cost assessment of BPMN diagrams, Sampathkumaran and Wirsing [8, 9] defined reliability as the rate at which a business activity is executed successfully, i.e., without errors. We follow this idea and adapt the concept of reliability for compliance activities. Since these are used to ensure compliance, reliability indicates the extent to which compliance violations, and thus monetary sanctions and fines, are prevented. The probability of a compliance violation decreases with increasing reliability. Similar to the determination of costs, reliabilities can also be determined for the basic control flow patterns. In a *Sequence*, all $ca_{i,j}$ are executed one by one, so that the overall reliability of the sequence pattern depends on the reliabilities of its activities. To calculate the pattern reliability $r_{p,j}$ of a sequence consisting of $i = 1 \ldots m$ activities $ca_{i,j}$, the $r_{i,j}$'s are simply multiplied [9]:

$$r_{p,j} = \prod_{i=1}^{m} r_{i,j} \qquad (3)$$

For the example of the sequence pattern shown in Table 1, the reliability is calculated by $r_{p,j} = r_{A,j} * r_{B,j} * r_{C,j}$. The calculation of the pattern reliability for the *Parallel Split* and *Synchronization* pattern follows the same calculation principle as shown in (3). Here, too, all $ca_{i,j}$ are executed so that the reliability of each individual activity influences the overall reliability of the pattern. For the example of the parallel split and synchronization pattern shown in Table 1, the reliability is calculated by $r_{p,j} = r_{A,j} * r_{B1,j} * r_{B2,j} * r_{C,j}$. In contrast, to calculate the pattern reliability for the *Exclusive Choice* and *Simple Merge* pattern, the branch probabilities b_k for $k = 1 \ldots v$

exclusive branches must be taken into consideration. The sum of the products of branch probabilities b_k and branch reliabilities $r_{k,j}$ results in the expected pattern reliability:

$$r_{p,j} = \sum\nolimits_{k=1}^{v} b_k * r_{k,j} = \sum\nolimits_{k=1}^{v} b_k * \prod\nolimits_{i=1}^{m} r_{k,i,j} \qquad (4)$$

For the example of the exclusive choice and simple merge pattern shown in Table 1, the reliability is calculated by $r_{p,j} = b_1 * \left(r_{A,j} * r_{B,j} * r_{D,j}\right) + b_2 * \left(r_{A,j} * r_{C,j} * r_{D,j}\right)$.

As mentioned above, a compliance risk CR_j is mathematically described as the product of an uncertain consequence c_j and its probability of occurrence p_j [3]. If a single pattern hedges CR_j, then $r_{p,j}$ describes the extent to which a compliance violation is prevented by this pattern, which is actually the opposite of p_j:

$$CR_j = p_j * c_j = (1 - r_{p,j}) * c_j \qquad (5)$$

3.3 Economic Benefit of Compliance Measures

Taking into consideration the formulas defined in the previous sections, the economic benefit per instance $eb_{P,j}$ can be determined for compliance measures whose process design is based strictly on the basic control flow patterns (see Table 2). The monetary damage prevented by a pattern is determined by multiplying c_j by $r_{p,j}$. The $tc_{p,j}$ are deducted from this value to determine $eb_{P,j}$. Of course, calculations can also be performed for more complex processes in which, for example, exclusive and parallel gateways are combined. To determine the economic benefit eb_j of such complex processes, the cost formulas of the patterns must be combined additively and the reliability formulas multiplicatively. A corresponding example is provided in Sect. 4.

Table 2. Formulas for calculating the economic benefit of patterns

Pattern	Economic benefit
Sequence	$eb_{P,j} = c_j * \left(\prod_{i=1}^{m} r_{i,j}\right) - \sum_{i=1}^{m} tc_{i,j}$
Parallel Split & Synchronization	$eb_{P,j} = c_j * \left(\prod_{i=1}^{m} r_{i,j}\right) - \sum_{i=1}^{m} tc_{i,j}$
Exclusive Choice & Simple Merge	$eb_{P,j} = c_j * \left(\sum_{k=1}^{v} b_k * \prod_{i=1}^{m} r_{k,i,j}\right) - \sum_{k=1}^{v} \sum_{i=1}^{m} b_k * tc_{k,i,j}$

4 Running Scenario

We illustrate the applicability of our approach using a simplified process model of an online retailer (see Fig. 2). In the business process, once a customer's order has been received, the order is processed, the goods are delivered, and an invoice is sent.

The process is affected by Article 32 (1) GDPR, which corresponds to the compliance source CS_1 (with $j = 1$). Article 32 (1) deals with data-processing security and

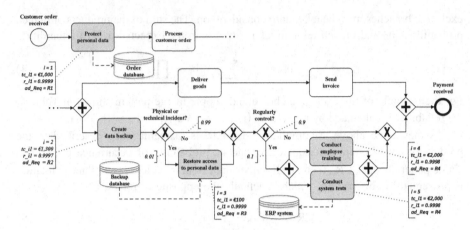

Fig. 2. Simplified BPMN process model of an online retailer

stipulates that controllers and data processors must take appropriate technical and organizational measures to adequately hedge data-protection risks. Table 3 provides an overview of the compliance requirements (*R1–R4*) applying to the process. These were derived directly from the legal provisions laid out in Article 32 (1) GDPR and must be fulfilled for preventing a compliance violation. To meet the requirements *R1–R4*, five compliance activities $ca_{i,1}$ (highlighted in gray in Fig. 2) were integrated into the process. To simplify matters, we assume that $w = 50$ instances of the illustrated ordering process are executed per year, whereby a total worldwide annual turnover of €500,000 is realized. The compliance risk CR_1 arises from a potential violation of *R1–R4* and is subject to an administrative fine of 2% of the total worldwide annual turnover of the financial year in accordance with Article 83 (4) GDPR. This results in an uncertain consequence per instance of c_j = €10,000.

Table 3. Compliance requirements pursuant to Article 32 (1) GDPR

#	Compliance requirement
R1	Before the customer's order can be processed, his or her personal data must be protected from unauthorized access
R2	To ensure the availability and security of personal data, a copy of the data record must be stored in a backup database after taking data-protection measures
R3	Given a physical or technical incident, the access to personal data must be restored
R4	The effectiveness of technical and organizational measures is ensured by regularly conducting employee trainings and system tests

In [7], it is shown that parameters such as costs and probabilities can be annotated to BPMN models for calculation purposes. Following this, we annotated an activity identifier i, total costs $tc_{i,1}$ and reliabilities $r_{i,1}$ as well as the addressed compliance requirements (ad_Req) to each $ca_{i,1}$ in Fig. 2. For example, $ca_{2,1}$ (i = 2) "create data backup" addresses requirement $R2$ with a reliability of $r_{2,1} = 0.9997$. In other words, the activity and thus the backup fails with a probability of 0.03%. If $ca_{2,1}$ generates variable personnel costs for operating the backup software of $cv_{2,1} = €100$ and fixed annual costs for the software license of $cf_{2,1} = €60,000$, the total costs are calculated by $tc_{2,1} = cv_{2,1} + cf_{2,1} * w^{-1} = €100 + €60,000 * 50^{-1} = €1,300$. For reasons of space and clarity, we refrained from differentiating between variable and fixed activity costs in Fig. 2 and simply annotated all as $tc_{i,1}$.

For the assessment of compliance measures, only the $ca_{i,1}$ and the gateways linking these $ca_{i,1}$ are taken into consideration. Accordingly, the business activities "process customer order", "deliver goods", and "send invoice" as well as the first splitting gateway can be disregarded. Since the process model contains two exclusive choices linking $ca_{i,j}$, there are four alternative paths. In accordance with formulas (2) and (4), the path probabilities b_k must be determined for each path. This is done using the transition probabilities that are annotated on the edges emanating from the exclusive gateways. The first path P_1 contains $ca_{1,1}$ "protect personal data" and $ca_{2,1}$ "create data backup", given *physical or technical incident = FALSE* and *regularly control = FALSE*. The associated path probability b_1 results from multiplying the transition probabilities of the exclusive choices: $b_1 = 0.99 * 0.9 = 0.891$. The other b_k are calculated analogously; the remaining results are shown in Table 4.

Table 4. Interim results of the calculation example

P_k	b_k	$tc_{k,1}$	$r_{k,1}$
$P_1 = \{ca_{1,1}, ca_{2,1}\}$	0.8910	€2,300	0.9996
$P_2 = \{ca_{1,1}, ca_{2,1}, ca_{3,1}\}$	0.0090	€2,400	0.9995
$P_3 = \{ca_{1,1}, ca_{2,1}, ca_{4,1}, ca_{5,1}\}$	0.0990	€6,300	0.9992
$P_4 = \{ca_{1,1}, ca_{2,1}, ca_{3,1}, ca_{4,1}, ca_{5,1}\}$	0.0010	€6,400	0.9991

After resolving the exclusive choices, $tc_{k,1}$ must be determined by summing $tc_{k,i,1}$, and $r_{k,1}$ must be determined by multiplying $r_{k,i,1}$ for each P_k. For $P_2 = \{ca_{1,1}, ca_{2,1}, ca_{3,1}\}$ the total path cost is $tc_{2,1} = \sum_{i=1}^{3} tc_{2,i,1} = €1,000 + €1,300 + €100 = €2,400$. The reliability of P_2 is $r_{2,1} = \prod_{i=1}^{3} r_{2,i,1} = 0.9999 * 0.9997 * 0.9999 = 0.9995$. The other $tc_{k,1}$ and $r_{k,1}$ are calculated analogously; the remaining results are shown in Table 4. Using these interim results, the economic benefit of the compliance measures contained in the process model can be determined. For reasons of traceability and clarity, we use mathematical vectors for the subsequent calculation:

$$eb_1 = \left(\left(\begin{pmatrix} 0.891 \\ 0.009 \\ 0.099 \\ 0.001 \end{pmatrix} * \begin{pmatrix} 0.9996 \\ 0.9995 \\ 0.9992 \\ 0.9991 \end{pmatrix} \right) * €10,000 - \begin{pmatrix} 0.891 \\ 0.009 \\ 0.099 \\ 0.001 \end{pmatrix} * \begin{pmatrix} €2,300 \\ €2,400 \\ €6,300 \\ €6,400 \end{pmatrix} \right) = €7,294.59$$

The economic benefit eb_1 is positive and amounts to a total of €7,294.59 per instance. The expected total cost of an instance is t_1 = €2,701 and results from the scalar product of path probabilities and path costs (both right vectors). The expected reliability of an instance is $r_1 = 0.9996$ and results from the scalar product of path probabilities and path reliabilities (both left vectors). This allows for the probability of occurrence to be determined by $p_1 = 1 - r_1 = 0.0004$. Accordingly, the probability of an instance violating Article 32 (1) GDPR is 0.04%.

5   Related Work

Magnani and Montesi [6, 7] developed an approach to evaluate the costs of business processes based on four selected classes of BPMN diagrams. Using BPMN annotations, business activities were expanded by costs, and cost models were derived considering the control flows of the four classes. Sampathkumaran and Wirsing [8, 9] introduced a construct called business cost, which allows the calculation of costs necessary for a business process to reach its goal. Similar to [6, 7], they define four process patterns, but the possibility of combining them opens up a wider range of application. However, the applicability of both approaches is somewhat limited by the fact that these are based on selected process templates and not on the frequently occurring workflow patterns of van der Aalst et al. [15]. In addition, both approaches disregard the compliance perspective.

Narendra et al. [16] analyzed the feasibility of continuous process monitoring at runtime considering economic criteria, thus building a bridge between process-based assessment and compliance. They are concerned with the selection of a reasonable number of policy clauses imposing requirements on process tasks that are referred to as control points. They define a two-tier optimization problem that identifies the minimum number of control points while taking costs into consideration. Bhamidipaty et al. [17] developed an integrated quantitative system for compliance management in IT service delivery (called Indra), which extends the optimization of control points by a subsequent analysis phase. Calculated optimal solutions are checked for real optimality and, in the event of deviation, a new optimization phase is initiated. However, both approaches deal exclusively with the optimization of control points but not with the economic assessment of compliance measures.

6   Conclusion

The approach presented in this paper is an initial attempt towards the economic evaluation of process-based compliance measures taking into account the monetary consequences of compliance violations. We introduced variables and formulas for calculating the costs and reliabilities of compliance activities and for quantifying compliance risks. In addition, for combinations of compliance activities that are based on the basic control flow patterns of van der Aalst et al. [15], we derived calculation rules for pattern reliabilities, pattern costs and economic benefit. The combinability of the pattern formulas allows for an assessment of even more complex processes, as has been shown by an exemplary ordering process affected by Article 32 (1) GDPR.

Our approach is subject to a number of assumptions, such as disjoint business and compliance activities, a one-period view or the focus on monetary consequences of compliance violations. Furthermore, we did not distinguish between different types of requirements, such as requirements depending on roles or data. The assumptions made are necessary for the applicability of our approach, but limit its scope. It takes future research to eliminate these assumptions and to expand the scope of application. Moreover, our approach is a pessimistic one that assumes that monetary damages increase proportionally with the number of compliance violations. This was intended to ensure that uncertain monetary consequences would not be underestimated. In practice, however, monetary damages are often not necessarily proportional to the number of violations but instead depend on the type and severity of the infringements. Furthermore, we derived formulas only for the basic control flow patterns. Although these are very common patterns, further research is needed to derive costs and reliability formulas for other patterns to broaden the scope of the approach. In addition, the applicability has been demonstrated only by way of an example. In the future, the usefulness and usability of our approach must be evaluated with the help of scientific methods, which will be the consequent next step of our research.

The assessment approach presented offers application potentials especially for situations in which the use of alternative compliance measures is conceivable, e.g., alternative compliance tools or different employees with varying qualifications and salaries. In this case, this approach can be used to identify the economically best alternative. It also opens up the potential for sensitivity analyses and, for example, the identification of fixed-cost degression effects of compliance measures.

References

1. Governatori, G., Hashmi, M., Lam, H.-P., Villata, S., Palmirani, M.: Semantic business process regulatory compliance checking using LegalRuleML. In: 20th International Conference on Knowledge Engineering and Knowledge Management, pp. 746–761 (2016)
2. Schultz, M.: Towards an empirically grounded conceptual model for business process compliance. In: 32nd International Conference on Conceptual Modeling, pp. 138–145 (2013)
3. Kühnel, S.: Toward a conceptual model for cost-effective business process compliance. In: Proceedings of Informatik 2017. Lecture Notes in Informatics (LNI), pp. 1631–1639 (2017)
4. Pham, H.C., Pham, D.D., Brennan, L., Richardson, J.: Information security and people. A conundrum for compliance. Australas. J. Inf. Syst. **21**, 1–16 (2017)
5. Kumar, A., Yao, W., Chu, C.-H.: Flexible process compliance with semantic constraints using mixed-integer programming. INFORMS J. Comput. **25**, 543–559 (2013)
6. Magnani, M., Montesi, D.: Computing the cost of bpmn diagrams. Technical report UBLCS-07-17, Bologna (2007)
7. Magnani, M., Montesi, D.: BPMN: how much does it cost? An incremental approach. In: Alonso, G., Dadam, P., Rosemann, M. (eds.) BPM 2007. LNCS, vol. 4714, pp. 80–87. Springer, Heidelberg (2007). https://doi.org/10.1007/978-3-540-75183-0_6

8. Sampath, P., Wirsing, M.: Computing the cost of business processes. In: Yang, J., Ginige, A., Mayr, H.C., Kutsche, R.-D. (eds.) UNISCON 2009. LNBIP, vol. 20, pp. 178–183. Springer, Heidelberg (2009). https://doi.org/10.1007/978-3-642-01112-2_18
9. Sampathkumaran, P.B., Wirsing, M.: Financial evaluation and optimization of business processes. Int. J. Inf. Syst. Model. Des. 4(2), 91–120 (2013)
10. Vom Brocke, J., Recker, J., Mendling, J.: Value-oriented process modeling: integrating financial perspectives into business process re-design. BPM J. 16, 333–356 (2010)
11. Hammer, M., Champy, J.: Reengineering the Corporation. A Manifesto for Business Revolution. HarperBusiness Essentials, New York (2003)
12. Seyffarth, T., Kühnel, S., Sackmann, S.: A taxonomy of compliance processes for business process compliance. In: Carmona, J., Engels, G., Kumar, A. (eds.) BPM 2017. LNBIP, vol. 297, pp. 71–87. Springer, Cham (2017). https://doi.org/10.1007/978-3-319-65015-9_5
13. Kirzner, I.M., Boettke, P.J., Sautet, F.E.: The Economic Point of View. An Essay in the History of Economic Thought. Liberty Fund, Indianapolis (2009)
14. Rastrepkina, M.: Managing variability in process models by structural decomposition. In: International Workshop on Business Process Modeling Notation, pp. 106–113 (2010)
15. van der Aalst, W.M.P., ter Hofstede, A.H.M., Kiepuszewski, B., Barros, A.P.: Workflow Patterns. Distrib. Parallel Databases 14, 5–51 (2003)
16. Narendra, N.C., Varshney, V.K., Nagar, S., Vasa, M., Bhamidipaty, A.: Optimal control point selection for continuous business process compliance monitoring. In: International Conference on Service Operations and Logistics, and Informatics, pp. 2536–2541 (2008)
17. Bhamidipaty, A., Narendra, N.C., Nagar, S., Varshneya, V.K., Vasa, M., Deshwal, C.: Indra: an integrated quantitative system for compliance management for IT service delivery. IBM J. Res. Dev. 53, 1–12 (2009)

Visual Representation of the TOGAF Requirements Management Process

Elena Kornyshova[1](✉) and Judith Barrios[2]

[1] CEDRIC, Conservatoire National des Arts et Métiers, Paris, France
elena.kornyshova@cnam.fr
[2] GIDyC, Systems Engineering School, University of Los Andes, Mérida,
Venezuela

Abstract. TOGAF Architecture Development Method (ADM) covers different aspects of Enterprise Architecture (EA) management as well as it provides textual guidelines to adapt and perform EA processes including the requirements management (RM) process. We observed that adopting ADM following these guidelines is an intricate task because the effort required to define a sequential interaction between related activities is meticulous and hard. During a real case experience we have formalized the ADM-TOGAF Requirements Management textual guidelines with the Business Process Model and Notation (BPMN) to facilitate its comprehension and usage. Afterwards, the usefulness of the proposed process models has been qualified with a group of engineering and master students.

Keywords: Requirements management · ADM-TOGAF · Process model
Visual notation · Industrial experience

1 Introduction

The Open Group and Gartner consider Enterprise Architecture (EA) as a discipline that assists an organization defining, developing and exploiting information flow capabilities in order to achieve organization´s strategies and intentions [1–3]. In order to really take advantage of an EA inside an organization or a business, many frameworks have emerged. Among the more known and used we have Zachman [4], FEAF [5], DODAF [6], TOGAF [3], and GEAF [1]. TOGAF (The Open Group Architecture Framework) is one of the most used in modern organizations [7, 8]. TOGAF framework also provides a dedicated method ADM (Architecture Development Method) for supporting the work of developing EA.

TOGAF covers all different aspects of the EA management work as well as it provides specific textual guidelines to adapt and complete the development processes and steps prescribed in its method. However, we have observed in practice that understanding, interpreting and implementing TOGAF processes through the ADM method and following its textual guidelines is not a simple task. It is reported in the literature that textual guidelines are heavy to read, to follow and to understand, even for domain users [9]; whereas, visual representations are easy to understand, to follow and they may provide a wider perspective of the available courses of action included into

© Springer Nature Switzerland AG 2018
C. Woo et al. (Eds.): ER 2018 Workshops, LNCS 11158, pp. 239–248, 2018.
https://doi.org/10.1007/978-3-030-01391-2_29

the ADM workflow. The research work of Kathrin Figl [9] expounds, among other significant issues, that visual process models meet better user preferences because process models provide a more complete and precise visual representation.

The main intention of this work is to provide a visual representation of the ADM-TOGAF Requirements (RM) textual guidelines using the Business Process Model and Notation (BPMN). This work was done in a commercial organization[1].

Based on our experience, we consider that, a holistic method view, built with visual process models, is easiest to read and to understand, thus to be followed than existing textual guidelines representations. We preserve the original meaning of method guidelines prescriptions by keeping consistency between textual and visual guidelines representations. Afterwards, the usefulness of the proposed process models has been qualified with a group of engineering and master students.

The remaining of this paper is structured as follows. Section 2 presents the problem and motivation of our work. Section 3 contains the BPMN models of the ADM-TOGAF Requirements Management activities. In Sect. 4 we present an analysis of the positive and negative aspects of using visual process models used to represent ADM guidelines and give results of the obtained models validation by students. Finally, Sect. 5 concludes the paper and outlines future work.

2   Problem and Motivation

Methods are usually accompanied with a set of rules of action, suggestions, strategies or practices attested to be effective to complete or to perform a prescribed step, to execute an action or a technical procedure, in turn. As mentioned earlier TOGAF is a framework for developing and managing enterprise architectures [3]. It has been developed by the members of the Open Group working within the Architecture Forum [10]. This means that experience and knowledge of enterprise architects are included at essential sections of the framework; thus, it increases the potential of the framework for its practical use. TOGAF considers three business perspectives building three complementary architectures: Business Architecture, Information Systems Architecture (Applications and Data Architectures), and Technology Architecture. The framework allows enterprise architects to develop a consistent, feasible and integrated model from the business, information systems and IT infrastructure perspectives of any enterprise.

However, the process of architecting an enterprise is technically complex and the design of heterogeneous, multi-perspective architectures is not easy, thus TOGAF provides the method ADM (Architecture Development Method) that aims to help users while designing and implementing an enterprise architecture. It is composed of a set of ten phases, considered as business processes, their corresponding steps or activities, and some guidelines to assist method customization. The phases are: Preliminary phase, Architecture Vision, Business Architecture, Information System Architecture, Technology Architecture, Opportunities and Solutions, Migration Planning, Implementation Governance, Architecture Change Management, and Requirements Management.

---

[1] The name of the organization was intentionally hidden in order to preserve confidentiality.

The Requirements Management (RM) phase is the central and key element of the ADM method. RM plays a particular and essential role in the entire EA development process and is strong related to all other ADM phases. The goal of the RM phase, as defined in TOGAF 9.1 document [3], is "To define a process whereby requirements for enterprise architecture are identified, stored, and fed into and out of the relevant ADM phases". It means that the requirements phase drives the whole process of development an Enterprise Architecture and, that current or new requirements for the EA under development, as well as any modification on their specifications, should be considered at any stage of an ADM cycle. For each new requirement, an impact analysis is prepared; thus its potential effect over current EA requirements as well as on EA development project plans is considered. The main result of the impact analysis is the identification of those ADM phases that should or need to be revised in order to tag subsequent implications and changes on current requirements. Finally, if a change on actual requirements is approved, it must be registered and scheduled in current project transition plan.

In order to assist architects, ADM-TOGAF provides textual guidelines to adapt and complete the prescribed development processes of the ADM and its steps. A guideline is a recommendation of what ought to be done in order to fulfill a specific method target or just the description of how to perform a particular technical task included as part of a method.

As mentioned earlier, textual guidelines are heavy to read, to follow and to understand, even for domain users [9]; additionally, we observed in practice that to follow, interpret and implement what is prescribed by TOGAF processes through the ADM method is an intricate task as well as it demands a significant amount of time and dedicated effort from architects and project leaders. This difficulty is due not only to the fact that guidelines are documented as connected text – hypertext, that can be routed according to the level of refinement demanded by architects; but, to the fact that it does not exist an entire method representation linking inputs/outputs to sources/destinations for every method phase and step. Consequently, the levels of details trailed by the architect to build a specific project workflow may yield him/her to lose the holistic perspective of current process, and the wider view of what has to be done or how setting-up current process inputs/outputs with other method phases sources/destinations.

To support our work, we have reviewed some domain literature and we found no research or practical work directly associated with providing supplementary help, support or guidelines about understanding, interpreting or following the ADM guidelines. However, we found some tips and informal advises about what to do or what to consider while developing an EA project in some enterprise architectures specialized blogs [11]. Nevertheless, and considering that the RM process is our instructive example, we search and analyzed some ADM-RM reported works and practices as well as some guidelines provided by some EA supporting tools. For instance, the framework TOGAF-SABSA [12] describes the integration of both perspectives TOGAF and SABSA® (Sherwood's Applied Business-driven Security Architecture) showing how the requirements management processes in TOGAF may be complemented with the SABSA concept of Business Attribute Profiling. The guidance support provided by this integration concerns enabling an efficient management of enterprise architecture security requirements: business and risk management and, secure architectures to support business outcomes.

In order to complete our work position, we also revised some EA modeling and management tools like Enfocus Solutions [13] which enables the capture, development and traceability of architecture requirements defined as IT service components; and, the EA (Enterprise Architecture) of Sparx Systems [13], that enables ADM requirement activities by providing facilities for manipulating and producing TOGAF inputs and outputs artifacts. There are some others tools like iServer for EA from Orbus Systems, Abacus, BiZZDesign, etc.[2], that put into service the OMG Archimate notation concepts and other visual notations for facilitating the production and management of artifacts used by ADM phases, including inputs and outputs of the Requirement Management phase. But, none of them provides complementary ADM adaptation guidelines or textual description of ADM guidelines.

Assenting with Zachman [14] and the work reported in [15], about the benefits of having a holistic vision of the information needed to define the architecture of an enterprise, we decided to represent ADM guidelines by using business process models. In addition, this decision is supported by the following arguments: (1) the fact that enterprise architects and EA team members have technical competences and conceptual fundaments for using, understanding, interpreting and exploiting business process models because they are used to work with business processes models[3]; (2) because graphical business process models have evidenced to be useful and pertinent for describing complex workflows as well as their inter-processes communications [16–18]; and, (3) because business process models are powerful tools for designing, analyzing, implanting, controlling and optimizing business processes as well as they offer a way for representing practical process knowledge which is acquired from performing EA projects with their corresponding method customization [19, 20].

Finally, it is important to mention that the process models presented here are independent of the graphical notation used to represent business process models. Architects may choose UML activity diagrams [21] or UML business [16], BPMN [17], Archimate [13], among the most used. In our case, we have selected BPMN because it is a standard and complete notation for representing business processes; it allows modelers to represent several and connected levels of processes detail as well as their inputs/outputs, dependencies and links and, because BPMN diagrams may be transformed into executable sets of workflows thus processes can be simulated, measured and enhanced, if necessary.

---

[2] https://www.capstera.com/enterprise-architecture-tools/.

[3] Process modelling is included in specialized domain curricula like EABoK, ACM curricula, SEBoK.

3   Process Model of the ADM Requirements Management Guidelines

In this section we describe process models that represent in detail the activities included as part of the RM phase. Process models have been built based on TOGAF 9.1 official documents and represented using BPMN (Figs. 1, 2, and 3)[4].

Considering that BPMN activity diagrams are all-inclusive, we will give just a brief explanation of each one. The most relevant feature of this process representation is that it allows to assure consistency between modeling levels.

Figure 1 represents the RM process at the high level. In order to keep correspondence with textual ADM guidelines, we preserved the same ADM activity′ numbers in the process model representations of the ADM guidelines. The RM process includes four composed activities detailed as sub-processes. Figures 2 and 3 offer two sub-processes of the RM phase namely "Determine baseline requirements" and "Identify changed requirements and record priorities", respectively. For the lack of space we do not present the detailed activities of the two other sub-processes which are included in other ADM processes ("Identify changed requirements" and "Assess and issue impact of changed requirements").

Following the ADM description, Fig. 1 includes two pools corresponding to the RM process and processes from other ADM phases, respectively. As mentioned before, the RM process is tightly related to other phases thus, it is mandatory to show what are its interactions with activities performed in other ADM phases.

The process is initiated from one of the other ADM phases by identifying and documenting new requirements which are transferred to the RM phase. The baseline requirements are determined. The sub-process of the activity "Determine baseline requirements" is given at Fig. 2. It includes the following activities: Determine priorities arising from current phase of ADM, Confirm stakeholder buy-in to resultant priorities, Record requirements priorities and place in Requirements Repository. All these activities are performed based on the information from the Requirements Catalog ("Stored requirements" from the Requirements Repository) and, the new requirements are also stored in the Requirements Repository at the end of this sub-process. Thereafter, the requirements are monitored.

If requirements changes are detected during any other ADM phase, they are correspondingly transferred to the RM phase to be analyzed. This is performed by the activity "Identify changed requirements and record priorities" which is detailed as a sub-process shown in Fig. 3. The sub-process includes the following activities: Identify changed requirements and ensure the requirements are prioritized, Record new priorities, Ensure that any conflicts are identified and managed through the rest of ADM phases and, Generate Requirements Impact Statement. This sub-process uses the stored requirements from the Requirements Repository, takes into account the changed requirements from the other ADM phases and analyses them to define priorities. At the end, it produces the Requirements Impact Statement, sends it to corresponding ADM phases and, stores the changed requirements with their priorities into the Requirements Repository.

---

[4] The processes in this paper are modeled with Bizagi Modeler (https://www.bizagi.com/en/products/bpm-suite/modeler).

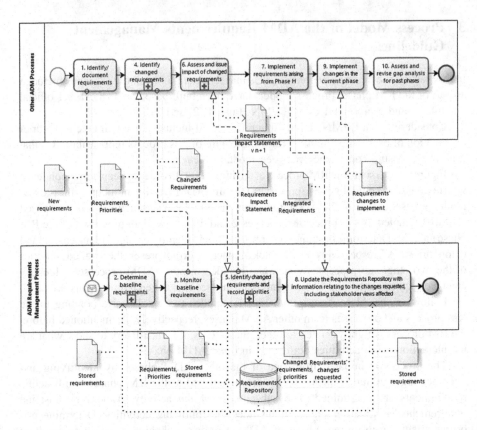

Fig. 1. Activities of the RM process at the high level. (We chose to represent ADM phases as pools to separate the RM team process responsibilities from others EA teams' process responsibilities.)

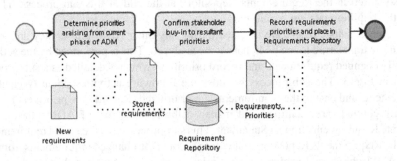

Fig. 2. Sub-process "Determine baseline requirements".

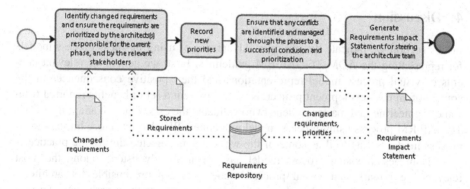

Fig. 3. Sub-process "Identify changed requirements and record priorities".

Compared to the textual description from the ADM-TOGAF documentation, we have completed these models with different links between activities (the BPMN message flows) and with documents which are not mentioned in the ADM textual description, but are useful to understand the corresponding processes. Concerning the links, the textual guidelines suggest only sequences naturally existing between steps (from 1 to 2, from 2 to 3, and so on). However, these links and activities are not sufficient. As sequences between steps were not explicit, we have completed them following our experience (for instance, the step 4 follows the step 1, and takes into account the results of the step 3).

Regarding documents, the ADM textual guidelines do not mention that activities of the RM process use the Requirements Catalog available in the Requirements Repository. However, the information about existing requirements (we call them "Stored requirements") is used at different steps as steps 2, 5, and 8 (at step 2 to compare with other existing requirements and define priorities; at step 5, to compare the changes with the previous version of the requirements and associated priorities; at step 8, to update information in the Requirements Repository). The ADM textual guidelines do not detail explicitly the input "Stored requirements" for these steps. We have added this input to three mentioned activities to have a more complete vision of the documents moving along the RM process. The same was done with the "Integrated requirements" transferred from the step 7 to the step 8 (the ADM text does not mention any document at this place).

Additionally, several amendments were made to transform the textual description into the process models due to the modeling constraints:

1. The step 2 was renamed from "Baseline requirements" to "Determine baseline requirements". It was done to have a verb in the activity name.
2. The name "Assess and issue impact of changed requirements" was given to the step 6. As this step in the ADM textual guidelines contains only different activities and does not have a name.

4 Discussion

In this section we indicate the observed advantages of using business process models for representing the ADM-TOGAF RM guidelines. First, we generalize relevant benefits of visual process model representations and their positive consequences in the context of an EA development project. Next, we present a set of benefits reported from some engineering and master students who validated the experience of adapting ADM-TOGAF RM process using process model guidelines. We also introduce limitations of process model use as well as some of the weak points detected during the practice.

Advantages of visual process model have been already listed; among the most relevant we remark that visual process models are comprehensible for architects because of their professional background and, that by using them architects can represent several levels of detail of an ADM process; architects can describe method processes from high and general level to low and specific detailed level without losing coherence and integrity between levels. Therefore, a visual process shows an ADM business process, its inputs, outputs, sources and destinations allowing architects to visualize, at a glance, the set of artifacts and documents that are implicated in the execution of process activities as well as the interactions with other ADM phases. Thus, architects can make associations according to current project factors, analyze alternative orders for actions that should be executed before setting what is the specific workflow that current project situation needs. Accordingly, architects' practical knowledge can be kept and reused in other similar projects, because EA project's experiences can be included as part of ADM existing process models; same way, contextual factors may also be added thus method management process can be enhanced and optimized, if needed.

For validating the usefulness of the proposed process model guidelines, we prepared an exercise with a group of 18 selected engineering and master students. As part of the practical work assigned in a requirements engineering course, students had to adapt the RM-ADM prescribed process for a specific project situation. A half of selected students (group A) had all information about ADM as textual TOGAF guidelines; the other half (group B) used the proposed process models guidelines. The main results of this experience, based on a qualitative observation, are that the students of group A asked for extra time for reading, understanding and tailoring once the requirement management process; whereas group B had the opportunity of rereading the documentation and adapt two times the tailored process. Group B students qualified the proposed process model representation as a valuable means because it shows, at a glance, the set of EA project elements to be considered at a given method phase. In consequence, this guidelines representation opens them the option to inspect method details like terms glossary, process' inputs and their sources, transforming tasks that should be executed and, project team responsibilities. One relevant point remarked by group B students is they have always the option of going back to the original RM process perspective, in one click, at any time. According to students, it was easier and faster to understand and follow graphical that textual guidelines; besides, a lower possibility of getting loss inside the vast ADM documentation.

Finally, we have detected two important weak points about taking advantage of process model guidelines: first, we corroborate that it is mandatory to have practical knowledge on process model notations as well as medium to high level of modeling abilities for well understanding and really take advantage of what it is represented by process model guidelines; second, we verified that practical exploitation of process models is strong limited by technology constraints. It is recommended to install a Business Process Management (BPM) tool not just to use any graphical representation software: a business process modeling tool has, among other user interface facilities, to simplify the work required to state process models consistency, to keep account of models updates as well as enabling their customization. Further, a BPM tool will facilitate the definition, analysis, store, enactment and automatic execution, with BPEL, of process models guidelines, if required.

5 Conclusions and Future Work

In this paper, we proposed to represent ADM-TOGAF guidelines by using process models because this kind of models seem to be suitable tools for describing business processes and activities as well as for detailing their dependence and sequence links.

A visual process model of the ADM-TOGAF RM textual guidelines offers a comprehensive high level representation containing the whole participant elements and activities. Along with the relevant benefits listed for any business process model like reuse, simulation, process automation and optimization, ADM guidelines could be completed with references coming from architects' practical work on EA projects improving method recommendations and enhancing future processes of tailoring method to specific project characteristics.

In addition, we observed, from a qualitative perspective, the usefulness of the results of our experience by performing a short validation exercise with a group of engineering and master students. Main result of this validation process was that the time needed for the group (A) of students to tailor ADM-RM by following textual guidelines almost doubled the time needed for the group (B) that used process model guidelines. It is important to observe that after students' experience, the time required for understanding what they might do before they have to do it, was a key factor for well adapting ADM RM phase to each of their particular project situations. Furthermore, knowing which documents, products and artifacts are in/out of each RM activity helped students to improve their understanding of when and how to perform a customized activity.

Future work will first concentrate on a formal validation study of the practical advantages of using process model representation for expressing ADM guidelines. After that, we look forward to deliver process model guidelines for the whole ADM process allowing architects to store and manage their own data from practical process model customized guidelines.

References

1. James, G.A., Handler, R.A., Lapkin, A., Gall, N.: Gartner Architecture Framework Evolution (2005)
2. Lapkin, A., et al.: Gartner clarifies the definition of the term "enterprise architecture". Gartner Research (2008)
3. TOGAF 9.1. The OpenGroup Architecture Framework (2011). http://www.opengroup.org/togaf/
4. Zachman, J.A.: About Zachman framework for enterprise architecture. Zachman International Enterprise Architecture Framework (2008)
5. A Common Approach to Federal Enterprise Architecture, Executive Office of the President of the United States (2012)
6. The DoDAF Architecture Framework Version 2.02. Department of Defense (2011)
7. A Comparison of the Top Four Enterprise-Architecture Methodologies. ObjectWatch, Inc. Microsoft Development Network (2007)
8. Cameron, B., McMillan, E.: Analyzing the Current Trends in Enterprise Architecture Frameworks (2013). http://ea.ist.psu.edu/documents/journal_feb2013_cameron_2.pdf
9. Figl, K.: Comprehension of Procedural Visual Business Process Models: A Literature Review. Bus. Inf. Syst. Eng. **59**(1), 41–67 (2017)
10. The Open Group working within the Architecture Forum (http://www.opengroup.org/architecture)
11. Enterprise Architecture Blogs. https://enterprisearchitectureblog.wordpress.com/top-enterprise-architecture-blogs/
12. The Open Group TOGAF-SABSA Integration Working Group, White paper, October 2015
13. The Open Group EA Standard Notation. Archimate 3.0. (http://www.opengroup.org/subjectareas/enterprise/archimate)
14. Lapalme, J., Gerber, A., Van der Melwer, A., Zachman, J., De Vries, M., HinKelmann, K.: Exploring the future of enterprise architecture: a Zachman perspective. Comput. Ind. **79**, 110–113 (2016)
15. Pinggera, J., Soffer, P., Fahland, D., Weidlich, M., Zugal, S., Weber, B., Reijers, H.A., Mendling, J.: Styles in business process modeling: an exploration and a model. Softw. Syst. Model **14**, 1055–1080 (2015)
16. Ericksson, H.-E., Penker, M.: Business Modeling with UML: Business Patterns at Work. Wiley, New York (2000)
17. Object Management Group (OMG): Business process model and notation (BPMN), version 2.0 (2011)
18. Moreno Montes de Oca, I., Snoeck, M.: Pragmatic Guidelines for Business Process Modeling. Technical Report. KU Leuven – FEB - Management Information Systems Group (2015). http://ssrn.com/abstract=2592983
19. Weske, M.: Business Process Management: Concepts, Languages, Architectures. Springer, Heidelberg (2007)
20. van der Aalst, W.M.P.: Business process management: A Comprehensive Survey. www.vdaalst.com
21. Object Management Group (OMG). Unified Modeling Language (UML) Version 2.5. OMG Document Number ptc./2013-09-05 (http://www.omg.org)

Technology-Transfer Requirements Engineering (TTRE) – on the Value of Conceptualizing Alternatives

Blagovesta Pirelli[✉] [ID] and Alain Wegmann

EPFL, Lausanne, Switzerland
{blagovesta.pirelli,alain.wegmann}@epfl.ch

Abstract. In this paper, we describe a requirements engineering method with a focus on the conceptualization of alternative service offerings. The practical context for our project is based on the first author's work in a startup. Our proposed method is suitable for exploring market opportunities while specifying a service offering. Our method helps requirements engineering practitioners understand the business and technology worlds by modeling business needs and technical capabilities in the same model.

Keywords: Technology transfer
Market-driven requirements engineering · Conceptualization
Modeling · Service design

1 Introduction

One of the key roles of requirements engineering (RE) during software development projects is to achieve consensus and a common understanding between business and technology. A major challenge of bringing business and technology together is the different conceptualization of the world that people from the two domains hold. The business domain requires understanding the inner workings of an enterprise, its strategy, operations, and organization. Technology, and in particular research technology, requires specialized knowledge to understand and use it. To find a solution to a business problem and an application for a piece of technology, RE practitioners need to elicit requirements from the business and understand the capabilities of the technology.

Technology-intensive startups (e.g., research spin-offs) have to undergo a technology-transfer process that is a type of market-driven RE (MDRE) process [17]. Startups scout the market space for opportunities that they can address with their technology, instead of producing a customized product for a single customer. They implement a technology push strategy, which involves collaboration with potential customers to find applications for the research technology. The collaboration and feedback from the interaction with potential customers

© Springer Nature Switzerland AG 2018
C. Woo et al. (Eds.): ER 2018 Workshops, LNCS 11158, pp. 249–259, 2018.
https://doi.org/10.1007/978-3-030-01391-2_30

is valuable, as service offerings require information to shape the technology in a usable and profitable way.

The RE community currently explores the area of RE for technology transfer [2,6]. Technology transfer projects lack direct applicability to an immediate market, unlike most traditional software-development projects. Yet, technology transfer projects have a potential for a high impact. For example, Popek and Golberg developed the hardware virtualization principles in 1974 [15]. It was only 30 years later that the founders of VMWare[1] used these principles to advance the research and to create the market for virtualization [4]. We believe that RE can accelerate the process of finding a technology-market fit for new research projects.

In this paper, we present the artifact from the first iteration of a design science research (DSR) [10] project - a proposal for a technology transfer RE method to systematically specify alternative service offerings based on research technology in the process of market exploration. The DSR project is based on the experience of the first author who worked at technology-intensive startup for over a year. The method is an explicit prescription [9] to the following question: *How to productize a research technology in a service?*

The structure of the paper is the following. In Sect. 2, we present our research methodology. In Sect. 3, we present the knowledge base of the DSR project. In Sect. 4, we present our RE method. In Sect. 5, we present the business environment with the example of a pilot project in secure data exchange for an e-health platform. In Sect. 6, we conclude and give an outlook for future work.

2   Research Methodology

In our research, we use the design science research (DSR) framework as described by Hevner et al. [10]. To both solve a practical problem and contribute an abstract theory, DSR gives guidelines to researchers on how to build socio-technical artifacts, such as system design, methods to design systems, and requirements of systems [14].

A DSR project has three main components: the environment, the knowledge base, and the IS research artifact. Our artifact, a requirements engineering method for early-stage technology-intensive ventures, falls under the category of methods for designing systems [9]. Our method for technology transfer RE is prescriptive: it gives explicit guidelines on how to achieve a task, rather than provide an analysis, an explanation, or a prediction [9,10]. Our contribution is a theory for design and action [9].

Environment: We obtained our data from the environment via a qualitative study that examines the real-life settings of a startup. The first author was the first employee of the startup (denoted as XYZ), and, as such, could follow the evolution of the company from the beginning. The observed period was fifteen months. Our practical question stems from the environment.

---

[1] www.vmware.com.

Qualitative research methods have a number of characteristics [13]: (1) the data are rich and contextual; (2) the emerging themes are validated often with informants; and (3) the researcher gains an integrated (holistic) overview of the study. This data collection method is an ethnographic method. We use the Gold's classification of roles: a complete observer, an observer-as-participant, a participant-as-observer, and a complete participant [7]. The first author assumed a participant-as-observer role, with the observer role being secondary to the participant role. It enables researchers to gain access to more information, as they become part of the environment they study [8]. The drawback is that the collected data are subject to interpretation, reflection and reflexivity [20].

Knowledge Base: The knowledge base consists of foundations and methodologies. The foundations are previously published theoretical contributions. The foundations for our method are requirements engineering theories, service science, and conceptual modeling literature. The methodologies from the knowledge base are formalisms such as models and code.

Artifact: Our artifact is a method for early-phase technology transfer RE. To refine and evaluate the artifact, we use concepts from the knowledge base and interact with the environment. Our assumptions are that with the help of existing RE methods, the conceptual modeling techniques, the service science lens, and technology road-mapping concepts, we can accelerate the technology-transfer process of technology-intensive startups, such as research spin-offs.

3   Literature Review

Our literature review presents the knowledge base of our DSR project. The different streams of literature are conceptual modeling, service science, and RE. In this section, we outline the main relevance of these theories and the portion of them that we use to design our DSR artifact.

3.1   Conceptual Modeling

We base our understanding – of how people conceptualize and model their observations – on the literature of conceptual modeling. A universe of discourse (UoD) is an observed portion of the "reality" during conceptual modeling [1]. A UoD is a set of objects, such as people, physical objects, and technology. The observer establishes semantics (i.e., gives meaning) to the UoD, and with a predefined syntax (formalisms, models) expresses observations [1]. Different purposes for modeling result in different levels of detail and of precision of the models.

An observer, after perceiving the UoDs, *conceptualizes* and *models* the observed portion of the reality. The conceptualization is possible after the observer is able to break down the observed UoDs into nameable, i.e., meaningful, objects (or entities), such as a service, a graphical interface, a data storage, a department [3]. Conceptualization gives meaning to the observed entities and their relationships, whereas modeling gives a structure to the conceptualization.

3.2 Service Science

We use service science to structure our observations. Service science recognizes the value of combining different theories to achieve a holistic view of the service environment [11]. Service-dominant logic provides a perspective over the combination of elements into service systems, and it is the philosophical foundation of service science [18,19]. Services are the application of competences for the benefit of others hence the means to create value for others [19]. Service systems are configurations of resources - people and technology [11]. Service science assumes a service system to be the basic unit of analysis of the value exchange [12]. In these value networks, the value added by the socio-technical elements determine their place in the configuration of components.

3.3 RE Methods

We use guidelines from the RE literature to understand how to conduct RE in a technology-intensive project. Some of the challenges that RE faces are to ground its findings in the reality in which the technical system will operate and instead of describing the system to be built [23]. There are differences between RE for market-driven and bespoke development, e.g. how to elicit, prioritize, negotiate and analyze the impact of requirements [5,17]. The main goal of market-driven RE (MDRE) is to gauge a product for a broad market [17].

We use SEAM, an RE method: it enables service designers to model and design systems [21]. SEAM recognizes the importance of an observer of the reality that is being modeled. SEAM includes methods for analyzing explicitly the relationship between different stakeholders [22] and to conduct goal-oriented RE [16]. This enables service designers to define terms concretely ("What is a market/benefit/value?") and to perform analysis of the stakeholders' goals in the environment in which an IT system would operate. The explicit positioning of technology within a service system model enables us to also reason about the benefits for service adopters, or the market position of new customers (either market re-segmentation, or customers of competitors within the same segment). The service models use the same ontology to represent different parts of the reality – both business and technology.

4 Technology Transfer Requirements Engineering (TTRE)

We choose to model the service offerings with the help of SEAM and to further develop its applicability to designing new service offerings based on research technology. We formulate explicit hypotheses about potential benefits directly stemming from the technology for service adopters, and we build proof-of-concept prototypes to test the assumptions we have regarding the market propensity. After multiple feedback cycles, we keep only the requirements for the features that show high impact for a broad market. The distillation from concrete steps to conduct such early-phase RE lead to TTRE:

1. Observe - Examine the existing technology and the potential business settings in details.
2. Conceptualize and model - Define **explicitly** a set of alternative services that emerge from the observations.
3. Construct - Build proof-of-concept prototypes for the current prospects of the company to validate that services are of actual value for them.
4. Contextualize - Collect feedback from the prospects to make adjustments to the service alternatives and to the market assumptions.

The TTRE method is iterative and the expected outcome is to be able to outline a general profile of a customer and extrapolate the industry and the market segments that the service addresses.

5   TTRE in Context

In this section, we present the environment of the DSR project and the way TTRE applies in it, based on the first author's experience. We describe the four different steps (Observe, Conceptualize and model, Construct, Contextualize) with the example of XYZ and a pilot project with E-Health Cisco Switzerland (Cisco for short). We show only one industry and illustrate TTRE with alternative technical service offerings. Another application for TTRE, and area for future work, is to show the same service alternatives in different industries.

5.1   Observe

Technology. The initial technology of XYZ enables operations such as search and share on end-to-end encrypted data. In this context, end-to-end refers to the notion of encryption at the point of creation of the data and storage of the data only in an encrypted form (cipher text) when it is outside of the boundaries of the creating unit. XYZ's technology consists of a set of various components, each of which has a different function. The purpose of the components is to provide searchable and sharable encrypted data. The features that the technology can offer are (1) libraries - encryption, searchable encryption, and sharable encryption libraries, (2) storage - data and search index storage, and (3) graphical user interface for end-user applications.

We stop making the observations on the research technology at the level where the components are meaningfully divided into different-purpose components. The components contribute to achieving the main goal of the technology: in XYZ's case, search and share end-to-end encrypted remotely-stored data.

Business. Our running example is a project between Cisco and XYZ. The current offering to the healthcare organizations is a health-data exchange platform. Cisco's role is to provide a secure medical-data exchange infrastructure that enables different stakeholders to communicate in a unified way.

While in the process of delivering treatment to patients, certain data is generated, for example, prescribed medications, test results, courses of treatment, and check-up dates. All of these data are associated with the patient and stored in their record. The electronic health record is sensitive information and requires a secure exchange process between different actors. Additionally, different stakeholders have access to only a subset of the records.

We stops the observations on the business domain at the level of service exchange between our potential customer (Cisco) and their customer(s) (hospitals, medical practitioners). These observations give us information on what value we can bring to the value network of our customer to enable them to offer a feature desired by their customer.

5.2   Conceptualize and Model

Observers form concepts about what they have observed, as a precursor for expressing ideas in concrete formal models. Conceptualization is an interpretative step of translating the observed signals into semantically coherent entities and relationship between those entities. Using these observations, we compose different configurations (alternatives) of the XYZ's technical components.

Technology. The observed features of the technology are grouped together in three possible services. There are different ways to group features. For example, at random, with the core functionality in mind (share data or search data), and based on the abstraction level (software libraries, development toolkit, end-user graphical interface). We choose to group the features of the alternatives service offerings, based on their abstraction level. Table 1 shows the grouping of features that form the three alternative services for XYZ's technology: an encryption package, a secure data-storage platform, and an encrypted cloud-storage system.

Table 1. Features and services.

Feature	Encryption package	Secure data-storage platform	Encrypted cloud-storage system
Encryption libraries	✓	✓	✓
Shareable encryption libraries	✓	✓	✓
Searchable encryption libraries	✓	✓	✓
Data storage		✓	✓
Search index storage		✓	✓
GUI			✓

Encryption Package. The shareable encryption package is a library that offers encryption that preserves operations (search and share) on the data. The user of this package is a software developer, who develops their own applications and already has mechanisms in place to interact with their storage servers.

Secure Data-Storage Platform. In this package, we propose a service: a platform that handles the interaction with the data and the search servers, as well as the data encryption and decryption. The user is a software developer who develops end applications for their users and makes use of the platform to enhance the security of users' data.

Encrypted Cloud-Storage Application. This service alternative bundles all the features of XYZ's technology. The user of this package is an end user.

Business. With the help of formalized models, we are able to reason about the different actors in the service system, the process in which they participate and the possible integration points for XYZ's service offerings. Figure 1 represents the model of the e-health data exchange ecosystem. We choose to model the ecosystem in three levels; each of them represents the grouping of features based on Table 1. On the topmost level, the model depicts the e-health data-exchange service system in Switzerland. The actors (medical practitioners, pharmacies, post-hospital facilities, and the Swiss Post e-health unit) all contribute their services to the process of "secure exchange of electronic patients data". These are the customers of our customer.

One of the alternative service offerings - the encrypted cloud-storage system - can take the place of the electronic medical record (EMR) system. This requires working with chief officers (CxOs) of the organizations to further define the requirements of the system and working with the end users from the first level. However, there is an EMR system provided by the "software vendor" already in place, hence the competition with the incumbent software vendor would be hard.

The next integration point for XYZ is within the software vendor's value network on Level 2. The software vendor's value network provides the EMR system as a service. This value network is on the same level of granularity as XYZ's platform. The EMR system of the "Software vendor" provides the infrastructure for storing and searching through medical records and the mechanisms to share them with other members of the system. XYZ's platform service offering has a limited but well-defined role. The main feature of the service is encrypting the most sensitive parts of the data that contains details about person's health, e.g., EMRs. XYZ has to communicate with the people (project managers and software developers) who choose and configure the different systems that will work with each other.

The last alternative, the encryption package, would require going to Level 3, where the system is built. On this level, the service offering is a software library for the software developers who build the EMR system. XYZ works with the developers to understand their requirements.

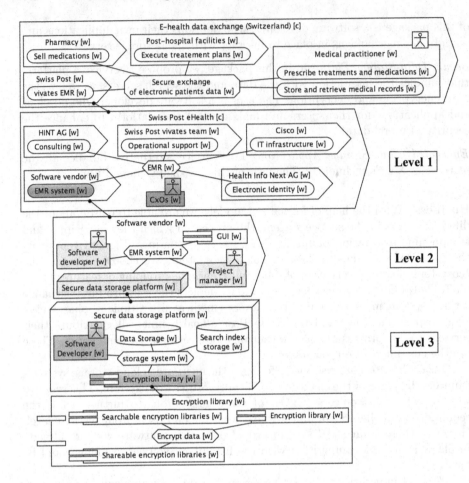

Fig. 1. E-health data-exchange service system (Switzerland)

5.3   Construct

During the construct, we take as input the models from the previous conceptualization step and build prototypes. The construction has a two-fold purpose: to test the market propensity towards the service offering by placing in the hands of prospective users a concrete implementation and to check the technical feasibility of the service offering.

In the case of XYZ, there were prototypes for all three alternatives. XYZ's software developers, guided by the identified dependencies between the service offerings, used the service offering alternatives to package the components.

5.4   Contextualize

The last step is to "place" the service offering in the environment of the customers. At this stage, the service offering prototypes serve both to sell to customers (generate revenue) and to orient the technology development during the next iteration. The service alternatives transfer a limited vocabulary from the technology to the business application domain. Little by little searchable and shareable encryption becomes a part of the business vocabulary. Also, the business vocabulary influences the technology UoD with terms such as electronic health/medical records. The RE practitioners' role is to carry these concepts from one UoD to the other and to show the potential to the other is crucial.

Also, the explicit position of the alternatives points towards a different approach in the service-integration process, depending on who the customer is and who the user is. For example, the CxOs are the customers of the storage system but the hospital practitioners are the users of it, hence the technical development has to satisfy the needs of both.

5.5 Findings

The value of the explicit conceptualization is that XYZ categorizes the user and the customer profile of each alternative and collects systematized feedback. This categorization serves to test the viability of alternatives by cross referencing feedback from different industries.

In the case of XYZ, the search for systematically conceptualizing alternative service offerings resulted in a catalog of opportunities that linked the market, the market segment and the layering of technology components. XYZ worked on projects for various industries, namely insurance, e-health, banking compliance. In each industry, XYZ had a pilot project using one of the service alternatives. The conceptualization of the services helped the development team develop a stack of services to offer to the different projects and to reuse as many components as possible.

6 Conclusion and Future Work

In this paper, we have addressed the question on how to transfer technology with RE methods. We have presented the results of a DSR project for technology transfer RE. We have described a method highlighting the value of conceptualizing alternative technical service offerings. We have presented a startup in its search for a service/market fit and a project that we observed.

Our future work is to show not only how technical alternatives, but also industry alternatives, influence the choice of technology and market evolution. For a next step, we will link TTRE with value-based RE to explicitly estimate market potential in monetary terms. Furthermore, we will expand our modeling tooling and use goal-oriented RE to delve into the motivational analysis. One of the limitations of our work is that it is based on one startup case. To test the generalizability of the TTRE method, we plan to evaluating the method in another context.

References

1. Boman, M., Bubenko Jr., J.A., Johannesson, P., Wangler, B.: Conceptual Modelling. Prentice-Hall, Inc., London (1997)
2. Cleland-Huang, J., Damian, D.: Ready-set-transfer! Technology transfer in the requirements engineering domain. In: 2011 IEEE 19th International Requirements Engineering Conference, pp. 327–328. IEEE, August 2011
3. Daft, R.L., Weick, K.E.: Toward a model of organizations as interpretation systems. Acad. Manag. Rev. **9**(2), 284–295 (1984)
4. Devine, S.W., Bugnion, E., Rosenblum, M.: Virtualization system including a virtual machine monitor for a computer with a segmented architecture, May 2002
5. Dos Santos, J.R.F., Albuquerque, A.B., Pinheiro, P.R.: Requirements prioritization in market-driven software: a survey based on large numbers of stakeholders and requirements. In: 2016 10th International Conference on the Quality of Information and Communications Technology (QUATIC), pp. 67–72. IEEE (2016)
6. Duarte, C.H.C., Gorschek, T.: Technology transfer-requirements engineering research to industrial practice an open (ended) debate (panel). In: 2015 IEEE 23rd International Requirements Engineering Conference (RE), pp. 414–415. IEEE (2015)
7. Gold, R.L.: Roles in sociological field observations. Soc. Forces **36**(3), 217–223 (1958)
8. Gray, D.E.: Doing Research in the Real World. Sage, London (2013)
9. Gregor, S.: The nature of theory in information systems. MIS Q. **30**(3), 611–642 (2006)
10. Hevner, A.R., March, S.T., Park, J., Ram, S.: Design science in information systems research. MIS Q. **28**(1), 75–105 (2004)
11. Maglio, P.P., Spohrer, J.: Fundamentals of service science. J. Acad. Mark. Sci. **36**(1), 18–20 (2008)
12. Maglio, P.P., Srinivasan, S., Kreulen, J.T., Spohrer, J.: Service systems, service scientists, SSME, and innovation. Commun. ACM **49**(7), 81–85 (2006)
13. Miles, M.B., Huberman, A.M., Saldana, J.: Qualitative Data Analysis: A Sourcebook. Sage, Beverly Hills (1984)
14. Offermann, P., Blom, S., Schönherr, M., Bub, U.: Artifact types in information systems design science – a literature review. In: Winter, R., Zhao, J.L., Aier, S. (eds.) DESRIST 2010. LNCS, vol. 6105, pp. 77–92. Springer, Heidelberg (2010). https://doi.org/10.1007/978-3-642-13335-0_6
15. Popek, G.J., Goldberg, R.P.: Formal requirements for virtualizable third generation architectures. Commun. ACM **17**(7), 412–421 (1974)
16. Regev, G., Wegmann, A.: Defining early it system requirements with regulation principles: the lightswitch approach. In: 2004 Proceedings of the 12th IEEE International Requirements Engineering Conference, pp. 144–153. IEEE (2004)
17. Regnell, B., Brinkkemper, S.: Market-driven requirements engineering for software products. In: Aurum, A., Wohlin, C. (eds.) Engineering and Managing Software Requirements, pp. 287–308. Springer, Heidelberg (2005). https://doi.org/10.1007/3-540-28244-0_13
18. Vargo, S.L., Lusch, R.F.: Evolving to a new dominant logic for marketing. J. Mark. **68**(1), 1–17 (2004)
19. Vargo, S.L., Lusch, R.F.: Service-dominant logic: continuing the evolution. J. Acad. Mark. Sci. **36**(1), 1–10 (2008)
20. Weber, R.: The reflexive researcher. MIS Q. **27**(4), v–xiv (2003)

21. Wegmann, A.: On the systemic enterprise architecture methodology (SEAM). In: International Conference on Enterprise Information Systems (2003)
22. Wegmann, A., Julia, P., Regev, G., Perroud, O., Rychkova, I.: Early requirements and business-IT alignment with SEAM for business. In: Proceedings of the 15th IEEE International Requirements Engineering Conference, RE 2007. pp. 111–114. IEEE (2007)
23. Zave, P., Jackson, M.: Four dark corners of requirements engineering. ACM Trans. Softw. Eng. Methodol. (TOSEM) 6(1), 1–30 (1997)

Conceptual Modeling to Support Pivoting – An Example from Twitter

Vik Pant[1(✉)] and Eric Yu[1,2]

[1] Faculty of Information, University of Toronto, Toronto, Canada
vik.pant@mail.utoronto.ca, eric.yu@utoronto.ca
[2] Department of Computer Science, University of Toronto, Toronto, Canada

Abstract. Pivoting is used by many startups and large enterprises to recon-figure their structures and relationships in line with their changing environments and requirements. However, pivoting is a non-trivial undertaking that has far reaching consequences for the focal organization. Conceptual models of actor intentionality can be used to design and analyze organizational pivots in a systematic and structured manner. Conceptual modeling is preferable to ad hoc evaluation as it can provide more detailed and systematic analysis of pivoting decisions. It can be used to uncover mistakes and gaps in reasoning that are missed or obscured via ad hoc evaluation. Actor- and goal-modeling can be used to differentiate among beneficial and deleterious pivoting options. Correctly designed and implemented pivots can avoid substantial value destruction from direct damages as well as opportunity loss for the focal organization. In this paper we present conceptual models of pivoting based on a retrospective case example of Twitter.

Keywords: Pivoting · Modeling · Entrepreneurship · Design · Analysis

1 Introduction

Pivoting refers to changing the structure and relationships of an organization to improve its overall viability and sustainability [1]. The business world offers numerous examples of organizations that have used pivoting succesfully to their advantage. For example, Microsoft pivoted away from protecting the market for its Windows operating system (OS), which was one product line, to capturing the market for its larger product portfolio including applications for Linux OS [2]. This allowed Microsoft to pursue a larger addressable market of application software for multiple OSs rather than the market for just one OS. Amazon pivoted away from being an online bookseller to one of the world's largest eCommerce websites [3] selling everything from avocados to zinc.

Such examples of successful pivots notwithstanding, pivoting remains a risky and complicated undertaking for many organizations. This is because it involves redesigning and reconfiguring an organization and its relationships. The case example of Twitter [4–8], a popular microblogging social network, shows that undertaking the wrong pivot can be harmful for an organization since it can lead to value destruction through direct damages as well as opportunity loss.

© Springer Nature Switzerland AG 2018
C. Woo et al. (Eds.): ER 2018 Workshops, LNCS 11158, pp. 260–270, 2018.
https://doi.org/10.1007/978-3-030-01391-2_31

Ries [1] popularized the notion of pivoting in the entrepreneurial context via his book titled "Lean Startup". He defined pivoting as the continual testing of the fundamental beliefs, logics, and assumptions underlying an organization's business model [1]. Ries proposed ten pivot archetypes [1] and other researchers have since proposed other pivot archetypes [9]. While Ries defined the pivot archetypes and highlighted numerous real-world examples of each archetype in action – he did not offer a structured or systematic methodology for designing and evaluating pivots.

As a consequence, many startup enterpreneurs and corporate decision-makers planned and analyzed pivots for their organizations in an ad hoc manner. This exposed their organizations to risks and uncertainties that could have otherwise been abated or mitigated through the application of a structured or systematic methodology for generating and discriminating organizational pivots [10].

Conceptual modeling provides a means for defining and exploring problem and solution spaces in an orderly and meticulous manner. It entails building abstract patterns and decontextualized representations of real-world phenomenon that may comprise descriptive, predictive, and prescriptive qualities. Many types of models (e.g., physical, conceptual, etc.) are used by humans to communicate ideas, explicate assumptions, codify designs, test configurations, and simulate outcomes.

Strategic modeling, with *actor-* and *goal-* models, is used to represent and reason about relationships in and between actors whose goal structures might be partially congruent and partially divergent [11]. Such techniques can be useful for reasoning about organizational pivoting because pivoting involves changing the configuration of an organization as well as its relationships with other organizations and individuals.

*i** (stands for "Distributed Intentionality" or "Intentional STrategic Actors Relationships") modeling is an *actor-* and *goal-* modeling language [12]. It comprises of two types of models – Strategic Dependency (SD) and Strategic Rationale (SR) diagrams [13]. SD diagrams portray dependencies between different *actors*. *Actors* depend on each other to take advantage of opportunities that they cannot pursue alone. However, dependencies also make the depending *actor* vulnerable to the actor on which it depends.

SR diagrams depict the inner intentional structure of *actors* to explicate the reasons for their dependence on other *actors*. *i** is useful for modeling organizational pivots and has been applied by Pant et al. in [10] for this purpose. This paper presents strategic models of pivoting based on a retrospective case example of Twitter. This example of pivoting by Twitter is used to illustrate the point that strategic modeling can provide more detailed and systematic analysis than an ad hoc approach to uncover mistakes in reasoning. A systematic analysis can potentially avert premature implementations, which, upon discovering that they don't work out, necessitate the focal organization to go back, or pivot again to something else.

In this paper we use a historic case to illustrate the application of conceptual modeling to support pivoting. In such a case the solution space (i.e., To-Be options) is already known to the modeler. In a real-world setting, subject matter experts (SMEs) and domain specialists would apply their contextual knowledge and situational awareness to generate a solution space with new alternatives incrementally, creatively, and iteratively.

2 Case Example: Twitter's Pivot from "Advertising" to "Commerce" and Back

Twitter is a popular social network for microblogging with an average of 330 million active users per month. Users publish posts of 140 characters or less and these "Tweets" can include video, audio, and images. Users subscribe to other users whose tweets they wish to view in their social feed. Twitter content is accessible via web browsers on desktop and laptop computers as well as on mobile devices such as smartphones and tablets.

The reach and span of Twitter makes it attractive for businesses to advertise and promote their offerings. Brands create Twitter accounts from which they tweet content related to marketing and public relations. Brands also hire influencers to promote their products overtly or covertly. The importance of Twitter for organizations is evidenced by the viral mainstreaming of new entrants (e.g., upstart startups) as well as large-scale boycotts of established incumbents.

Advertising revenue is the original source of income for Twitter. Monetizing of Ads was Twitter's first income model and was first implemented in 2010. Organizations pay Twitter for displaying their tweets in the feeds of Twitter's users. A Twitter user can click on a link and is redirected to the marketer's website. Marketers target users based on demographic and psychographic attributes and Twitter displays the marketer's tweets (marking them as advertisements) in feeds of its users that meet the marketer's targeting criteria. Twitter leverages an auction system for ad-buying wherein marketers compete to place their advertisements on Twitter users' feeds. Twitter offers CPI (cost per impression) and CPC (cost per click) pricing options to marketers. Fees are based primarily on targeting criteria and quantity of ad placements.

In 2015, Twitter pivoted to a different income model which was based on monetization of Commerce revenue stream. Twitter introduced a "Buy Now" button that allowed its users to buy items directly in a Twitter storefront. Twitter assumed that its users would welcome the ability to conduct transactions directly from their Twitter feeds. Twitter wanted to be a first mover and market maker by popularizing social commerce in the industry since the advertising income model was already in use by other social networks such as Facebook and LinkedIn.

By contrast, Social Commerce was not well established in the industry in 2015 and Twitter attempted to change that by pivoting its income model to monetize this revenue stream. Twitter assumed that an innovative income model could help it to better achieve its objective of revenue growth. It presumed that by processing transactions in its storefronts it could increase its revenues by earning sales commissions rather than charging fees on ad placements. Sales commission would be based mainly on the price and quantity of the item purchased by a Twitter user. However, Twitter's attempts at monetizing the Commerce revenue stream was unsuccessful and, in 2017, Twitter abandoned it by reverting to its advertising-based business model.

3   Conceptual Modeling of Twitter's Pivoting

Conceptual modeling with concepts of goals and agents can be used to answer several important questions to guide decision-making about organizational pivoting. These include questions such as 'is a pivoting option attractive?' as well as 'what is the effect of a specific pivoting option on certain objectives?'. We note that the content of the models in this paper is based on details about the case example of Twitter that are obtained from secondary sources including [4–8]. Therefore, some of these details may be incomplete or erroneous and are used in this paper just for illustration.

3.1   Rationale for Pivoting Away from "Advertising"

Figure 1 presents an *i** SR diagram showing the Advertising income model in the Twitter case. "Twitter", "Twitter User", and "Vendors" are portrayed as *actors*. In *i**, *actors* are intentional and strategic. They reason about their strategic interests within a network of interdependent social actors. Twitter's highest-level objective mandated "revenue growth for the corporation". This is depicted as a *softgoal* which is a quality criterion that is judged to be sufficiently satisfied or denied from the subjective perspective of its encompassing *actor*.

Twitter's top-level objective was supported by its mid-level aim of "revenue growth via advertising". This mid-level aim is denoted as a *softgoal* and its relationship with Twitter's top-level objective is shown via a 'Help' *contribution link*. *Softgoal contribution* (SC) *links* can be of different types (e.g., help, hurt) and are used to show the impact of a *softgoal* (or *task*) on other *softgoals*. SC links can be used to depict hierarchies of quality criteria with positive, negative, and neutral impacts going up as well as across different branches of *softgoals*. The top-most 'Help' *contribution link* in Twitter connects the *softgoal* corresponding to the Advertising income model with the *softgoal* corresponding to its superordinate objective (i.e., "revenue growth for the corporation"). Similarly, the Advertising income model is itself supported by contributions from "Ad Sales of placement".

When softgoals have been refined to become sufficiently specific, they can be operationalized. The softgoal "Ad Sales from Placement" is operationalized into the task "Auction of impressions". This lowest-level *softgoal* is operationalized via an "Auction of impressions" which corresponds to bidding for ad placements by Vendors. This operationalization allows Twitter to satisfy its functional *goal* of "Content be served" to Twitter Users wherein the content refers to ads mixed in with "Social Network Updates". Operationalizations are represented in *i** as *tasks* since a *task* is a specific way of achieving a *goal*. *Actors* can have *goals* which are states of affairs that they wish to achieve in the world. *Task* are connected to goals via *Means-Ends* (ME) *links* whereby the completion of any *task* leads to the fulfilment of the *goal* to which it is associated. SC and ME links jointly help an analyst to perform trade-off analysis among multiple competing options with respect to the quality criteria (*softgoals*) that each satisfies or denies.

Twitter must "Track" click-throughs on ads as part of its "Auction of impressions" task otherwise neither it nor Vendors would be able to measure the performance of their ad campaigns. Therefore, "Track click-throughs" is depicted as a sub-*task* of the *task*

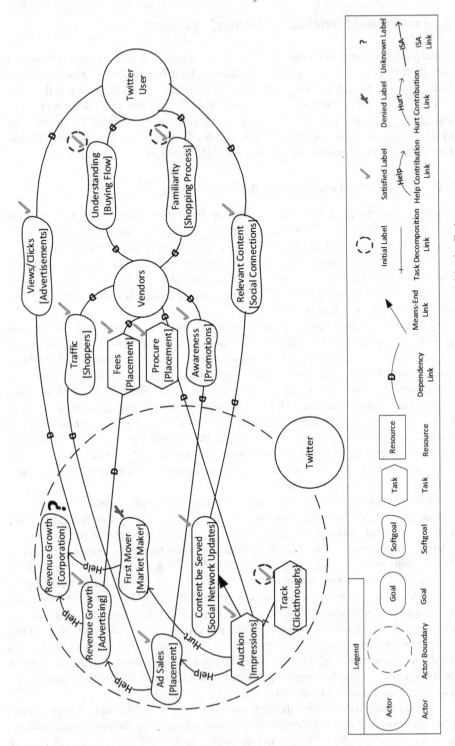

Fig. 1. *i** SR diagram showing Advertising income model in the Twitter case

"Auction of impressions". *Tasks* can be decomposed via *Task-Decomposition* (TD) *links* to show their constituent parts wherein each constituent part must be successfully performed for the *task* to succeed. In evaluating the viability of goal achievement, ME links between *goals* and *tasks* behave as "OR" links whereas TD links between *tasks* and other elements act as "AND" links.

Twitter Users depend on Twitter for "Relevant Content generated by their social connections" and Twitter depends on Twitter Users for "Views/Clicks on ads in their social feeds". In *i**, a *depender* is an *actor* that depends on another *actor* while a *dependee* is an *actor* on which another *actor* depends. A *dependum* is the thing for which the *depender* depends on the *dependee*. When *dependums* are regarded as being sufficiently fulfilled or denied solely from the perspective of the *depender* then they are shown as *softgoals*.

Initial labels denote starting points in the analysis and are shown as dashed circles around evaluation labels. Twitter's Advertising income model was predicated on Vendors "Procuring ad placements" and paying Twitter "Fees for ad placements". Vendors placed ads and paid fees to Twitter to direct "Traffic of shoppers" (Twitter Users) to their websites and to generate "Awareness about their promotions". This arrangement only benefited any Vendor if Twitter Users were "Familiar with the shopping processes" and "Understood the buying flows" on their eCommerce website. If these preconditions were not satisfied – then ad placements and fees payments to Twitter by a Vendor would not translate into additional revenues for that Vendor. Therefore, these preconditions are denoted with initial labels such that if they are denied then Twitter's Advertising income model would be unsuccessful.

Twitter's Advertising income model was established in the industry and was already in use by other social networks including Facebook and LinkedIn. Twitter Users were accustomed to seeing ads in their social feeds because this was commonplace in other social networks such as Facebook and LinkedIn. Figure 1 shows that Twitter's Advertising income model was reasonably successful because all but one of the model elements were satisfied. The objectives of each *actor* other than Twitter were fulfilled. However, Twitter's quality criterion of being a "First Mover and market maker" was denied because it did not innovate the Advertising income model. Therefore, top-most *softgoal* is labeled as unknown in the Advertising income model. In 2015 Twitter pivoted its income model from Advertising to Commerce. It did so to better achieve its top-most *softgoal* of "Revenue Growth for the corporation".

3.2 Rationale for Pivoting to "Commerce"

Figure 2 presents an *i** SR diagram showing the Commerce income model in the Twitter case. Original model elements corresponding to the Advertising income model from Fig. 1 are greyed-out to simplify visual presentation of the model. *Actors* as well as dependencies between Twitter and other *actors* related to the Advertising income model have been updated in Fig. 2 to reflect the changes resulting from the switch to a Commerce income model.

Twitter's highest-level *softgoal* of "revenue growth" for the corporation was supported by its pivot to a mid-level *softgoal* of "Net revenue via Commerce". Twitter intended to switch from its income model of charging fees for placement of ads to

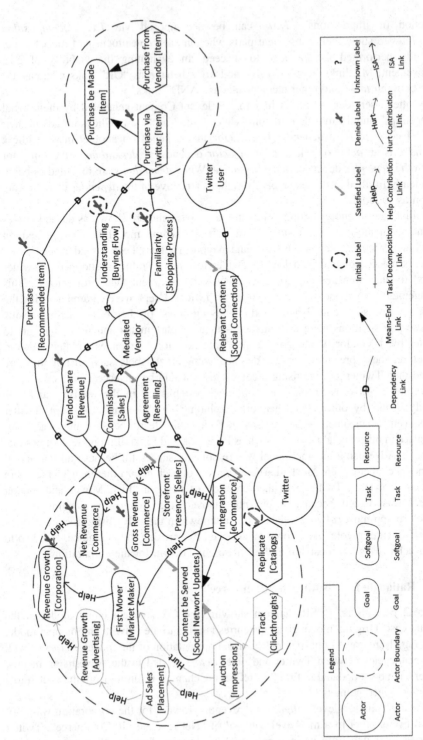

Fig. 2. *i** SR diagram showing Commerce income model in the Twitter case

earning commissions on items sold in its storefronts. Therefore, this Commerce income model is supported by an intermediary objective of "Gross revenue from Commerce" and this is supported by the lowest-level *softgoal* of "Storefront presence for Vendors" on Twitter. Twitter mediated the relationship between Vendors and Twitter Users.

The "Storefront Presence" softgoal, having been sufficiently refined, is operationalized into the task of "Integration" with eCommerce systems of Vendors. This necessitated "Replication of Vendor catalogs" by Twitter so that they could be accessed by Twitter Users in storefronts on Twitter. Compared to the Advertising-based business model, this operationalization is an alternate way for Twitter to satisfy its goal of "Content be served" to its users wherein the content now includes "purchase recommendations" mixed in with "Social Network Updates". Clicking on a "purchase recommendation" would take a Twitter User to the Twitter storefront of the Vendor whose item was recommended for purchase. That Twitter User could then buy that item directly on that Twitter storefront.

As in the Advertising income model, Twitter Users in the Commerce income model also depended on Twitter for "Relevant Content" generated by their Social Connections. However, unlike with Advertising, in Commerce – Twitter depended on Twitter users to "Purchase the recommended item" in their social feed from a Twitter storefront. This was different from the case with Advertising where Twitter depended on Twitter Users for "Views/Clicks on Advertisements" in their social feeds.

The Commerce income model necessitated Twitter to establish different relationships with Mediated Vendors. Twitter depended on Mediated Vendors for "Commissions" on Sales and for "Agreements to Resell" Mediated Vendors' items in Twitter's storefronts. Mediated Vendors depended on Twitter for "Vendor's Share" of Revenue. Twitter Users' relationships with Mediated Vendors also changed across Advertising and Commerce income models because in the former Twitter User's bought items directly from a Vendor while in the latter Twitter Users bought items from Mediated Vendors indirectly via Twitter. Twitter earned "Commissions" on Sales and Mediated Vendor earned "Vendor's Share" of Revenue only when Twitter Users made purchases directly in a Twitter Storefront.

By performing goal modelling evaluation as shown in Fig. 2, one could conclude that Twitter's Commerce approach would be unsuccessful. The business logic underlying the Commerce approach was beneficial for any Mediated Vendor only when Twitter Users were "Familiar with the shopping processes" and "Understood the buying flows" on the Twitter storefront of that Mediated Vendor. However, most Twitter Users were neither "Familiar with the shopping processes" nor "Understood the buying flows" on Twitter because unlike eCommerce websites, social networks were not regarded by users as marketplaces. Twitter was unable to convince Twitter Users to make "Purchases" of Recommended Product directly on Twitter.

Despite success with some of the elements in the model, the overall objective was not achievable. Consequently, Twitter was unable to generate "Gross Revenue" or "Sales Commissions" because Mediated Vendors were unable to generate "Vendor Share of Revenue" from orders in Twitter storefronts by Twitter Users. Revenues earned via "Sales Commissions in Commerce" were lower than Revenues accruing via "Placement Fees in Advertising". This meant that the larger *softgoal* of Twitter (i.e., "Revenue Growth for the corporation") was denied because of the pivot from

Advertising to the Commerce. Analysis of this $i*$ model indicates that Twitter's pivot to a Commerce approach was unsuccessful because its central premise was faulty. This is depicted by denied initial labels over *dependums* between Mediated Vendors and Twitter Users. Twitter abandoned its unsuccessful Commerce approach and reverted back to selling Advertisements in 2017 [4–8].

4   Related Work

This paper focused on the strategic modeling of pivoting in organizations. We presented strategic models of pivoting based on a retrospective case example of Twitter. Our work is related to the strategic modeling literature because researchers in that discipline express and analyze organizations and their relationships using visual models. In this respect our work is related to research by Samavi et al. [14] that proposes a conceptual modeling approach to reason about strategies and business models. Our work is also related to research by Giannoulis et al. that describes a language for modeling strategy maps [15]. Kim et al. posit a modeling technique to depict a value chain of virtual enterprises [16]. Cardoso et al. [17] offer architectural guidelines and methodological recommendations for modeling different types of goals in organizations. This paper also builds upon our prior work [10] in which we presented models that were abstract generalizations and decontextualized representations of the ten pivot archetypes proposed by Ries in [1].

5   Conclusion and Future Work

This paper contributes to conceptual modeling research by applying $i*$ to analyze pivoting based on a retrospective case example of Twitter. $i*$ modeling allowed us to express and assess pivoting in a structured and systematic manner. However, in doing so we also encountered certain limitations of $i*$ for modeling this case example. These limitations pertain to lack of support in $i*$ for temporal and conditional reasoning as well as expression of negative dependencies and quantities. These are open topics that are being studied by strategic modeling researchers.

$i*$ does not support the analysis of temporality and this is a hinderance for designing organizational pivots because pivots entail shifting the structural configuration of an organization (and its relationships) from a past or present state to a future state. Currently, a sequence of $i*$ models are needed to depict structural configurations at different time steps. $i*$ also does not support the analysis of conditionality which is necessary for understanding inclusivity and exclusivity relationships between elements. Some options may be viable only when certain other options are deactivated while some options may be viable only when certain other options are activated. Currently, different $i*$ models are needed to portray such conditionality among elements. $i*$ does not support the depiction of negative *dependencies* which makes it difficult to separate those *dependencies* which are left out of the model because they are unimportant from those that are left out of the model because they must be avoided. $i*$ also does not support the depiction of quantities which are necessary for assessing attainment of numerical targets and for comparing among tactics using objective indicators. However, extensions of $i*$ such as Goal-oriented Requirements Language (GRL) support the depiction of quantities.

Overcoming these inadequacies of $i*$ for modeling organizational pivots satisfactorily shall comprise the core of our future work. Some of these gaps can be partially filled by providing tool support for $i*$ modeling. Such a tool can present $i*$ models to humans in a way that is explainable and interpretable. Examples of functions and features in such a tool could include coloring/discoloring, revealing/hiding, expanding/collapsing, pulsing/fixing and enlarging/shrinking model elements.

References

1. Ries, E.: The Lean Startup: How Today's Entrepreneurs Use Continuous Innovation to Create Radically Successful Businesses. Crown Publishing Group, New York (2011)
2. Townsend, K.: Why Microsoft's Linux lovefest goes hand-in-hand with its Azure cloud strategy, 15 July 2016. https://www.techrepublic.com/article/why-microsofts-linux-lovefest-goes-hand-in-hand-with-its-azure-cloud-strategy/
3. Penenberg, A.L.: Amazon's Pivot, 02 May 2017. https://www.fastcompany.com/1842702/amazons-pivot
4. Guynn, J.: Buh bye Twitter 'buy' button, 20 January 2017. https://www.usatoday.com/story/tech/talkingtech/2017/01/18/goodbye-twitter-buy-button/96726766/
5. McCarthy, A.: Twitter Unclicks Buy Button, 19 January 2017. https://www.emarketer.com/Article/Twitter-Unclicks-Buy-Button/1015070
6. Gagliordi, N.: Twitter kills the buy button, ends focus on ecommerce, 19 January 2017. https://www.zdnet.com/article/twitter-kills-the-buy-button-ends-focus-on-ecommerce/
7. Lunden, I.: Twitter is phasing out the "Buy" button, will continue to offer donations, 18 January 2017. https://techcrunch.com/2017/01/17/bye-buy-on-twitter/
8. Rey, J.D.: Twitter's 'Buy' button is officially dead, 18 January 2017. https://www.recode.net/2017/1/18/14311230/twitter-buy-button-dead-killed-shuts-down
9. Bajwa, S.S., Wang, X., Duc, A.N., Abrahamsson, P.: "Failures" to be celebrated: an analysis of major pivots of software startups. Empir. Softw. Eng. **22**(5), 2373–2408 (2017)
10. Pant, V., Yu, E., Tai, A.: Towards reasoning about pivoting in startups and large enterprises with i*. In: Poels, G., Gailly, F., Serral Asensio, E., Snoeck, M. (eds.) PoEM 2017. LNBIP, vol. 305, pp. 203–220. Springer, Cham (2017). https://doi.org/10.1007/978-3-319-70241-4_14
11. Pant, V., Yu, E.: Modeling simultaneous cooperation and competition among enterprises. Bus. Inf. Syst. Eng. **60**(1), 39–54 (2018)
12. Yu, E.S.K., Giorgini, P., Maiden, N., Mylopoulos, J. (eds.): Social Modeling for Requirements Engineering. MIT Press, Cambridge (2011)
13. Pant, V., Yu, E.: Modeling strategic complementarity and synergistic value creation in coopetitive relationships. In: Ojala, A., Holmström Olsson, H., Werder, K. (eds.) ICSOB 2017. LNBIP, vol. 304, pp. 82–98. Springer, Cham (2017). https://doi.org/10.1007/978-3-319-69191-6_6
14. Samavi, R., Yu, E., Topaloglou, T.: Strategic reasoning about business models: a conceptual modeling approach. IseB **7**(2), 171–198 (2009)
15. Giannoulis, C., Petit, M., Zdravkovic, J.: Towards a unified business strategy language: a meta-model of strategy maps. In: IFIP Working Conference on the Practice of Enterprise Modeling (PoEM), pp. 205–216 (2010)

16. Kim, C.H., Son, Y.J., Kim, T.Y., Kim, K., Baik, K.: A modeling approach for designing a value chain of virtual enterprise. Int. J. Adv. Manuf. Technol. **28**(9–10), 1025–1030 (2006)
17. Cardoso, E., Mylopoulos, J., Mate, A., Trujillo, J.: Strategic enterprise architectures. In: Horkoff, J., Jeusfeld, Manfred A., Persson, A. (eds.) PoEM 2016. LNBIP, vol. 267, pp. 57–71. Springer, Cham (2016). https://doi.org/10.1007/978-3-319-48393-1_5

Modeling and Management of Big Data (MoBiD) 2018

Preface to MoBiD 2018

Seventh International Workshop on Modeling and Management of Big Data (MoBiD 2018) at the 37th International Conference on Conceptual Modeling (ER 2018)

Xi'an, China
October 22–25, 2018

Organized by
Il-Yeol Song, Jesús Peral and Alejandro Maté

In recent years, the quick adoption of Big Data technology has led to the apparition of a wide landscape of technologies and applications. Together with a wide variety of devices, including sensor-enabled smart devices, and all types of wearables, connected to the Internet, vast amounts of data are powering newly connected applications and solutions. From real-time information to purely digital environments, like social media where all information is stored in the Cloud have led to the apparition of new expressions such as Internet of Things, Industry 4.0, Big Data, Cloud Computing or Fog Computing among others.

Given the necessity to obtain faster and deeper insights related to business objectives from these data, there is a special interest in reducing the effort of integrating, processing, and understanding Big Data. In this sense, conceptualization and methods to effectively model and manage Big Data have led to the apparition of new ways of analyzing Big Data. Thus, the objective of MoBiD'18 is to be an international forum for exchanging ideas on the latest and best proposals for modeling and managing big data in this data-driven paradigm. The workshop is a forum focused on the use of the conceptual modeling approaches for the development of new applica-tions based on Big Data, where researchers and practitioners are invited to exchange ideas and present their novel methodologies.

The workshop has been announced in the main announcement venues and attracted papers from four different countries distributed all over the world: China, United States, Finland, and Spain. We have finally received 5 papers and the Program Committee has selected 3 papers, making an acceptance rate of 60%.

The first paper by Jie Lian et al. shows a comparison between SQL and NoSQL systems for storing big spatio temporal data. The authors propose a general architecture oriented to the storage and retrieval of high-dimensional spatio-temporal data. Using this architecture, the authors compare the performance of MongoDB against PostgreSQL, showing how they behave in terms of space and insertion time as the volume of data increases. The architecture allows an easy comparison of technologies oriented to big spatio temporal data and highlights the most important factors to take into account when selecting a technology. The second paper by Guangxuan Zhang et al. presents an approach to conceptualize conflict models from unstructured information stored in emails and online discussions. The approach uses categories of argumentation and features from text mining to detect argumentative messages, obtaining better results than the baseline models that relied solely on classifiers. In this way, the approach allows developers and online communities to better understand the nature of the discussion and improve conflict management in virtual collaborative work environments. The third paper, by Jiaheng Lu et al. presents a vision to manage the variety in Big Data by means of a unified architecture that covers both relational and NoSQL systems. The architecture makes use of a two-layer system. The first layer, Core layer, is dedicated to query processing and schema managing, using a model-agnostic query language that makes use of indexes dedicated to each technology. The second layer, Storage layer, is dedicated to the extraction and storage of information through a model-agnostic data storage interface that interacts with different datasource technologies. Finally, the keynote by Tok Wang Ling, *Conceptual Modeling Views of Relational Databases vs Big Data*, discusses the importance of conceptual modeling and semantics in relational databases and Big Data environments, and which set of characteristics can help us decide between one or the other technology.

Acknowledgments. We would like to express our gratitude to the Program Committee members for their hard work in reviewing papers, the authors for submitting their papers, and the ER 2018 organizing committee for supporting our workshop. We also thank ER 2018 workshop chairs Carson Woo and Jiaheng Lu for their direction, guidance and support. MoBiD'18 was organized within the framework of the following project SEQUOIA-UA (TIN2015-63502-C3-3-R) from the Spanish Ministry of Economy and Competitiveness (MINECO).

October 2018

Il-Yeol Song
Jesús Peral
Alejandro Maté
Program Co-chairs

Conceptual Modeling Views of Relational Databases vs Big Data

Tok Wang Ling

Department of Computer Science, School of Computing,
National University of Singapore, Singapore, Singapore
lingtw@comp.nus.edu.sg

The concepts of object class, relationship type, and attribute of object class and relationship type, are the three basic concepts in Entity Relationship Model. They are termed ORA-semantics. Without knowing the ORA-semantics in the databases, the quality of some database areas are low. Both Relational Model and Big Data Model do not capture ORA-semantics.

We first give an introduction to the current 3 Vs + 2 Vs characteristics of big data applications and the types of big data applications and their characteristics. The outline of the talk is as follows:

- We recall the requirements and criteria for traditional database applications in RDBMS using SQL and highlight some limitations and performance issues of RDBMS. We review the use of ACID for handling concurrent update transactions, join of relations, redundancy and updating anomalies, normal forms, performance issues in parallel and distributed databases, etc.
- Briefly, we present the basic data models of the 4 major categories of NoSQL databases for big data applications, namely, key-value store, wide-column store, document store, and graph database. We further compare the expressive power of the NoSQL databases with relational databases for capturing the ORA-semantics in the data, such as object class, relationship type, redundancy and normalization, recursive relationship type, multiple relationship types between/among same object classes, ISA relationship type, etc.
- Next, we compare the relational model and big data model using a set of characteristics. We describe some of the existing techniques which can be used to improve the performances of certain database applications, such as materialized view design, the concepts of strong FD/MVD and weak FD in physical database design, horizontal and vertical partitioning of data, etc.
- We further present some seldom mentioned but very important issues in data and schema integration, such as entity resolution vs relationship resolution, primary key vs object identifier (OID), local OID vs global OID, system generated OID vs manually designed OID, local FD/MVD vs global FD/MVD, semantic dependency vs FD/MVD, etc. All these concepts are related to ORA-semantics and they have a significant impact on the quality of the integrated relational databases and big data. We need to discover these ORA-semantics in the data in order to perform data integration with good value quality.
- Lastly, we give conceptual modeling views on RDBMS vs Big Data Computing using a set of requirements and characteristics of the applications. This set of characteristics helps us to decide when to use SQL or NoSQL for big data applications.

SQL or NoSQL? Which Is the Best Choice for Storing Big Spatio-Temporal Climate Data?

Jie Lian[1]([✉]), Sheng Miao[2], Michael McGuire[2], and Ziying Tang[2]

[1] Shanghai Normal University, 100 Guilin Road, Shanghai, China
lianjie@shnu.edu.cn
[2] Towson University, 8000 York Road, Towson, MD, USA

Abstract. Management of big spatio-temporal data such as the results from large scale global climate models has long been a challenge because of the sheer vastness of the dataset. Although different data management systems like that incorporate a relational database management system have been proposed and widely used in prior studies, solutions that are particularly designed for big spatio-temporal data management have not been studied well. In this paper, we propose a general data management platform for high-dimensional spatio-temporal datasets like those found in the climate domain, where different database systems can be applied. Through this platform, we compare and evaluate several database systems including SQL database and NoSQL database from various aspects and explore the key impact factors for system performance. Our experimental results indicate advantages and disadvantages of each database system and give insight into the best system to use for big spatio-temporal data applications. Our analysis provides important insights into the understanding of performance of different data management systems, which is very useful for designing high dimensional big data applications.

Keywords: Spatio-temporal database · NoSQL
Big spatio-temporal data · Performance

1 Introduction

Big data is not a new term in the field of climate science. Advances in various technologies including high precision remote sensing systems, distributed data computation have resulted in a large accumulation of high dimensional data. Over time, the management of such a dataset becomes unwieldy and difficult to access, explore, and analyze. For example, in climate science, the dataset is multi-dimensional and multi-variable, which is computational prohibitive to find patterns. To complicate this problem, a legacy exists where the data are managed using array-based file formats such as the popular NetCDF format [1] where scientists have to first download the entire dataset and run manual software to

© Springer Nature Switzerland AG 2018
C. Woo et al. (Eds.): ER 2018 Workshops, LNCS 11158, pp. 275–284, 2018.
https://doi.org/10.1007/978-3-030-01391-2_32

decompress and preprocess the data to access a single time step. Recently, with improved capabilities in distributed computing, using database technology to manage this data has become a reality. Recent research has investigated the use of NoSQL databases such as Hadoop [2] and MongoDB [3] for managing spatial and temporal data. There are many attributes to NoSQL databases that make them attractive for this application including scalability, flexibility, and the ability to deal with unstructured data. In this paper we investigate the performance of a document-oriented NoSQL database in its ability to store and query a multivariate global climate dataset. We compare this performance to a traditional relational database management system, and analyze the results with theoretic explanations. The research presented in this paper is largely motivated by the high dimensional spatio-temporal datasets that result from large-scale simulations of the Earth's Climate. Consider applying a basic data mining operation such as clustering on a dataset that have millions of data points, where the clustering needs to be done for each time step. Under the current data management method, one would have to first download all of the required NetCDF files, decompress, and extract them before applying any clustering operations. This process alone is very time consuming. Furthermore, the entire file must be loaded into main memory for processing no matter user's interest is only two months, which wastes a lot of memory space. Because of this, we are motivated to apply new technology in cloud-based NoSQL databases.

The focus of this research is to study and understand the data storage and management requirements for high dimensional datasets by implementing and analyzing database solutions for big spatio-temporal climate data, which compares and evaluates the performance of the SQL and NoSQL database systems. With this in mind, we make the following contributions. First, we design a document-oriented schema to manage multivariate climate data and implement the design using a NoSQL database in both single node and multi-nodes environments. Second, we implement a similar database design in PostgreSQL and compare the performance of each implementation in terms of disk usage, memory usage, insert time, and query time. Third, we provide analysis and insights on different types of database system, which can be used as references for any high dimensional database applications. The rest of this paper is organized as follows. After reviewing related works in Sect. 2, we discuss three different data models in Sect. 3. Section 4 presents experimental results and discussions. Section 5 concludes this paper with limitation and future works.

2   Related Work

A revolution in distributed computation is currently underfoot driven by cloud computing and virtualization solutions. This emerging technology enables the analysis of big data. One of the important aspects of climate data research is that the spatial and temporal domain of the climatic system is multi-scale in nature. The inclusion of multiple scales of spatio-temporal data is therefore very computationally intensive. With this in mind, using a distributed cloud-based architecture scales well to support these types of analyses. Cloud-based data warehouses have been shown to be efficient on big data processing [4] and it is possible to conduct aggregation queries using the MapReduce framework [5]. Recently, there has been some researches on processing spatial queries using the MapReduce framework [6] and a grid-based architecture has been implemented on Amazon EC2 cloud storage to perform hydrologic simulation [7]. Document databases such as MongoDB have shown to be very effective in the distributed management of spatial data [8] and geoanalytic frameworks [9]. Also, the cloud architecture has proven to be effective for visual analytics of spatio-temporal data [10].

There have also been a number of studies who propose custom database solutions for the storage, manipulation, and query of multi-dimensional arrays. Early solutions have proposed a database and query language for the management of multi-dimensional arrays [11]. This research eventually evolves into the RasDaMan or raster data manager array database system which is implemented as an extension to PostgreSQL [12]. Bimonte [13] et al. presented a UML profile for spatial data warehouse, and a efficient ROLAP architecture. Tang [14] developed a parallel spatial computing framework for vector-based geo-spatial data. However, to our knowledge, there has yet to research that tests the performance of a multi-node NoSQL database for the use of storing high dimension spatio-temporal data. This paper adds to the state of the art in cloud computing and NoSQL databases in that we test the performance of a distributed database in its ability to store and retrieve multidimensional data such as that associated with global climate simulations.

3 System Architecture

The data models presented in this paper are candidate databases to be used as the core database for the spatio-temporal climate data analysis. This work will contribute to the overall design of this system in that it will guide the choice for the central data model. As such, one of the main requirements of the system is efficient data insert and query performance and because of commodity hardware, less of a focus is placed on data storage size and memory usage.

3.1 NoSQL Data Model

MongoDB is a document-based NoSQL database system that uses flexible JSON-style documents. It can host many databases in one server and each database owns a number of collections. Every collection stores any numbers of documents that are grouped by a set of key-value pairs. To put this in climate data, a single MongoDB database would consist of data for a given climate model, e.g. the Reanalysis I dataset. Then within this database, there would be a set of collections that would represent a given set of variables such as air temperature. Then within each collection, a set of documents would exist for each time step in the model where the document in its most simple form would store both a time step and a multi-dimensional array.

Single-Node MongoDB. In our framework, data is four dimensional (latitude, longitude, time stamp, values) spatio-temporal data which will be queried by users during data processing. To store such data in a database system, there are two approaches. The first approach is to convert the n-dimensional array to a number of individual documents where each document stores a tuple consisting of latitude, longitude, time stamp, and value. However, doing so needs to store the same information repeatedly and therefore is not efficient. The second approach is to directly load the n-dimensional array into the database system. It improves disk usage and reduces I/O operation times. This method which named as "multi-dimensional array framework" has been widely used, especially in relational database systems. This framework is based on array abstraction which can be mapped in a relational database. Whereas, retrieving one record from the n-dimensional array is similar to search for the desired value directly from a flat file. Both of these scenarios need to load entire array into memory and only choose the desired subset for processing. Therefore, finding one appropriate data model is based on the requirements for the specific application. Since our application is to cluster every point in the global dataset into six climate types based on six geographical attributes daily, we choose to store the data using two key/value pairs, time and space. By doing so, query time is reduced and disk space is minimized. Moreover, matrix operations could also be used directly at the application level in this method. In MongoDB database, array object could be encoded in BSON (Binary JSON). BSON is a binary-encoded serialization of JSON-like documents. It supports the embedded documents and arrays. After retrieving data from database, the BSON documents could be decoded to array for data processing. In MongoDB data model, we have a database at the topmost level that stores all of the climate data. Then we have collections for each climate attribute. Documents in every collection are key-value pairs; key is set by 'date' which is also index, and value is a two dimensional array.

Multi-node MongoDB. Big data often requires the scaling out of a database where a single database is distributed across a number of nodes. With respect to climate data, as more and more data is produced, this becomes a big data problem and thus the ability to scale out is required. MongoDB has the capability to scale out horizontally across a number of nodes. Horizontal scaling is different from vertical scaling where more resources are added to a single node; horizontal scaling adds more nodes to an existing framework. This approach is not only easier for setup also exceed one machine's power limit. In the distributed implementation, there are generally three types of components, including the Mongos query router, config server, and slave nodes, also known as shards. First, data are divided into shards based on shard key which is an index or composite index. The chosen shard key determines on which node each document is stored. As such, performance will vary based on the shard key and it is optimal to keep data that is typically queried together on the same node. Therefore, choosing an ideal shard key is important in big data system. Another component of the system is the Mongos Query Router which is responsible for processing queries and locating on which shard a particular document exists. Moreover, all the other database operations are executed in the Mongos Query Router. To improve high frequency operations, more Mongos Query Routers could be added to share the load. The last component is Config Server which is responsible for serving as a connection between mongos and shard clusters. The Config Server also stores metadata of sharded clusters. It only does read-write operations when there is a new Mongos Query Router instance initializing and data chunks are splitting or migrating between shard clusters. And for this data model, time is used as both the index and the shard key.

3.2 PostgreSQL Data Model

Another option for the database system is PostgreSQL which is a relational database management system. PostgreSQL has been used to manage spatial data using the PostGIS extension. PostgreSQL also provides data storage and database operations and supports storage of array objects. In the PostgreSQL model, there are six tables for each variable and a Date table. The six variable tables store the multi-dimensional arrays and include a primary key and foreign key that will be related to the Date table as well as an array object. The Date table as the center in data model stores date_id which is the primary key and also the other six tables foreign key as well as the date value. D_id is set for index in Date table and every attribute id is set for their own index.

4 Expriment Results

4.1 Environment Setting and Testing Data

The goal of the experiments presented in this paper is to evaluate the three data models presented in Sect. 3 to determine which should be used for storing and retrieving big climate data. With this in mind, we first test the performance of the single node MongoDB implementation versus the PostgreSQL implementation. Then, a more detailed test was performed to compare the single node MongoDB versus the multi-node MongoDB. In particular, tests were performed to evaluate insert time, disk space usage, memory usage and query time. In order to do so, a VM Hypervisor (VMware vSphereESXi 5.1) was employed to build our testing platform on a PC with a single Intel Core i5-3570@3.40 GHz processor, 16 GB RAM and a 2 TB SATA hard drive. For the experiments, the system resources were partitioned into four nodes. The first three nodes each had 4 GB of RAM and a 400 GB disk and were used to house the single node MongoDB implementation, PostgreSQL server, and the Mongos Query Router for the multi-node MongoDB implementation. The fourth node was given 2 GB of RAM and a 100 GB disk and was used to house the multi-node MongoDB config server. This VM was given less resource because the MongoDB config server generally needs less system resources than the other nodes. In addition to the hypervisor PC, three more PCs were used as shard nodes in the multi-node MongoDB system.

They each have 8 GB RAM and 2 TB disk. The NOAA Reanalysis I Dataset [15] is used in this experiment. This dataset consists of daily modeled results based on in situ readings for air temperature, pressure, relative humidity, precipitation, u-wind, and v-wind from 1948 to 2012. The data is distributed on a 73 by 144 latitude longitude grid.

4.2   Performance Metrics

We evaluate the performance of the database implementations using four metrics: inserting time, disk size, querying time and memory usage. Inserting time and querying time mean the time to do insert and query operations on database system. Disk size is defined as the storage usage of database. Memory usage is recorded during querying data. In general, query time and memory usage are the most important metrics. In the experiment, insert time was measured by recording the time to insert data into the single node MongoDB, multi-node MongoDB and PostgreSQL. To show the result of increasing data size, the dataset used for testing was increased from one to thirty years. After the insert operation, disk size was calculated as the dataset size increased. We recorded the disk size from the database configurations.To record the query time, we used a script to retrieve the same dataset from three database systems separately. The running time of the script was recorded as the query time. Memory usage was calculated by capturing the peak value and steady value from the system monitor when the query operation was being performed. For the multi-node MongoDB implementation, Mongos who handles most of database operations was chosen for record memory usage instead of recording all the nodes memory usage.

4.3   Results

In this section, we present the results of our experiments.

The results of insert time between PostgreSQL and the single node MongoDB database are shown in Fig. 1(a). The blue dotted line represents the insert time for PostgreSQL and red solid line represents the insert time for single node MongoDB, which is the same in subsequent figures. It is shown that the single node MongoDB system uses less time to insert the same data than PostgreSQL. The insert time of the PostgreSQL increases linearly with the size of the dataset and the insert time of the single node MongoDB shows an inflection point when inserting 26 years of data. The reason is that the system has reached the physical memory limit and has begun to use swap space. It is known that frequent swapping between physical and virtual memory will lead to poor efficiency. Figure 1(b) shows the comparison between the single node in red line and multi-node MongoDB in green line. From the figure, we can see that the multi-node MongoDB takes more time to insert data than the single node MongoDB at the beginning. However, the multi-node MongoDB system does not have the inflection point because it distributes data to three shards and each shard uses its own memory to insert data. As a result, the multi-node MongoDB will face memory bottleneck much later. The experiment dataset was not large enough to reach

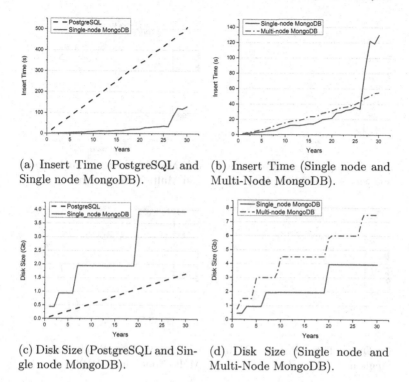

(a) Insert Time (PostgreSQL and Single node MongoDB).

(b) Insert Time (Single node and Multi-Node MongoDB).

(c) Disk Size (PostgreSQL and Single node MongoDB).

(d) Disk Size (Single node and Multi-Node MongoDB).

Fig. 1. Insert time and disk size comparison of PostgreSQL and MongoDB.

the total physical memory size for the multi-node MongoDB but this result suggests that this is an ideal database for big data. All in all, both single node and multi-node MongoDB outperform the PostgreSQL in insert time.

Figure 1(c) shows disk usage for PostgreSQL and the single node MongoDB. For PostgreSQL, the storage size rises linearly as the data size increases. For the single node MongoDB, the disk size increases in steps. It is because MongoDB uses a disk pre-allocation algorithm where disk space is pre-allocated to support data of a predetermined size. The purpose of this mechanism is to reduce disk fragmentation and to assure that documents are stored in a contiguous block on the hard disk. As documents are inserted, the pre-allocated space fills up. However when there is not enough free space for a new record, MongoDB will pre-allocate disk space that is twice the size of the previous pre-allocation size. As the data further increases, MongoDB will continue to pre-allocate more and more double-sized data files. This mechanism will result in an inefficient use of disk space. Another reason is that MongoDB does not use any data compression but PostgreSQL implements TOAST (The Oversized-Attribute Storage Technique), which is a form of LZ compression. Disk usage of the single node and multi-node MongoDB implementations are shown in Fig. 1(d). Multi-node MongoDB system has a similar behavior in terms of the pre-allocation of disk space yet the growth

282   J. Lian et al.

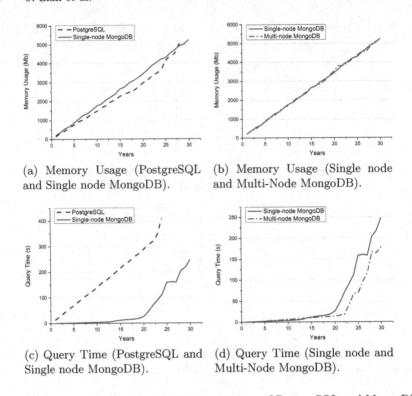

(a) Memory Usage (PostgreSQL and Single node MongoDB).

(b) Memory Usage (Single node and Multi-Node MongoDB).

(c) Query Time (PostgreSQL and Single node MongoDB).

(d) Query Time (Single node and Multi-Node MongoDB).

Fig. 2. Memory usage and query time comparison of PostgreSQL and MongoDB.

in disk space is more pronounced than the single node implementation, which is because the multi-node MongoDB implementation uses three shards and the disk space for each shard is pre-allocated. As a result, there is more unused space totally. Thus for the disk size comparison, MongoDB should be used in situations where disk size is not a major concern. Figure 2(a) shows the memory usage of PostgreSQL and the single node MongoDB during query. From the figure, it is evident that the growth trends of two database systems are similar. But it is still notable that the single node MongoDB takes up more memory than PostgreSQL for two reasons. First, MongoDB adopts memory-mapped file mechanism, which associates data files with memory. It makes querying data as if it were directly located in memory. But the fixed memory size will generate some data files that are not mapped into memory which results in a page fault and thus deteriorates performance. In general, this memory mapping mechanism brings out an efficient query time at the expense of using more memory. Second, the MongoDB database employs as much memory as it can for query operations. But its memory usage is dynamic; when other processes need memory, it will yield cached memory. The memory usage of the single node and multi-node MongoDB are shown in Fig. 2(b). It is understandable that these two systems have similar memory usage as they have the same data model and the same memory-mapping mechanism. And in multi-node, rather than recording the performance of three nodes, the Mongos Query Router is used to calculate memory usage. In general, our results show that all the three database systems occupy similar memory usage with PostgreSQL being slightly more efficient.

Figure 2(c) shows the query performance of the single node MongoDB and PostgreSQL. It is obvious to see that single node MongoDB performs much better than PostgreSQL. When querying small amounts of data, PostgreSQL has a linear trend in query time, but after querying more than 24 years of data, the query time increased sharply. The exponential growth is due to frequent memory swapping. The MongoDB shows an amazing speed before querying 20 years data because of the memory-mapping mechanism referred to the previous section. After that, the query speed also increases rapidly which is also because of memory swapping. However, even with virtual memory, it is evident that the speed of the single node MongoDB is still extremely faster than PostgreSQL. The single node and multi-node MongoDB query time results are shown in Fig. 2(d). The two systems perform in a similar fashion. However, after querying 20 years data, both of the two database systems need more time to do querying which is due to memory swapping but the multi-node outperforms the single node because the overall amount of memory available in the system is much larger. The results of query performance show that MongoDB has advantage over PostgreSQL.

5 Conclusion

In this paper, we have proposed a novel data management platform to study and understand the data storage and management requirements for big high dimensional dataset. We analyze the performance of document-oriented NoSQL database system from various aspects, and compare it with traditional SQL database system. The focus of our work is to understand the key impact factors of a system's performance and the theoretic reasons behind them. Our experimental results are satisfactory and clearly indicate advantages and disadvantages of each database system. We believe that our analysis and insights are very useful for designing high dimensional data applications. In the future, we will include other database types such as scalable multi-dimensional array database such as the RasDaMan array database. We will also consider comparing performances of other NoSQL databases such as Hadoop, Spark as well as content-based data storage and object-based data storage.

References

1. Unidata: NetCDF. http://www.unidata.ucar.edu/software/netcdf/
2. Apache: Hadoop (2011). http://hadoop.apache.org/
3. MongoDB: Mongodb. http://www.mongodb.org/
4. Cuzzocrea, A., Moussa, R.: A cloud-based framework for supporting effective and efficient OLAP in big data environments. In: IEEE/ACM International Symposium on Cluster, Cloud and Grid Computing, pp. 680–684 (2014)

5. Brezany, P., Yan, Z., Janciak, I., Chen, P., Ye, S.: An elastic OLAP cloud platform. In: IEEE Ninth International Conference on Dependable, Autonomic and Secure Computing, pp. 356–363 (2011)
6. Singh, H., Bawa, S.: A survey of traditional and MapReducebased spatial query processing approaches. ACM SIGMOD Rec. **46**(2), 18–29 (2017)
7. Chiang, G.T., Dove, M.T., Bovolo, C.I., Ewen, J.: Implementing a grid/cloud eScience infrastructure for hydrological sciences. In: Yang, X., Wang, L., Jie, W. (eds.) Guide to e-Science. CCN, pp. 3–28. Springer, London (2011). https://doi.org/10.1007/978-0-85729-439-5_1
8. Ameri, P., Grabowski, U., Meyer, J., Streit, A.: On the application and performance of MongoDB for climate satellite data. In: IEEE International Conference on Trust, Security and Privacy in Computing and Communications, pp. 652–659 (2014)
9. Jern, M., Franzen, J.: "GeoAnalytics" - exploring spatio-temporal and multivariate data. In: Tenth International Conference on Information Visualization, pp. 25–31 (2006)
10. Lian, J., Mcguire, M.P., Moore, T.W.: FunnelCloud: a cloud-based system for exploring tornado events. Int. J. Digit. Earth **10**, 1–25 (2017)
11. Baumann, P.: Management of multidimensional discrete data. VLDB J. **3**, 401–444 (1994)
12. Baumann, P., Dehmel, A., Furtado, P., Ritsch, R., Widmann, N.: The multidimensional database system RasDaMan. In: ACM SIGMOD International Conference on Management of Data, pp. 575–577 (1998)
13. Bimonte, S., Zaamoune, M., Beaune, P.: Conceptual design and implementation of spatial data warehouses integrating regular grids of points. Int. J. Digit. Earth **10**, 1–22 (2017)
14. Tang, W., Feng, W.: Parallel map projection of vector-based big spatial data: coupling cloud computing with graphics processing units. Comput. Environ. Urban Syst. **61**(11), 187–197 (2014)
15. Arndt, D.S., et al.: State of the climate in 2011 special supplement to the bulletin of the American meteorological society. Bull. Am. Meteorol. Soc. **93**(7), S1–S263 (2012)

UDBMS: Road to Unification
for Multi-model Data Management

Jiaheng Lu[1]($\boxtimes$), Zhen Hua Liu[2], Pengfei Xu[1], and Chao Zhang[1]

[1] University of Helsinki, Helsinki, Finland
jiaheng.lu@helsinki.fi
[2] Oracle, Redwood Shore, CA, USA

Abstract. One of the greatest challenges in big data management is the *"Variety"* of the data. The data may be presented in various types and formats: structured, semi-structured and unstructured. For instance, data can be modeled as relational, key-value, and graph models. Having a single data platform for managing both well-structured data and NoSQL data is beneficial to users; this approach reduces significantly integration, migration, development, maintenance, and operational issues. Therefore, a challenging research work is how to develop an efficient consolidated single data management platform covering both NoSQL and relational data to reduce integration issues, simplify operations, and eliminate migration issues. In this paper, we envision novel principles and technologies to handle multiple models of data in one unified database system, including model-agnostic storage, unified query processing and indexes, in-memory structures and multi-model transactions. We discuss our visions as well as present research challenges that we need to address.

1 Introduction

As data in all forms and sizes are critical to making the best possible decisions in businesses, we see the continued growth of demands to manage and analyze massive volume of different types of data. The data may be presented in various types and formats: structured, semi-structured and unstructured [14,16,22]. In the case of structured data, data might be structured as relational, key-value, and graph models [15]. In the case of semi-structured data, data might be represented as XML and JSON documents [19]. Consequently, beyond the traditional relational database (RDBMS), there has been a blooming of different big data management solutions, specialized for different kinds of data and tasks, to name a few, distributed file systems (e.g. Ceph and HDFS), NoSQL data stores (e.g. Bigtable, Redis, Cassandra, Mongodb, Neo4j), and distributed data processing frameworks (e.g. MapReduce, Spark). This is also known for *"one size does not fit all"* argument. However, managing heterogeneous data sources across the systems imposes a big challenge for practitioners in that the deployment of multiple systems has led to a wide diversification of data store interfaces and the loss of a common programming paradigm.

C. Woo et al. (Eds.): ER 2018 Workshops, LNCS 11158, pp. 285–294, 2018.
https://doi.org/10.1007/978-3-030-01391-2_33

Let us consider three application scenarios to illustrate the variety of data. First, consider an application called *customer-360-view*, which often requires to aggregate multiple data sources, including graph data from social networks, document data from product orders and customer information in a relation database. Second, in Oil & Gas industry, a single oil company can produce more than 1.5 TB of diverse data per day. Such data may be structured or semi-structured and come from heterogeneous sources, such as sensors, GPS devices, and other measuring instruments. Third, in health-care: North York hospital needs to process 50 diverse datasets, including structured and unstructured data from clinical, operational and financial systems, and data from social media and public health records. *These emerging applications clearly demand the need to manage multiple-model data in complex, modern applications.*

There exist two approaches to address the challenge of multi-model data management: (i) *polyglot persistence* [10,13] and (ii) *multi-model database*. The first solution uses multiple databases to handle different forms of data and integrates them to provide a unified interface, whereas the second supports multiple data models against a single, integrated backend while meeting the growing requirements for fault tolerance, scalability, and performance. We discuss the issues of both solutions below.

The history of polyglot persistence may trace back to the federation of relational engines [11], or distributed DBMSs, which were extensively studied in depth during the 1980s and early 1990s. Polyglot persistence approach is similar to the use of *mediators* in early federated database systems. For example, Musketeer [10] provides an intermediate representation between applications and data processing platforms. DBMS+ [13] aims at embracing several processing and storage platforms for declarative processing, and BigDAWG [8] has recently been proposed as a federated system that enables users to run their queries over multiple vertically-integrated systems such as column stores, NewSQL engines, and array stores. Overall, the polyglot persistence solution needs to integrate multiple systems to provide a unified interface, which imposes operational complexity and cost, because the integration of multiple independent databases imposes a significant engineering and operational cost. Further, in order to answer a global query, all of the sub-systems need to remain up, which makes the fault tolerance of the application equal to the weakest component in the stack.

The second approach is to develop a new database to support multiple data models against a single, integrated backend, while meeting the growing requirements for scalability and performance. We observe a recent trend among relational and NoSQL databases in moving away from one single model to multimodel databases. For example, OrientDB is extending graph database to support key-value and json models. ArangoDB is moving from document model to support key-value and graph models. AgensGraph simultaneously supports both the property graph model and the relational model based on a PostgreSQL kernel. Compared to the polyglot persistence, a single multi-model database can reduce the cost of integration, migration, development and maintenance on multiple systems. Furthermore, during query execution, it can leverage a unified query language directly to access the data stores without query decomposition because of its native store nature.

This paper envisions a multi-model database system, called UDBMS (Unified DataBase Management System), that provides several new components and functions to enable a unified and efficient management of multi-model data. The contributions of this paper are summarized as follows. (1) We envision the architecture of UDBMS system to enable the unified data access and management; (2) We show five model-agnostic properties of UDBMS on data storage, query processing, index structure, in-memory structure and transaction; and (3) we illustrate three new research challenges of UDBMS on schema discovery, model evolution and multi-model data sharding.

In the sequel, we introduce UDBMS's architecture, and walk the readers through the main technical elements of our solutions. Finally, we discuss related works, then conclude.

2 Architecture

The overview of UDBMS architecture is illustrated in Fig. 1.

The main component of the system includes two layers: the *core layer* receives queries from the client and returns the unified results; and the *storage layer*

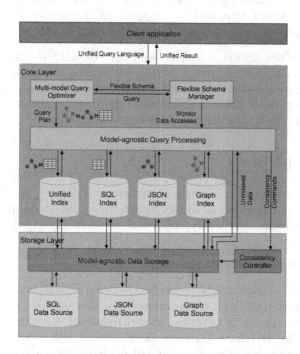

Fig. 1. Architecture of the UDBMS.

loads data with diverse models and stores them in a unified storage platform. We depict the important components in these two layers below:

(i) *Model-agnostic query processing* is responsible for validation and compilation incoming unified queries, as well as query plan generation and optimization. The validation and compilation are supported in cooperation with the flexible schema manager. Moreover, multiple types of *indexes* are developed to efficiently answer ad hoc queries.

(ii) *Flexible schema manager* is to handle the diverse schema of multi-model data. Relational DBMS works according to *"schema-on-write"* principle, which pre-defines the schema prior to feeding data into system. Whereas NoSQL usually agrees *"schema-on-read"*, where schema is needed only when reading data from system. To gain a high level of schema flexibility, we continue to use the "schema-on-read" principle by employing a schema manager which tracks data accesses (including read and write) and automatically generates schema and refines them when necessary. By doing this we are able to give an answer without any input schema - we can generate one by ourselves.

(iii) *Model-agnostic data storage* is to provide a model-agnostic abstraction of multi-model data to the accesses of data. The physical data can be stored in different formats and dispersed on a distributed platform, e.g. Hadoop and Spark.

(iv) *Consistency controller* controls the level of data consistency for a single query by using multi-model locks on data. The locks have various types for different models of data, and they support the fine-grained management for better efficiency, e.g., record-, document-, and node/edge-level.

From the above description, it can be seen that UDBMS is different from the existing wrapper-mediator systems [10,13], where data resides in various stores and query execution is divided between the mediator and the wrappers. Instead, UDBMS imports data across the different-data model stores and employs the same abstraction and approach to access and manage them.

3   Under the Hood

UDBMS aims at efficiently managing heterogeneous datasets through a unified set of interface and abstraction. In this section we lay out several model-agnostic properties and discuss how we approach each of them respectively.

3.1   Model-Agnostic Data Storage

Classical RDBMS makes a tight connection between logic data model and physical storage so that the storage engine assumes that data is physically stored in a particular sequence of bytes to support relational access pattern. Though Object-Relational Database Management System (ORDBMS) can be considered as an early version of multi-model DB via handling non-relational data in RDBMS, its starting point is that relational data is the first class citizen. But in

UDBMS, relational data is just one kind of data models, and there is no specific bias towards relational data. Therefore, the storage layer of UDBMS makes no assumption of how multi-model data is internally laid out. It provides a collection of abstraction API that holds a set of objects. Each of the object is accessed by an object id with version based time stamp. There is a key-value interface that supports generic *put()/get()/replace()/delete()* APIs for an object collection. From storage layer perspective, the value is a sequence of bytes which are not interpreted by the storage engine. The sequence of bytes can be a relational row with well-defined schema, or a piece of graph, or an XML document etc.

Specifically, the *get(objectId, timeStamp, objectBytes)* interface returns a sequence of bytes from the storage given an object id. In addition to the object id, the *get()* interface accepts a time stamp snapshot parameter. This parameter identifies the version of the object bytes for that object id that the data storage engine will provide. The *put(objectId, objectBytes)* interface returns a new object id for a sequence of bytes for storage. The *replace(objectId, objectBytes)* interface replaces the content of the object with object id and a new sequence of bytes and returning the new time stamp for the data. The *delete(objectId)* interface deletes the object with the id. However, *get(objectId, oldTimestamp)* may still be able to retrieve the object with the old time stamp. In addition, to support both ACID, BASE or hybrid transactional semantics, the object collection can be declaratively specified as ACID or BASE. The transaction layer then enforces different transaction semantics for the model-agnostic data.

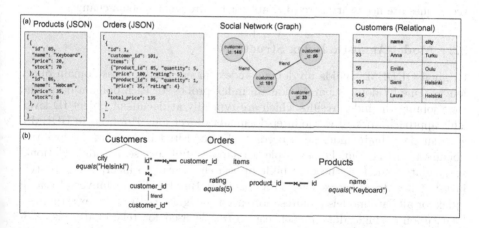

Fig. 2. Example of (a) multi-model datasets and (b) query represented with a forest pattern. The elements marked with asterisk (*) are returned in query results.

3.2 Multi-model Query Processing

The original ORDBMS does not embrace a language to process multi-model data, nor does it address the idea of doing inter-model compilation and optimization. To develop a unified query to accommodate all the data, there are several existing works towards providing a global language to query multi-model data simultaneously. For instance, SQL++ [21] is proposed to query both JSON native stores and relational data. ArangoDB AQL can be used to retrieve and modify both document and graph data. We use the following example to demonstrate the core structure in a multi-model query.

Example 1. Consider an application involving JSON documents, a relational table and a graph data in Fig. 2(a). One example query is to return *the friends of the customers in Helsinki who bought a keyboard and gave a five-star feedback.* This query can be used for product recommendation. Note that there are three types of joins in the query of Fig. 2(b): *graph-relational* ($\bowtie_a$), *relational-JSON* ($\bowtie_b$) and *JSON-JSON* ($\bowtie_c$) joins. The answer of this query is two pairs of customer IDs: (101,145) and (101,56).   □

To process the above query efficiently, as the order of joins can significantly affect the execution time, a query optimizer evaluates available plans and selects the best one. For example, how to decide the join orders among $\bowtie_a$, $\bowtie_b$ and $\bowtie_c$ in Example 1. Therefore, one challenge is to develop new algorithms to select the best query plan for a multi-model query. In addition, statistics, such as *histogram* or *wavelet* can be used to provide detailed information about data distribution for query optimization. The existing statistics techniques on RDBMS are developed based on the static relation schema, but multi-model data requires the diverse and flexible schema. Therefore, it starts to crystallize that dynamic statistics techniques are necessary here to adapt for the frequent schema changes.

3.3   Model-Agnostic Index Structure

Original ORDBMS builds up domain index for each data, cross-domain query join is done by doing separate domain index probes for each domain data, and then joining the index results which are typically at document object ID level. This approach works if we do inter-document object join. However, in UDBMS, such single domain index idea needs to be re-visited if we want to do intra-document object join. For example, to support full text search and relational scalar data search, we need to built up search indexes to incorporate IR-style inverted lists to index various data together. But building a universal search index for all data models requires more deep thoughts. Existing index structures focus upon a single data model, e.g. *B-tree* is used for relational joins, *XB-tree* [4,20] is developed for XML data, and *gIndex* [23] is applicable for graph queries, our visioned system, however, executes queries on more than one data model. Therefore, *how to index multiple data models to accelerate operations such as cross-model filtering and join?* For example, how to support $\bowtie_a$ and $\bowtie_b$ operations in Example 1 for graph-relation and JSON-relation joins efficiently?

In general, we envision two types of auxiliary structures. The first is using inverted index based search index for full-text search. This index integrates domain context aware inverted index for XML and JSON, full text, and leaf scalar relational data together as one unified index. The second is building ad-hoc global indexes to capture the structural feature in multi-model data to speedup structural query processing.

3.4 Multi-model In-Memory Structure

Classical RDBMS uses buffer cache and assumes that the data layout on disk is the same as what is cached in memory. Nevertheless, UDBMS data storage engine makes no assumption of the domain specific data layout, nor it makes any assumption on how the domain specific data format is when it is cached in memory for fast query. Each domain specific data model may cache the data in different format from what is stored on disk.

Multi-model in-memory cache merely uses callback to delegate to each data-model to get its specific in-memory format. For example, the proposed in-memory data structures for relational database include, to name a few, Oracle columnar layout, IBM DB2 BLU Acceleration and SQL Server In-Memory Columnstore. Through adopting the idea of decoupling the storage format with in-memory query friendly format, UDBMS will provide fast in-memory access for multi-model data using its domain specific query language without worrying about finding the optimal storage model for multi-model data. In fact, there is probably no single best storage format for a domain specific data to satisfy all the workload requirements. For example, columnar layout of the relational data might not always be faster than row layout of the relational data for OLTP query. Another aspect of in-memory cache for domain specific model data is to enable efficient intra-object navigation and traversal when domain specific query language is being evaluated.

3.5 Model-Agnostic Transaction

RDBMS supports ACID guarantee, while NoSQL employs BASE pertaining as the ways for scaling and workaround on CAP theorem. We envision a per-query choice of consistency between ACID and BASE for multi-model data, which is flexible so that the user has a clear understanding and control over the performance as well as the consistency guarantees. To support a hybrid transactional semantics, the model-agnostic object collection can be declaratively specified as ACID or BASE. The transaction layer then enforces different transaction semantics for the data.

Further, to boost the performance of transaction execution, a fine-granular isolation at different levels in multi-model data can achieve the flexibility and performance benefit. For example, objects can be isolated in the forms of subtree locks, subgraph locks, path locks and neighbor node locks. Further, an effective global node labeling scheme can be developed to enable the quick jump to a particular inner data node as required in the lock manager (e.g. to support *getParent* operations in a tree).

3.6   Other Research Challenges

Schema Discovery. Original ORDBMS assumes the perfect schema based world. Semi-structure data and unstructured data challenge ORDBMS with schema-less design. We understand the value of NoSQL point of schema-less DB development, but further enhances it to argue that schema-less for write is half of the story, schema-rich for query is the other half of the story via auto-schema discovery. Therefore, UDBMS is expected to support schema discovery interface. Enhancing schema discovery for multi-model data is a new challenge which involves knowledge and techniques from neighboring research communities such as data mining, machine learning and human-computer interaction.

Model Evolution is a unique challenge in a multi-model database [17]. With the increasing maturity of NoSQL databases, a plethora of applications turn to store data with JSON documents or graph representations. But the legacy data are still stored in the traditional RDBMS. Thus, the model evolution may affect the usability of queries and applications developed on the RDBMS. There-fore, a research challenge is how to perform model mapping and query rewriting to automatically handle such model evolution. Note that model evolution is a more complicated than schema evolution on RDBMS, because it raises the issues involving both the attribute difference and the structural difference.

Multi-model Sharding is a method for distributing data across multiple machines. A relational database shard is a horizontal partition of data and each shard is held on a separate database server instance to spread load. But in the scenario of multi-model data management, do we support inter-object or intra-object sharding ? It is easy to support inter-object sharding. But if a graph or a tree is a big object, then we need to consider about intra-object sharding. Therefore, the distributed data sharding technology needs further investigation in UDBMS.

4   Related Work

Heterogeneous Query Processing. The interests of providing a unified data access and processing interface for heterogeneous data sets have been extensively explored in previous systems and prototypes [6,9,18]. For example, BigIntegra-tor [24] supports SQL-like queries that combines data in Bigtable stores in the cloud and data in relational stores; Forward presents SQL++ [21], an SQL-like language designed to unify the data model and query language capabilities of NoSQL and relational databases; Dremel [2] uses semi-structure data model, however, it can execute queries based on a flattened columnar storage. Vertexica [12] runs graph queries in a relational database; Asterix follows the motto *"one size fits a bunch"* and has built a data model-agnostic query compiler substrate, called Algebricks [3], and has used it to implement three different query lan-guages: HiveQL, AQL, and XQuery, on different format of data. Compared to the previous works which focus on specific models, our approach is more generic, with several model-agnostics principles (for data storage, query processing and index structures) that enable the efficient heterogeneous query processing for multiple data stores.

Tightly-Coupled Multistore System. Recent works (e.g. Polybase [7], HadoopDB [1] and Estocada [5]) also demonstrate the performance benefits of using seamless integrated multi-stores, these systems aim at efficient management of structured and unstructured data for big data analytic, particularly for integration of HDFS and RDBMS data. Different from the purpose of combining relational and distributed unstructured stores, UDBMS focus on the unification of relational and transactional NoSQL stores for on-line hybrid OLTP and OLAP data processing.

5 Conclusion and Future Works

"The intellect seeking after a unified theory cannot rest content with the assumption that there exist two distinct fields totally independent of each other by their nature."
— Albert Einstein in his Nobel lecture in 1923

In this paper, we present our visions to build a system to query, index and update multi-model data in a unified fashion. While the road to unification is full of challenges, this paper has laid down our visions to build a UDBMS with model agnostic properties and structures. In the future work, we shall investigate the following three categories of challenges on UDBMS: diversity, extensibility and flexibility.

(1) **Diversity:** The first challenge is the *"diversity"* of multi-model data. The existing results for query optimization and transaction model mainly work on a single model, either structured or semi-structured data. The highly diverse nature of multi-model data makes a unified system complicated and fascinated.

(2) **Extensibility:** The second challenge is to identify the boundary of UDBMS. In this paper, we envision a unified system for several types of data, i.e. relation, JSON, XML and graph. A further question is how to adopt more types of data such as streaming data and time series data. This calls for the future research on the extensibility of a multi-model system.

(3) **Flexibility:** Finally, NoSQL DBMS can support schema-less for storage, and schema-rich for query according to the automatically discovered schema. It would be interesting to explore the model-agnostic storage and query processing in UDBMS – in particular, we call this *"what we store is **not** what we get"*.

Acknowledgment. Contact email: Jiaheng.Lu@helsinki.fi. This work is partially supported by Academy of Finland (Project No. 310321).

References

1. Abouzeid, A., Bajda-Pawlikowski, K., Abadi, D.J., Rasin, A., Silberschatz, A.: HadoopDB: an architectural hybrid of MapReduce and DBMS technologies for analytical workloads. PVLDB **2**(1), 922–933 (2009)
2. Afrati, F.N.: Storing and querying tree-structured records in Dremel. PVLDB **7**(12), 1131–1142 (2014)
3. Borkar, V.R., et al.: Algebricks: a data model-agnostic compiler backend for big data languages. In: ACM SoCC, pp. 422–433 (2015)
4. Bruno, N., Koudas, N., Srivastava, D.: Holistic twig joins: optimal XML pattern matching. In: ACM SIGMOD, pp. 310–321 (2002)
5. Bugiotti, F., Bursztyn, D., Deutsch, A., Ileana, I., Manolescu, I.: Invisible glue: scalable self-tunning multi-stores. In: CIDR (2015)
6. Chen, J., et al.: Big data challenge: a data management perspective. Front. Comput. Sci. **7**(2), 157–164 (2013)
7. DeWitt, D.J., et al.: Split query processing in polybase. In: SIGMOD, pp. 1255–1266 (2013)
8. Elmore, A.J., et al.: A demonstration of the BigDAWG polystore system. PVLDB **8**(12), 1908–1911 (2015)
9. Franklin, M.J., Halevy, A.Y., Maier, D.: From databases to dataspaces: a new abstraction for information management. SIGMOD Rec. **34**(4), 27–33 (2005)
10. Gog, I., et al.: Musketeer: all for one, one for all in data processing systems. In: EuroSys, pp. 1–16 (2015)
11. Heimbigner, D., McLeod, D.: A federated architecture for information management. ACM Trans. Inf. Syst. **3**(3), 253–278 (1985)
12. Jindal, A., et al.: VERTEXICA: your relational friend for graph analytics!. PVLDB **7**(13), 1669–1672 (2014)
13. Lim, H., Han, Y., Babu, S.: How to fit when no one size fits. In: CIDR (2013)
14. Lin, C., Lu, J., Wei, Z., Wang, J., Xiao, X.: Optimal algorithms for selecting top-k combinations of attributes: theory and applications. VLDB J. **27**(1), 27–52 (2018)
15. Liu, Y., Lu, J., Yang, H., Xiao, X., Wei, Z.: Towards maximum independent sets on massive graphs. PVLDB **8**(13), 2122–2133 (2015)
16. Liu, Y., et al.: ProbeSim: scalable single-source and top-k simrank computations on dynamic graphs. PVLDB **11**(1), 14–26 (2017)
17. Lu, J.: Towards benchmarking multi-model databases. In: CIDR (2017)
18. Lu, J., Holubová, I.: Multi-model data management: what's new and what's next? In: EDBT, pp. 602–605 (2017)
19. Lu, J., Ling, T.W., Bao, Z., Wang, C.: Extended XML tree pattern matching: theories and algorithms. IEEE Trans. Knowl. Data Eng. **23**(3), 402–416 (2011)
20. Lu, J., Ling, T.W., Chan, C.Y., Chen, T.: From region encoding to extended dewey: on efficient processing of XML twig pattern matching. In: VLDB, pp. 193–204 (2005)
21. Ong, K.W., Papakonstantinou, Y., Vernoux, R.: The SQL++ semi-structured data model and query language: A capabilities survey of SQL-on-Hadoop, NoSQL and NewSQL databases. CoRR abs/1405.3631 (2014)
22. Xu, P., Lu, J.: Top-k string auto-completion with synonyms. In: Candan, S., Chen, L., Pedersen, T.B., Chang, L., Hua, W. (eds.) DASFAA 2017. LNCS, vol. 10178, pp. 202–218. Springer, Cham (2017). https://doi.org/10.1007/978-3-319-55699-4_13
23. Yan, X., Yu, P.S., Han, J.: Graph indexing: a frequent structure-based approach. In: SIGMOD, pp. 335–346 (2004)
24. Zhu, M., Risch, T.: Querying combined cloud-based and relational databases. In: CSC, pp. 330–335 (2011)

Extracting Conflict Models from Interaction Traces in Virtual Collaborative Work

Guangxuan Zhang[1], Yilu Zhou[2], Sandeep Purao[3]([⊠]), and Heng Xu[4]

[1] Microsoft, Redmond, USA
[2] Fordham University, New York City, USA
[3] Bentley University, Waltham, USA
spurao@bentley.edu
[4] American University, Washington, D.C., USA

Abstract. This paper develops a model of conflicts that relies on extracting text and argument features from traces of interactions in collaborative work. Much prior research about collaborative work is aimed at improving the support for virtual work. In contrast, we are interested in detecting *conflicts* in collaborative work because conflict undetected can escalate and cause disruptions to productive work. It is a difficult problem because it requires untangling conflict-related interactions from normal interactions. Few models or methods are available for this purpose. The extracted features, interpreted with the help of foundational theories, suggests a conceptual model of conflicts that include categories of argumentation such as reasoning and modality; and informative language features. We illustrate the extraction approach and the model with a dataset from Bugzilla. The paper concludes with a discussion of evaluation possibilities and potential implications of the approach for detecting and managing conflicts in collaborative work.

Keywords: Conflict · Conflict detection · Argumentation
Online collaboration

1 Introduction

Much collaborative activity is now conducted online, within and across organizations, with geographically dispersed teams that rely on email, instant messaging, and social media (Hinds and Mortensen 2005). In addition, online communities enable mass debate (Somasundaran and Wiebe 2010) and collaboration for complex tasks such as software development (Wang et al. 2015) and product design (Brabham 2008). Improving support for collaborative work continues to be an important research direction in this setting for tasks such as progress reporting, team awareness, agenda keeping and others. Platforms for collaboration are also increasingly sophisticated and track collaborative activity. As a result, the research community now possesses new opportunities to make sense of collaborative activity. We address one specific problem in this context: understanding and extracting models of conflict. An online discussion can contain multiple participants and overlapping threads. Detecting and modeling

© Springer Nature Switzerland AG 2018
C. Woo et al. (Eds.): ER 2018 Workshops, LNCS 11158, pp. 295–305, 2018.
https://doi.org/10.1007/978-3-030-01391-2_34

conflicts for the purpose of learning from them as well as managing them is, therefore, a difficult problem (Barcellini et al. 2005; Bex et al. 2013).

Some contemporary scholars are responding to the challenge (e.g. Bex et al. 2013; Lippi and Torroni 2015; Schneider 2014). However, few methods and tools are directly aimed at extracting arguments from the data available via online discussions. One reason for this may be that online communication tools, including group decision support systems, are rarely equipped with capabilities necessary for processing argumentative content (Zhang and Purao 2014; Zhang et al. 2016). Another is that the existing tools are focused on analyzing the argument structure in articles or essays (Moens et al. 2007; Ozyurt 2012), which tend to be different from online discussions in vocabulary and structure (Yates 1996). Further, these models tend to be inconsistent in feature selection and lack robustness due to the lack of strong theoretical foundations. To address the contemporary problems outlined above, the objective of this research is to design a conceptual model that can serve as the foundation for tools and methods for conflict detection in online discussions.

In this paper, we develop a theory based conceptual model that combines textual and argument features to effectively detect argumentative messages in online discussion. In particular, the model reflects five categories of argumentation functions: announcement, reasoning, modality, transition, and affect, and textual features that are informative for recognizing argumentative language. The paper develops the conceptual model and demonstrates, with the help of an example from the Bugzilla platform how it can be used for detecting conflicts. The model is evaluated with a dataset form Bugzilla that provides an indication of the effectiveness of the model compared to baseline models. The paper concludes with a brief discussion of implications for using the conceptual model for building tools, which may be useful for practice; and for future research related to extension of the model.

2 Related Work

We begin with a motivating example, followed by a review of relevant prior work.

2.1 A Motivating Example

Consider the following scenario faced by a small team of software designers, including John and Mary. The team is engaged in an effort to develop a systems integration solution for one of their clients.

The work involves constructing a catalog of services from a legacy scheduling application so that contractors and customers can use these, and outlining a set of procedures to work with other legacy applications in a similar manner. Their concerns revolve around scalability and security but they are also unclear about how they might go about aligning their current business processes. John and Mary propose different solution approaches. Because they work remotely on the project, their proposals are brought up and discussed on the email threads. The proposals rely on different technology platforms, one more mature and secure; the other more cutting-edge. The team engages in a discussion that includes a number of concerns such as characteristics of the platforms, resource constraints, and expertise available. The email discussion spawns several threads and sub-groups, cordial as well as argumentative. It is clear to some of the individuals in the group that at least some of the discussions have the potential to degenerate into conflicts that may be more harmful than helpful. They foresee posturing and arguments that include both substantive as well as affective concerns.

The problems faced by the integration team are not new. With online collaboration, the individuals in teams often express their points of view and preferences about a project work in public discussion threads. These contributions can include argumentative elements as well as uses of language to explain, convince and persuade. In teams that either work across projects or in long-term projects, the individual proclivities are sometimes well-known, allowing individuals to understand and predict conflicts. However, without the benefit of such expertise conflicts can be difficult to identify. We take first steps towards such proactive approaches to conflict detection by developing a theory-driven conceptual model (similar to Purao et al. 2018). The theory we draw upon is Argumentation Theory, outlined next.

2.2 Argumentation Theory

The labels Argument and Argumentation are often used loosely because they significantly overlap in meaning. For example, they have been used interchangeably in recent research (Lippi and Torroni 2015; Mochales and Moens 2011). In this study, we distinguish the two by first defining an argument as an utterance (expressed as a text fragment). An argument, then, contributes to the process of argumentation, which we define as the process of reasoning, systematically organized in support of an opinion. This distinction is important for online discussions and email threads that capture the traces of collaboration because an argumentation process may be spread across several messages (unlike, say, a scientific communication where all arguments may be part of the argumentation expressed within a single document). This dispersal of arguments across messages can also make it challenging to detect the progression of conflicts.

A complete argumentation consists of three elements: a proposition, a standpoint, and reasoning. A proposition is the matter of interest, which can be a description, a prediction, a judgment, or a suggestion. One can then take a positive standpoint to justify the proposition, a negative standpoint to refute the proposition, or a neutral standpoint. An argumentation is strengthened when some kinds of reasoning are provided to justify the standpoint (Eemeren et al. 2002). In a conversation, the three components may not be included in a single statement. For example, reasoning might not be included at the beginning of a discussion, while the proposition and parties' standpoints are often left out as the conversation moves forward. Eemeren et al. (2002) stress that verbal expressions provide strong indicators of argumentation and its elements. They provide an initial list of vocabularies that signal standpoint and reasoning, which however, need to be enriched and refined for the purpose of automatic argument detection and processing.

Eemeren et al. (2002), describe argumentation as the means of resolving a difference of opinion. It is performed when one party's opinion meets with objections from the other party and both parties have intentions to discuss the matter and address the difference (Thomas 1992). Argumentation theory arose in response to the need of understanding the actual argumentation practice in conversation (Eemeren et al. 2002). In contrast to the logician's view, which focuses on abstract forms and patterns of argumentation (Toulmin 2003; Walton et al. 2008), argumentation theory concentrates on rules and guidelines in constructing, detecting, and analyzing arguments. It has been employed to promote critical thinking and scientific argument (Bricker and Bell 2008), and recognize effective communication patterns in online environments (Clark and Sampson 2008).

2.3   Argument Detection Models

Argument detection in online discussion remains understudied. Relevant studies are mostly found in the field of argumentation mining, aimed at detecting, decomposing, and reconstructing argumentation in textual materials (Lippi and Torroni 2015; Palau and Moens 2009; Schneider 2014). To achieve this, a few models have been proposed (see Table 1). Most are developed to analyze documents like articles and essays. Their applicability for online discussion is, therefore, questionable because of differences in vocabulary and sentence structures (Yates 1996). One exception is the work by Levy et al. (2014). Their model aims to detect claims in online debates. Another important aspect of existing models is that they heavily rely on classic features for text representation, such as bag-of-words, word bigrams/trigrams, part-of-speech information, and text statistics (Conrad et al. 2012; Palau and Moens 2009; Stab and Gurevych 2014). There are, however, a few exceptions that incorporate sentiment analysis indicators (Conrad et al. 2012) and subjectivity scores of sentences (Levy et al. 2014). A third observation is that while word indicators have been recognized and applied in these models, there is little to no consistency across features that should be included for analyses. Lexicons in these models vary with regard to size, part of speech, and composition. Finally, features included in these models mainly represent researcher experience and peculiarities of the corpora, instead of foundational theories about arguments and argumentation.

3   A Conceptual Model for Conflicts in Online Collaborations

Building on this prior work, we develop a conceptual model for representing conflicts. The conceptual model builds on prior theories. We emphasize that this does not represent an exercise in YAML: yet another modeling language (see Purao and Woo 2014). Instead, our work represents a theory-driven approach similar to Purao et al. (2018), where the authors develop a modeling language for systems integration by extending coordination theory (Malone and Crowston 1994). In this paper, we focus on such a conceptual model that can capture conflicts in online collaborations. The paper does not elaborate the mechanics for extracting the components in such a model. Instead, our focus is on outlining the model itself. The model consists of elements that are inspired by two sources: foundational theories about argumentation, and features of the language that participants in an online setting use to express these arguments.

Table 1. Argument detection models in prior work

Works	Corpus	Machine learning features
Moens et al. 2007; Palau and Moens 2009; Mochales and Moens 2011	News articles, Legal text, Reports	Generic Features: *punctuation, text statistics, n-grams.* Special Lexicons: *verbs, adverbs, modal auxiliary, rhetorical relations*
Ozyurt 2012	Biomedical literature	Generic Features: *POS, position of sentence, semantic network.* Special Lexicons: *predict noun, head word*
Conrad et al. 2012	Blogs and editorials	Generic Features: *sentiment.* Special Lexicons: *relation phrases, noun, verb*
Levy et al. 2014	Debate forum	Generic Features: *cosine similarity between sentences, subjectivity score, sentiment, name entity.* Special Lexicons: *claim words*
Stab and Gurevych 2014	Persuasive essays	Generic Features: *text statistics, location of sentences, punctuation marks, n-grams, POS n-grams.* Special Lexicons: *verbs, adverbs, pronoun, reasoning markers*

3.1 Elements Inspired by Argumentation Theory

Drawing on prior work (e.g. Eemeren et al. 2002), we identify five elements of argumentation: announcement, reasoning, modal qualifier, transition and affect.

Announcement. Typing to speak is constrained by space, time pressure, and available attention. Participants in online collaboration, therefore, tend to express their ideas in a clear, straightforward language (Yates 1996), e.g., one often starts with an announcement of his/her standpoint with phrases such as: I think, I assert, I vote, and others. Negative standpoints can be announced with phrases like I debate, I doubt, I disagree and others as well as verb indicators that appear as nouns, e.g., My assertion is and The suggestion is.

Reasoning. Participants often use reasoning to support announcements, e.g. providing evidence, making cause-effect analysis, delimiting the problem scope, and making comparison/evaluation. This provides grounds, premises, warrants that can be used independently or combined with others (Toulmin 2003). Indicators of reasoning includes phrases such as for example, demonstrate, because, so that, respect, circumstance, more, than, superior and others.

Modal Qualifier. An argumentation may be made with total conviction or it may be uncertain. An experienced participant may present a mix of arguments that vary in force to highlight his/her firmness on some points (Joni and Beyer 2009). The strength of an argument may be expressed with phrases such as: absolute/absolutely, always, maybe, must, possible/possibly, perhaps, sometimes, and so on. They reflect one's confidence in the stated evidences and warrants.

Transition. Transitions are essential for organizing and connecting elements of argumentation. As an example, one technique, concession, relies on admitting some points claimed by opponents to build a cooperative relationship and promote reciprocity (Johnson and Cooper 2009). Transitions can be indicated by coordinators and conjunct adverbs, e.g. and, or, first, finally and others to establish links among clauses and argument components. Subordinators, such as although and despite, can indicate concession.

Affect. Conflicts may sometimes include heated debate and strong language. In online collaborations participants are often less constrained by social norms (Johnson and Cooper 2009). Offensive and abusive languages, therefore, can emerge when a debate escalates indicated by words such as f***, damn, stupid, hell and others. Other indicators may include words such as annoying and upset, and emoticons to convey one's anger, disappointment, or other negative attitude.

An argumentation lexicon can be created based on these categories, and drawing on prior work (e.g. for announcement and reasoning indicators, relying on Eemeren (2002), model qualifiers relying on Knott and Dale (1993), affective content based on Chen et al. (2012) and others).

3.2 Elements Based on Contextual and Language Features

Drawing on prior work related to linguistic features (Chen et al. 2012; Mochales and Moens 2011; Stab and Gurevych 2014), we identify additional features.

Contextual Lexicon. Similar to other text mining tasks (Chen et al. 2012), understanding context is important for argument detection. A contextual lexicon can help based on identification of vocabularies specific to the context. In our illustration that follows, this is the Bugzilla platform with vocabularies about topics such as usability, performance, and easy use; competitors such as other browsers Chrome, Google, and IE, and abbreviations such as IMHO and FWIW.

Sentence Style. In online discourse, it is a common practice that speakers use special punctuation and word formats to indicate feelings or speaking volume. Punctuations, such as repeated exclamation marks, or questions marks, or a combination of both, and words with all uppercased letter often appear in arguments of strong tone.

Sentence Structure. Sentence function indicates a speaker's purpose in uttering a specific sentence, phrase, or clause. In arguments, combinations of various sentence functions yield different intentions or different argumentation levels. For example, intensive use of interrogative sentences is observed when one refutes others' arguments. Explanative and imperative sentences often appear in strong persuasive arguments. These represent a third feature.

Sentiment. Sentiment analysis has been widely used to understand attitude (negative, neutral, and positive) to an idea. To incorporate sentiment information, we incorporate both positive score and negative score of a message' sentiment. The rationale is that an argument can be made either in favor of or in conflict with statements made in previous messages.

System Information. To facilitate communication and reduce confusion, commutation platforms provide functions like specifying reply-to messages, quoting previous messages, adding attachments, and so on. These functions have been intuitively employed in online discussion. For example, by explicitly indicating the reply-to message, one can deliver a targeted argument that is to refute a previous argument or provide supplement argument. Automatically generated system information on Bugzilla, including reply-to indicator, quotation, create attachment, and comment on attachment, are thus also included in TAD.

4   An Illustrative Example

To illustrate the conceptual model, we use a corpus of email threads from the online collaboration platform Bugzilla, selecting one email thread for the illustration. Figure 1 shows a visual representation of the thread with several sub-threads. From this figure, we investigate eight sub-threads. Table 2 provides a summary of each.

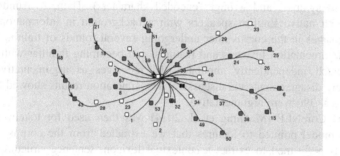

Fig. 1. An example bug discussion on Bugzilla: a visualization

An examination of the argument visualization, combined with\investigation of the data showed that a modeling exercise can, in fact, provide the ability to see the structure of an argument – and provided the foundation to create mechanics to distinguish argumentative and non-argumentative messages.

4.1   The Email Corpus, Initial Annotation and Feature Extraction

Annotated corpora for argument detection are still in short supply (Lippi and Torroni 2015). The few existing corpora are mainly based on articles and essays. To address this challenge, we created a corpus using email data collected from Bugzilla. We selected this data source for three reasons. First, arguments are a common phenomenon in its emaillists (Wang et al. 2015). Second, it represents a sample from a much larger data source that can be further employed to create a larger argumentation corpus. And third, it represents a corpus that is relatively easy to access. Based on email threads identified in Wang et al.'s study (Wang et al. 2015), we selected 7 email threads, consisting of 671 messages that constitute the corpus. The length of one message

Table 2. Sub-threads in the example discussion

#	Sub-threads	Description
1	5–6*–15*	Clarify what to do to unify the browser behaviors
2	18–21*	Propose a patch code
3	24*–25–26	Discuss a proposed patch code
4	28–43–45–48	Debate on the add-on solution to be used by those against
5	29–33	Remind to update user document
6	37–38	Debate on the rationale of the change
7	49–50	Argue against the change - breaking previous use patterns
8	51–52	Argue against the change – losing flexibility

Notes: Messages 6, 15, 21, and 24 mistakenly classified as argumentative.

ranges from 1 to 734 words (median 50 words). The time fame of a thread ranges from 1 month to 4 years (median 10 months).

During the initial annotation, we first removed stop words, and noise (such as unrecognized symbols and wrongly encoded characters). Then, two undergraduate students (both native English speakers with a background in information systems) labeled messages in the corpus after undergoing several rounds of training to understand the argumentation lexicons and features, and becoming familiar with topics on Bugzilla. Each independently annotated the messages as argumentative or not, resolving any disagreements via discussion. The annotation results showed that out of 671 messages, 396 were argumentative.

Stanford CoreNLP (Manning et al. 2014) was then used for tokenization. The conceptual model pointed to features that we extracted from the corpus. The same practice was performed to extract a contextual lexicon, sentence structure and style features. To extract sentiment feature, we adopted opinion lexicon provided by Hu and Liu (2004) and generated a positive and a negative sentiment score for each message. All these features are then used in text classification algorithms.

4.2 Feasibility Evaluation

To evaluate feasibility, we compared the possibilities offered by the conceptual model against two baseline models, which have primarily focused on text classification: Bag-of-Words (BoW) and 2-g. The Bag-of -Words approach uses each word as a representation, disregarding grammar and order. Word frequency, excluding stop words, is then used for classification. The 2-g approach captures all sequences of 2 words in a text to some certain contextual information. However, similar to Bag-of-Words, it does not concern grammar or orders between the tokens. Frequency of each 2-g token is used as a feature value in classification. The evaluation then consisted of a comparison between feature extraction based on the conceptual model against the baseline models. Four text-mining algorithms were used to make the comparisons: Decision Tree (J48), Naïve Bayes, Logistic, and Support Vector Machine.

In line with previous studies in text mining (Abbasi et al. 2012; Chen et al. 2012), standard evaluation metrics were used. Overall accuracy is captured as the percent of messages, including both argumentative and non-argumentative ones that are correctly classified. Precision is captured as the percent of identified messages that are truly argumentative messages. Recall is captured as the percent of actually argumentative messages that are correctly identified. F-measure (Chen et al. 2012) is used to represent the weighted harmonic mean of precision and recall, which is defined as: $F = 2*$ (Precision * Recall)/(Precision + Recall). ROC area (Abbasi et al. 2012) represents the ability of the model to correctly classify argumentative and non-argumentative messages.

We first tested baseline models with different classification algorithms. Both of them achieved best performance with the Support Vector Machine (SVM). The baseline models with SVM Bag-of-Words (baseline 1) performed better than 2-g model (baseline 2) on all measures, achieving a 0.77 F-Measure. The conceptual model (CML) achieved best performance with SVM, too. It achieved 82.7% accuracy, 87.9% recall and 0.86 F-measure. The F-measure is an 11.6% improvement from the best performing baseline model (0.77). The improvements in these four metrics are tempered by different learning mechanisms. The Accuracy scores improved from 76.3% (baseline 1) to 82.7% attributable to CML + SVM. Corresponding improvements for Recall scores show a move from 66.9% (baseline 1) to 87.9%, also attributable to CML + SVM; for the F-measure score, a move from 0.77 (baseline 2) to 0.86, again attributable to CML + SVM; and for the ROC area measure, a move from 0.78 (baseline 1) to 0.88, attributable to CML + Bayes.

Thus, we found that all extraction approaches that relied on the conceptual model (CML) achieved similar or better performance than the two baseline models. Although the two baseline models obtained better precision rates, they had low recall rates. Use of the CML improved recall for the two baseline models: by 31.3% and 60.4%. In other words, the baseline models missed a number of argumentative messages. The CML, thus, helped improve recall consistent with the strategy of maximally argumentative interpretation (Eemeren et al. 2002).

5 Conclusions and Next Steps

In this paper, we proposed a conceptual model that captures conflicts (as argumentative messages) in online discussions. As a part of this project, we hope to continue clarifying how the notion of conflicts (which we see as incompatible argumentation from different parties) can provide the foundation for future work. The model we build, therefore, relies on (a) categories of argumentation functions and (b) language features from classical text mining. The feasibility evaluation results show that the use of the conceptual model achieves higher Accuracy and Recall rates in detecting argumentative messages, compared to baseline models. The illustrative example shows the application and usefulness of the conceptual model for making sense of conflicts in online discussions. To the best of our knowledge, this is a first such conceptual model built upon features of argumentation theories, combined with classical text mining functions. There are some limitations to our work. Our corpus is comparatively small. It prevents us from conducting extensive evaluation (similar to challenges faced in other studies (Lippi and Torroni 2015)). The model we have proposed has the potential to parse an argument message further into claims and premises. Our illustrative example shows that argument information is more informative when integrated with other interaction data. Further work is needed to explore innovative use of argument information with respect to sense making, collaboration, and conflict management.

Acknowledgements. The work reported has been funded by the National Science Foundation under award number CNS 1551004. Any opinions, findings and conclusions or recommendations expressed in this material are those of the author(s) and do not necessarily reflect the views of the National Science Foundation (NSF). We also acknowledge the commentary from the review team that has helped us refine the paper.

References

Abbasi, A., et al.: Metafraud: a meta-learning framework for detecting financial fraud. Mis Q. **36**(4), 1293–1327 (2012)

Barcellini, F., et al.: A study of online discussions in an open-source software community. In: Van Den Besselaar, P., et al. (eds.) Communities and Technologies 2005, pp. 301–320. Springer, Netherlands (2005). https://doi.org/10.1007/1-4020-3591-8_16

Bex, F., et al.: Implementing the argument web. Commun. ACM **56**(10), 66–73 (2013)

Brabham, D.C.: Crowdsourcing as a model for problem solving an introduction and cases. Convergence **14**(1), 75–90 (2008)

Bricker, L.A., Bell, P.: Conceptualizations of argumentation from science studies and the learning sciences and their implications for the practices of science education. Sci. Educ. **92**(3), 473–498 (2008)

Chen, Y., et al.: Detecting offensive language in social media to protect adolescent online safety. In: International Conference Social Computing (SocialCom), pp. 71–80 (2012)

Clark, D.B., Sampson, V.: Assessing dialogic argumentation in online environments to relate structure, grounds, and conceptual quality. J. Res. Sci. Teach. **45**(3), 293–321 (2008)

Conrad, A., et al.: Recognizing arguing subjectivity and argument tags. In: Proceedings of ExProM 2012, Stroudsburg, pp. 80–88 (2012)

Eemeren, F.H., et al.: Argumentation : Analysis, Evaluation, Presentation. Routledge, Mahwah (2002)

Hinds, P.J., Mortensen, M.: Understanding conflict in geographically distributed teams. Organ. Sci. **16**(3), 290–307 (2005)

Hu, M., Liu, B.: Mining and summarizing customer reviews. In: Proceedings of the 2004 ACM SIGKDD International Conference on Knowledge Discovery and Data Mining, KDD 2004, vol. 04, p. 168 (2004)

Johnson, N.A., Cooper, R.B.: Power and concession in computer-mediated negot: an examination of first offers. Mis Q. **33**(1), 147–170 (2009)

Knott, A., Dale, R.: Using Linguistic Phenomena to Motivate a Set of Rhetorical Relations Human Communication Research Centre. University of Edinburgh, Scotland (1993)

Levy, R. et al.: Context Dependent Claim Detection. In: Proceedings of COLING 2014, the 25th International Conference on Computational Linguistics, Dublin, pp. 1489–1500 (2014)

Lippi, M., Torroni, P.: Argument Mining: A Machine Learning Perspective. In: Black, E., Modgil, S., Oren, N. (eds.) TAFA 2015. LNCS (LNAI), vol. 9524, pp. 163–176. Springer, Cham (2015). https://doi.org/10.1007/978-3-319-28460-6_10

Malone, T.W., Crowston, K.: The interdisciplinary study of coordination. ACM Comput. Surv. **26**(1), 87–119 (1994)

Manning, C.D., et al.: The stanford CoreNLP natural language processing toolkit. In: ACL (System Demonstrations), pp. 55–60 (2014)

Mochales, R., Moens, M.F.: Argumentation mining. Artif. Intell. Law **19**(1), 1–22 (2011)

Moens, M.-F., et al.: Automatic detection of arguments in legal texts. In: Proceedings of ICAIL 2007, New York, pp. 225–230 (2007)

Ozyurt, I.B.: Automatic identification and classification of noun argument structures in biomedical literature. IEEE/ACM Trans. Comput. Biol. Bioinform. **9**(6), 1639–1648 (2012)

Palau, R.M., Moens, M.-F.: Argumentation mining: the detection, classification and structure of arguments in text. In: Proceedings of ICAIL 2009, pp. 98–107. New York (2009)

Purao, S., Woo, C.: Conceptual modeling: going beyond the stigma of YAMA. SIGSAND Workshop, May 2014, St. Louis, MI (2014)

Purao, S., et al.: A modeling language for conceptual design of systems integration solutions. ACM Trans. Mis. Forthcoming (2018, Forthcoming)

Schneider, J.: Automated argumentation mining to the rescue? Envisioning argumentation and decision-making support for debates in open online collaboration communities. In: Proceedings of the First Workshop on Argumentation Mining, pp. 59–63 (2014)

Somasundaran, S., Wiebe, J.: Recognizing stances in ideological on-line debates. In: Proceedings of the NAACL HLT 2010 Workshop on Computational Approaches to Analysis and Generation of Emotion in Text, pp. 116–124, June 2010

Stab, C., Gurevych, I.: Identifying argumentative discourse structures in persuasive essays. In: Proceedings of the 2014 Conference on Empirical Methods in Natural Language Processing (EMNLP), pp. 46–56 (2014)

Thomas, K.: Conflict and negotiation process in organizations. In: Dunnette, M.D., Hough, L.M. (eds.), Handbook of Industrial and Organizational Psychology, vol. 3, pp. 651–717. Consulting Psychologists Press (1992)

Toulmin, S.: The Uses of Argument. Cambridge University Press, Cambridge (2003)

Walton, D., Reed, C., Macagno, F.: Argumentation Schemes. Cambridge University Press, Cambridge (2008)

Wang, J., Shih, P.C., Carroll, J.M.: Revisiting Linus's law: benefits and challenges of open source software peer review. Int. J. Hum. Comput. Stud. **77**, 52–65 (2015)

Yates, S.J.: Oral and written linguistic aspects of computer conferencing. In: Herring, S.C. (ed.) Computer-mediated Communication: Linguistic, Social, and Cross-cultural Perspectives, pp. 29–46. John Benjamins Publishing Co. (1996)

Zhang, G., Purao, S.: CM2: a case-based conflict management system. In: Tremblay, M.C., VanderMeer, D., Rothenberger, M., Gupta, A., Yoon, V. (eds.) DESRIST 2014. LNCS, vol. 8463, pp. 257–272. Springer, Cham (2014). https://doi.org/10.1007/978-3-319-06701-8_17

Quality of Models and Models of Quality (QMMQ) 2018

Preface to QMMQ 2018

5th Workshop on Quality of Models and Models of Quality (QMMQ 2018) at the 37th International Conference on Conceptual Modeling (ER 2018)

Xi'an, China
October 22–25, 2018

Organized by
Samira Cherfi, Beatriz Marín and Oscar Pastor

The 5th Workshop on Quality of Models and Models of Quality (QMMQ), organized by Samira Si-Said, Beatriz Marín, and Oscar Pastor has focused on the information systems development practices that are continuously evolving in the internet age. In a world in which we observe traditional methods that depends on heuristics in decision making and new data driven approaches, we believe that conceptual models of quality have a crucial role to play. Quality assurance has been and is still a very challenging issue within the Information Systems (IS) and Conceptual Modeling (CM) disciplines.

It is now more and more agreed that there is a need to develop the capacity to understand how the quality of data affects the quality of the insights that we derive from it. However, this quality, to be ensured, requires reliable IS that can only be designed with a precise ontological commitment. Moreover, research on quality needs more contributions based on experimentation to provide empirical evidences of successful IS design in different, challenging working domains to show how modeling provides benefits, and how assessing their quality becomes a major need. Empirical Software Engineering techniques and protocols should be followed in the CM discipline in order to provide reliable and useful results to assess IS quality.

With all these aspects in mind, for its 5th edition the QMMQ workshop has had a special focus on data quality. From a total of 6 submissions, up to 3 papers have been finally selected, which are related to data quality, data management in big data, and the integration of systems by following a model-driven approach. All of them provide a significant set of insights related to the problem under investigation, and they confirm an interesting technical program to stimulate discussion and going one step forward according to the QMMQ workshops series goals.

October 2018

Samira Cherfi
Beatriz Marín
Oscar Pastor
Program Co-chairs

Data Quality Evaluation in Document Oriented Data Stores

Emilio Cristalli[✉], Flavia Serra, and Adriana Marotta

Universidad de la República, Montevideo, Uruguay
{emilio.cristalli,fserra,amarotta}@fing.edu.uy

Abstract. Data quality management in document oriented data stores has not been deeply explored yet, presenting many challenges that arise because of the lack of a rigid schema associated to data. Data quality is a critical aspect in this kind of data stores, since its control is not possible and it is not a priority in the data storage stage. Additionally, data quality evaluation and improvement are also very difficult tasks due to the schema-less characteristic of data. This paper presents a first step towards data quality management in document oriented data stores. In order to address the problem, the paper proposes a strategy for defining data granularities for data quality evaluation and analyses some data quality dimensions relevant to document stores.

Keywords: Document store · Data Quality · Schema-less
Data quality dimensions · Data granularities

1 Introduction

The great popularity of document oriented data stores, a kind of NoSQL data stores, is mainly due to the flexibility they present for storing information, which is achieved through the absence of a rigid schema that imposes many restrictions to data (as there is in RDs). In document stores, data is organized in a semi-structured way, where the users are allowed to vary the data organization for each piece of data they store.

On the other hand, in this kind of data stores, data quality (DQ) is much more exposed and vulnerable than in structured databases, where the existence of a schema often enforces a lot of properties on data, such as "not null" or "unique" restrictions and more complex ones. Moreover, data manipulation for evaluating and improving DQ becomes much more difficult in "schema-less" contexts, due to the absence of a fixed schema that determines the way to access data items and groups of data items.

The main contributions of this work are the following: (i) a set of data granularities in document stores (it will be defined in Sect. 4), so that data can be manipulated in order to measure their quality, (ii) the identification and analysis of a set of DQ dimensions particularly relevant to this context, and (iii) an example of application of DQ metrics to a document store, defining a DQ model

© Springer Nature Switzerland AG 2018
C. Woo et al. (Eds.): ER 2018 Workshops, LNCS 11158, pp. 309–318, 2018.
https://doi.org/10.1007/978-3-030-01391-2_35

for a specific dataset. It is important to note that this proposal accomplishes document stores DQ management without the need to apply schema discovery techniques, which are extremely expensive. It shows that this kind of data can be manipulated for managing their quality while respecting the schema-less philosophy, fact that avoids costly data preprocessing, optimizing DQ management processes.

The rest of this document is organized as follows. Section 2 presents related work, Sect. 3 presents some preliminary concepts that are necessary to understand the proposal, Sect. 4 focuses on the problem of defining granularities for data manipulation, Sect. 5 presents the analysis of DQ dimensions, and Sect. 6 presents the application of the proposal to example data. Finally, Sect. 7 presents conclusions and future work.

2   Related Work

Recently, many works started to consider four "V" dimensions for characterizing Big Data, adding *veracity* to the three "Vs" *volume, velocity* and *variety* [3,11]. Although *veracity* refers to the same abstract concept in all these works, they present many differences at the time of exactly defining what *veracity* means and to which DQ dimensions it corresponds [4,5]. In some cases, authors focus on data integration contexts and they relate *veracity* to the quality of the sources (*coverage, accuracy*, and *timeliness*) [3]. In other cases, some DQ dimensions are selected as the main quality indicators. For example, in [11] *veracity* is *reliability, uncertainty, incompleteness*, in [6] the *consistency* and *completeness* dimensions stand out as the main quality indicators in Big Data analytics. In general, the most mentioned quality dimensions are: *accuracy, completeness*, and *consistency*. None of these works focuses on defining interesting DQ dimensions for document oriented data stores, although these are a very relevant kind of data stores among Big Data alternatives.

To the best of our knowledge, the problem of data manipulation for managing DQ in Big Data was not addressed by any work in the literature, neither the problem of DQ management in the specific context of document oriented data stores. If we consider semi-structured data in XML format, we find the proposal of [9], where a model for associating quality values to XML data is proposed. However, they do not deal with the problem of defining quality metrics and measuring quality of XML data solving the problem of structure heterogeneity. Our proposal focuses on document data stores, and aims to identify a set of DQ dimensions relevant to this specific context and to give solutions for measuring DQ in this kind of data, solving the problem of structure heterogeneity.

3   Preliminaries

In this section we present an example that will be used throughout the paper, some basic concepts and definitions about DQ that are necessary for the following sections, as well as the main characteristics of document oriented data stores.

3.1 Running Example

We introduce an example that will be used throughout the paper: A company selling TV cable subscriptions is interested in keeping track of their customers and the subscription plan they are paying for. The company needs to store personal information for each customer, like name, addresses and phones, and each customer is subscribed to one of the many subscription plans the company offers (basic plan, family plan, VIP, etc) which have different prices and descriptions.

3.2 Data Quality

DQ is a multi-faceted concept, which is represented by DQ dimensions that address different aspects of data. At the same time, the approach of DQ as *fitness for use* has been widely adopted by the research community. A huge set of DQ dimensions have been defined in the literature in the last 20 years, existing a subset that is used by most of the authors, with similar notions and/or definitions. Some works that gather definitions of DQ and DQ dimensions [8,10], show that there is not a standard or agreement about the set of dimensions that characterize DQ. Managing DQ in a dataset involves the complex tasks of evaluating, improving and monitoring DQ. To carry out these tasks it is necessary to define a *DQ Model* that serves as the basis and conduct all the involved processes. In this work we define a *DQ Model* as a set of *DQ dimensions* and *DQ metrics*, where the former represent general quality aspects and the latter define the way the DQ dimensions are measured. The same DQ dimension can be measured through many different DQ metrics, which address specific DQ problems. The DQ Model also includes the information of which DQ metrics will be applied over which data items. The DQ metrics are defined considering the following characteristics: Name (metric name), Description (what is measured by the metric and how), Measurement unit (how the result is represented; boolean, grade, enumeration), and Granularity (level of detail of the dataset at which the metric is applied, e.g., in relational model this would be cell, row, column, table or database). In the DQ model definition task, quality aspects and data items are prioritized, since all aspects cannot be measured for all data items. Therefore, the most relevant data items are selected and for each one, the most relevant quality aspects are chosen.

3.3 Document Oriented Data Stores

Document oriented databases store data as documents, usually in a format similar to JSON or XML. As explained in [7], a document can encapsulate multiple related objects that are to be treated as a unit. Documents can contain a complex structure of fields and embed a collection of related objects that altogether represent the unit. Whereas in a relational database (RD) these related objects would usually be stored in separate tables and referenced using foreign keys, in a document store they can be embedded in the same document to provide better performance when fetching an entity and its related objects (Fig. 1). This is known as an embedded data model, and even though it is particularly useful for sharding purposes (because the document can be stored as a unit in one node of the cluster, so the entity can be fetched with all its related data from the same node in one operation), it is also possible to use a more normalized data model referencing other documents, or even a combination of both models embedding part of the referenced document and also a reference to it. As mentioned in [2], each data model approach has its benefits and drawbacks in terms of performance, data duplication, consistency and other aspects, so choosing a data model is a trade-off decision and can vary for each attribute.

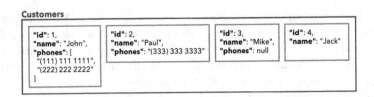

Fig. 1. Example documents in a document oriented data store.

Most document stores group documents in collections. Even though documents belonging to the same collection are not required to have the same structure, usually documents that represent the same type of entities are stored in the same collection to keep data organized and facilitate queries. For example, a collection might be used for all customer documents, and a different one for all the subscription plans. Documents are composed of field (*id, name* and *phones*) and value (*1, John* and *[(111) 111 1111, (222) 222 2222]*) pairs. In addition, document stores support a variety of data types for field values, including numbers, strings and arrays as well as nested documents. Documents that represent the same type of entities can have different structures, fields and data types for their values. In the example of Fig. 1, the field *phones* has an *array* value for the customer John, a *string* value for Paul, *null* for Mike, and the document for Jack does not include the field at all.

4   Dealing with Data Granularities

When evaluating the DQ of a dataset, it is extremely important to understand the underlying data model (e.g., relational, document oriented, key-value), so that different granularities for the DQ metrics can be properly defined. For example a customer with multiple phone numbers associated to him might be represented in two tables using foreign keys in a RD. But this same scenario might be modeled as a document in the customers collection with an embedded array of phones as shown in Fig. 1. To evaluate the quality of each phone in the document dataset, it would be necessary to look at each element of the embedded array, instead of looking at each row of the phones table in the relational model.

In Sects. 4.1 and 4.2, we propose six different granularities to measure DQ at different detail levels of a document dataset. Moreover, we discuss the potential recursiveness of these granularities due to the likelihood of having embedded documents.

4.1 Granularities

Database. In this case, the metric is applied over the database. A metric with this granularity evaluates the dataset as a whole, and as a result it provides only one DQ measure for the dataset. This is the granularity with the least specificity.

Example 1. In our running example, a completeness metric with Database granularity can be evaluated as the ratio between *null* values and non-*null* values.

Collection. Databases group documents in collections, being usual to store similar documents or documents representing the same type of records in the same collection. Collections can be compared to tables of RD. To store DQ measures of this granularity, the collections can be identified by their names, given that collection names are unique in document databases.

Example 2. Accuracy of the customers collection can be measured as the ratio of correct documents (for example verifying them with an external source) by the total amount of documents in the collection. This provides an overall DQ measure for the correctness of the collection.

Document. Documents are the main unit to store data in document stores, and therefore it is relevant to define document as another possible granularity. Albeit being comparable to a row of a relational table, a document can include multiple related objects. This feature allows evaluating the quality of the entity it represents as a whole, without having to analyze data from other documents (which would be different tables in a RD). Most document data stores require each document to have a unique identifier. This identifier, alongside the name of the collection, can be used to associate the obtained DQ measures to the corresponding documents.

Example 3. With this granuarity each customer document is evaluated. Figure 1 shows how very different groups of data would be measured through a metric with this granularity. Therefore, each document would have an associated DQ measure. One advantage of this granularity is that it allows identifying the documents that require data cleaning actions.

Field Across Documents. One of the traditional DQ granularities in RDs is the column granularity, which evaluates the quality of an attribute across all rows in a table. Even though the column concept that would cross all documents doesn't exist in document stores, the values for a field across all documents in a collection can be evaluated to measure the DQ of that field regardless of whether the field exists in all documents or not. We call this granularity "Field across documents".

Example 4. In our running example, the completeness of the address can be analyzed calculating the amount of empty (i.e., *null* value or field not present) values for that field, divided by the total amount of documents in the collection.

Field Value. This granularity captures the value of a specific field for a given document (analogously to cell granularity in RDs). Regardless of the data type, metrics with this granularity evaluate quality of the value as a unit, even if it is an array or nested document. Each field can be uniquely identified by its name inside a document, so this can be used to associate a DQ measure to a field value.

Array Element. Most document stores support having arrays as values for any field. The quality for the field value granularity can be evaluated considering the array as a whole, but in some cases it may be important to measure the quality for each element of the array. Using a combination of the field name and array element index, each element can be uniquely identified and associated with its DQ measure.

Example 5. In Fig. 1, the *phones* field for the first document is represented as an array of three values. The syntactic accuracy can be evaluated for each phone number and identified by *phones[0], phones[1], phones[n]*.

4.2   Nesting

The definition of the granularities mentioned in Sect. 4.1 can be also applied to nested documents. We propose using a "dot notation" and indexes to construct a path to identify the subject for each DQ measurement at different nested granularities. A dot indicates going one level deeper to a nested document, an index *[n]* references the n-th element of an array, and an empty index (*[]*) references a field across multiple documents of an array. For example, *field1.field2[3]* would reference the third element of the *field2*'s array value, of the document nested in *field1*. *Field1[].field2* would reference the field *field2* across all documents in *field1*'s array value.

Example 6. The customer document of Fig. 2 contains an array of address documents. We could measure: ***addresses[0]***, DQ for the first address document (Document granularity); ***addresses[].state***, DQ for the state field across all addresses (Field across documents granularity); ***addresses[0].state***, DQ for the state field of the first address (Field value granularity); ***addresses[0].phones[0]***, DQ for the first phone of the first address (Array element granularity);

```
"id": 1,
"name": "John",
"addresses":
                  "street": "1 Broadway",      "street": "2 Houston",
                  "city": "New York",          "city": "New York",
                  "state": "NY",               "state": "NY",
                  "zip": "10019",              "zip": "10001",
                  "phones": [                  "phones": [
                   "(111) 111 1111",            "(222) 222 2222"
                   "9999",                     ]
                  ]
```

Fig. 2. Document containing embedded documents.

5   Relevant DQ Dimensions

We propose the following group of DQ dimensions that we think are particularly relevant and useful when defining the DQ model for a document store dataset. We put special focus on a specific engine, *MongoDB*, so we can take advantage of its features and implementation details. We chose *MongoDB* since it is the most popular document store reported by [1].

5.1   Consistency and Redundancy

Document stores usually lack support for foreign keys. In order to be able to associate different documents, users have to manually reference them using a custom field with the target document's *id* for example. Because the system does not perform any checks on these manual references, it can lead to potential reference consistency issues, ending up with broken associations and orphan documents.

On the other hand, it is a common practice to embed documents instead of referencing them to avoid doing an extra query to get the referenced documents, due to the lack of support for *join* operations. In some cases it is useful to use a combination of both the embed and reference strategies (despite the risks it carries) depending on the frequency at which these documents are updated and read frequency. For instance, in our running example it might be better to store just the subscription plan's name and id embedded in the customer document and keep the rest of the fields in a different collection as shown in Fig. 3, because the subscription plan name is not likely to change and it is needed each time the customer information is shown. On the other side, the subscription plan price and description might change more often so it is good to keep them in another collection and avoid repeating this information for every customer.

5.2   Entity Heterogeneity

Most document stores do not enforce documents to comply to a predefined schema, and it is a big benefit when looking for flexibility or having heterogeneous data. However, it represents a problem for applications that query the

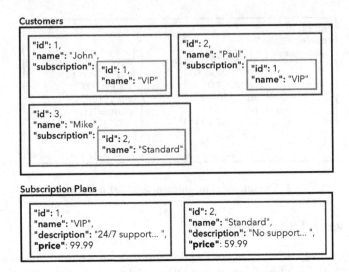

Fig. 3. Embedded and referenced documents.

data assuming certain fields or some kind of structure (implicit schema [7]). Entity heterogeneity in terms of data structure and data formats can be an important DQ dimension, in particular when applications expect some structure that some documents do not have.

5.3 Completeness

One of the most common strategies to evaluate data completeness is analyzing *null* or empty values. In document stores with flexible schema this requires more attention because values can also be empty arrays, empty documents, documents with empty values, or the field can even be completely missing.

5.4 Freshness

The usage of *ObjectId*s as *_id* in *MongoDB* (usually recommended [2]) can simplify the calculation of Freshness, eliminating the need of any extra data. Given a document that is using the default auto-generated *ObjectId* for the *_id* field value, the date and time of creation can be extracted using its first 4 bytes [2], and it can be used to calculate how fresh this object is in terms of creation date without explicitly storing any creation date information. This way of implementing a Freshness metric has the limitation that it does not allow calculating when the document was last updated.

5.5 Uniqueness

MongoDB has support for unique indexes which prevents the user from inserting two documents with the same value for the indexed field. In this scenario, uniqueness DQ dimension might be as important as in RDs. However, the existence of an index in this kind of data store is not often a suitable option when huge volumes of data are being managed.

6   Defining a DQ Model

A DQ model that specifies the dimensions, metrics and data to which they apply, should be defined. In Table 1 we present an example of a DQ model for evaluating DQ of a document store corresponding to the running example.

Table 1. DQ model.

Dimension	Metric	Measured On
Consistency	**Subscription Reference Integrity** Verify the existence of the referenced subscription plan for each customer document **Granularity:** Nested Field Value	**Collection:** customers **Nested Field:** subscription.id
	Subscription Reference Consistency Verify the consistency between the subscription data embedded in each customer document and the complete subscription document in the subscriptions collection **Granularity:** Field Value	**Collection:** customers **Field:** subscription
Heterogeneity	**Customer Structure Heterogeneity** Analise the percentage of documents with each field **Granularity:** Collection	**Collection:** customers
	Customer Address Data Type Heterogeneity Analise different data types used to represent customer addresses **Granularity:** Field value	**Collection:** customers **Field:** address
Completeness	**Empty addresses** Verify if a costumer document's address field is empty (either it doesn't exist, is *null*, an empty array or an empty document) *0* if address is empty, *1* otherwise **Granularity:** Field Value	**Collection:** customers **Field:** address
Accuracy	**Telephone Format** Verify each telephone number for each customer is composed of 9 digits *0* if telephone is has wrong format, *1* otherwise **Granularity:** Array element	**Collection:** customers **Array Field:** telephones
Uniqueness	**Duplicate telephones** Count the amount of duplicate telephone numbers in the customers collection **Granularity:** Field across documents	**Collection:** customers **Field:** telephones

7   Conclusions

With the rising popularity of NoSQL databases, and document oriented stores in particular, it is surprisingly hard to find literature and studies focusing on DQ aspects for this specific type of databases. In this paper we show a first approach to analyze DQ on document oriented stores taking into account their nature, features and limitations that can determine the quality of the data and increase the risk of potential DQ issues. We are currently working on applying these concepts, defining a DQ model and measuring DQ of a real-world dataset, to demonstrate the feasibility of our approach and the type of DQ problems it can discover. At the same time, a tool is being developed to be able to define a DQ model and evaluate DQ of document stores in a more automated way.

References

1. Db-engines ranking of document stores. https://db-engines.com/en/ranking/document+store. Accessed 03 Feb 2018
2. Chodorow, K.: 50 Tips and Tricks for MongoDB Developers: Get the Most Out of Your Database. O'Reilly Media, Sebastopol (2011)
3. Dong, X., Srivastava, D.: Big Data Integration. Synthesis Lectures on Data Management. Morgan & Claypool Publishers, San Rafael (2015)
4. Firmani, D., Mecella, M., Scannapieco, M., Batini, C.: On the meaningfulness of "big data quality" (invited paper). Data Sci. Eng. 1(1), 6–20 (2016). https://doi.org/10.1007/s41019-015-0004-7
5. Juddoo, S.: Overview of data quality challenges in the context of big data. In: 2015 International Conference on Computing, Communication and Security (ICCCS), pp. 1–9, December 2015. https://doi.org/10.1109/CCCS.2015.7374131
6. Kwon, O., Lee, N., Shin, B.: Data quality management, data usage experience and acquisition intention of big data analytics. Int. J. Inf. Manag. 34(3), 387–394 (2014). https://doi.org/10.1016/j.ijinfomgt.2014.02.002
7. Sadalage, P.J., Fowler, M.: NoSQL Distilled: A Brief Guide to the Emerging World of Polyglot Persistence. Addison-Wesley Professional, Upper Saddle River (2012)
8. Scannapieco, M., Catarci, T.: Data quality under a computer science perspective. Arch. Comput. 2, 1–15 (2002)
9. Scannapieco, M., Virgillito, A., Marchetti, C., Mecella, M., Baldoni, R.: The daquincis architecture: a platform for exchanging and improving data quality in cooperative information systems. Inf. Syst. 29(7), 551–582 (2004). https://doi.org/10.1016/j.is.2003.12.004
10. Shankaranarayanan, G., Blake, R.: From content to context: the evolution and growth of data quality research. J. Data Inf. Qual. 8(2), 9:1–9:28 (2017). https://doi.org/10.1145/2996198
11. Storey, V.C., Song, I.Y.: Big data technologies and management: what conceptual modeling can do. Data Knowl. Eng. 108, 50–67 (2017). https://doi.org/10.1016/j.datak.2017.01.001

Genomic Data Management in Big Data Environments: The Colorectal Cancer Case

Ana León Palacio[1][✉] ⓘ, Alicia García Giménez[1],
Juan Carlos Casamayor Ródenas[1],
and José Fabián Reyes Román[1,2] ⓘ

[1] Research Center on Software Production Methods (PROS),
Universitat Politècnica de València, Valencia, Spain
{aleon,jreyes}@pros.upv.es, algargi2@posgrado.upv.es,
jcarlos@dsic.upv.es
[2] Department of Engineering Sciences, Universidad Central Del Este (UCE),
San Pedro de Macorís, Dominican Republic

Abstract. If there is a domain where data management becomes an intensive Big Data issue, it is the genomic domain, due to the fact that the data generated day after day are exponentially increasing. A genomic data management strategy requires the use of a systematic method, intended to assure that the right data are identified, using the adequate data sources, and linking the selected information with a software platform based on *conceptual models*, which allows guaranteeing the implementation of genomic services with *quality, efficient* and *valuable* data. In this paper, we select the method called "SILE" –for Search, Identification, Load and Exploitation-, and we focus on validating its accuracy in the context of a concrete disease, the *Colorectal Cancer*. The main contribution of our work is to show how such methodological approach can be applied successfully in a real and complex clinical context, providing a working environment where Genomic Big Data are efficiently managed.

Keywords: SILE · Genomics · Big Data · Data quality · Colorectal cancer

1 Introduction

Since the completion of the Human Genome Project and the appearance of the next-generation sequencing (NGS) techniques, the sequencing cost per genome (i.e. reading the DNA of a person) has been exponentially decreasing, as well as the speed of the sequencing has been increasing [1]. Technological innovation combined with automation are responsible for the explosion in data production [2]. The postgenomic era in which we currently are, requires specific strategies for the interpretation, analysis and exploitation of large amounts of data.

Veracity is an important factor due to the fact that data comes –in some cases- from uncontrolled environments because of the lack of rules to submit the information. Moreover, in genomics, data *heterogeneity*, *variability* and *variety* are the main

C. Woo et al. (Eds.): ER 2018 Workshops, LNCS 11158, pp. 319–329, 2018.
https://doi.org/10.1007/978-3-030-01391-2_36

problems because they can derive in inconsistencies of the information managed. Thus, it is fundamental to identify the accurate and relevant data. The variability dimension is especially relevant for our work, as the problem of identifying the right data in the genomic domain depends on the disease under analysis. This is why applying the appropriate data selection method for one disease can be substantially different from applying it to other diseases. To understand analogies and differences, experiences from real cases related to different diseases are required. In this context, this paper reports the *colorectal cancer* (CRC) case, analyzing the methodological background that has been applied to identify those variations (changes in the DNA) that are considered relevant for the genomic diagnosis of CRC.

The challenge of our genome-oriented data management is to discard useless data, collect relevant information and turn it into knowledge. By identifying significant variations and loading them in the corresponding database, our work is intended to allow an early detection of the selected disease, as well as help the development of specific treatments to reduce health care costs and motivate one of the potential uses of Big Data in health care, i.e. *Personalized Medicine* [3].

There are multitude of public databases where scientists can extract information to validate the results of their experiments. For example, in its last update, the NAR Online Molecular Biology Database Collection summarizes information about 1,737 repositories [4]. This huge number of potentially valid data sources conforms a first problem: how to select those data sources that are considered relevant for the disease under investigation. Some criteria are strictly required: this is why we are going to focus on the *currency, accuracy, completeness* and *reliability* of the available information. The main contribution of this paper is to provide a systematic approach to assist the population of a genomic database with *useful, curated, verified* and *structured* data by applying a method called SILE and validating its accuracy in this specific context. As an additional value related to further work, doing this task in detail for a particular disease and reporting accurately the subsequent process details, it will be possible to analyze how the method works by comparing different instantiations of the used method when applied to different diseases.

With all this in mind, the following section presents a general view of the SILE method, that conforms the selected methodological background. In Sect. 3 the genetic causes of the CRC will be introduced. Next, in Sects. 4 and 5, the steps of the SILE method will be followed for the CRC case: searching relevant data sources, identifying the significant variations, loading the data into a specific database and using the *GenesLove.Me* (GLM) [5] platform for its interpretation. Finally, lessons learned, and future work are presented in the last section.

2 The SILE Method

The SILE method, developed within PROS Research Group[1], is composed by the following four steps: *search, identification, load* and *exploitation*. The aim of this method is to identify efficiently relevant genetic information (variations[2]) about the risk of suffering a certain disease, in this case CRC, and load them into the internally developed Human Genome Database (HGDB) [6]. Next, each step of the method is briefly explained.

2.1 Search

The first step of this method is based on exploring the different scientific sources (*articles, databases,* etc.) related to the disease. The goal is to select those repositories containing relevant information for the disease under study. In order to determine the optimal ones to obtain information from, the sources must be analyzed, according to some quality criteria, called dimensions. Three main dimensions are considered in this step:

- **Currency:** the extent to which data is sufficiently up to date for the task at hand. Genomics domain is in constant evolution, so the information stored in the databases must be frequently updated in order to maintain the data quality (less than one year before the last update).
- **Accuracy:** the extent to which data correctly describes the "*real world*" objects. For example, the DNA sequence is a four-letter code which contains A, C, T or G, no other letters. It is important that the selected databases use standards to represent the information, such as the HGVS nomenclature[3] for the name of a variation.
- **Completeness:** the extent to which data are in sufficient depth and breath for the task at hand. In order to obtain information as complete as possible, different databases must be checked, such as those specific of a disease, as the Colorectal Cancer Atlas[4], and more general ones such as the data sources of the National Center for Biotechnology Information (NCBI[5]).

The application of these criteria guides the process of selecting those data sources that will allow us to proceed with the next step: the precise identification of relevant variations for clinical genomic diagnosis purpose.

---

[1] Research Center on Software Production Methods (PROS), http://www.pros.webs.upv.es/.

[2] Variation: Naturally occurring genetic differences among organisms in the same species [*Scitable by Nature Education*].

[3] HGVS Nomenclature: http://varnomen.hgvs.org/.

[4] Colorectal Cancer Atlas: http://colonatlas.org/index.html.

[5] The National Center of Biotechnology Information: https://www.ncbi.nlm.nih.gov/.

2.2   Identification

Using the data sources selected in the previous step, the relevant variations must be identified. One of the problems to face when dealing with genomic information is the lack of a standard to define the core concepts, which leads to inconsistencies and discrepancies between data sources. The *Conceptual Schema of the Human Genome* (CSHG) [7, 8] was specifically developed to describe in a unified and structured way the main concepts of the domain. The use of the CSHG helps to describe these concepts in a non-ambiguous way and to determine the main attributes to consider when searching for information about a certain topic. In addition, it is important to highlight that the application of conceptual modeling techniques in this domain (complex) is essential to achieve information management based on data quality.

The meaningful information about the relevant variations is gathered from the selected databases according to the specification of the CSHG. The use of the conceptual schema determines which attributes are required, and it guides the mappings that must be specified to link conceptual schema attributes with their data sources attribute counterpart.

Once the information required from each source is identified, specific quality criteria controls must be proposed in order to assure a correct integration of the information. Three additional criteria are used in our work in this step: accuracy, completeness and reliability.

- **Data Accuracy** means that the data of a field is represented using the correct format and its value is correct. For example, it cannot be possible the existence of a chromosome 50 because there is a specific number of chromosomes from 1 to 22 and the sexual ones (X and Y).
- **Completeness** is related to the presence of missing values. Some attributes can be missing, but others are necessary to provide a consistent representation of the data, e.g. the position in the DNA where the variation occurs.
- **Reliability** is another important quality control. The reliability of a variation is related to the type of reports associated with it and the status of review. This is fundamental criteria to determine the significance of a variation when talking about its clinical application.

The aforementioned criteria are assured by the restrictions described in the conceptual schema, which guarantees the consistency of the data representation

2.3   Load

The Load step implies the extraction of information from the databases and its storage into the *Human Genome Data Base* (HGDB), the database that corresponds to the CSHG. In order to accomplish this task, three more sub steps are performed: *data extraction, transformation* and *load*.

First of all, the information of interest is extracted from the databases. Then, that information is transformed by making the correspondence between the fields in the datasets and the fields in the HGDB. Besides the controls applied in the identification step, other quality controls must be taken into account:

- **Consistency** assess if that data coming from the implied data sources are consistent and represented in the same format. It is not an easy task because most genomic databases where developed in an ad-hoc way to solve a specific research requirement and the use of ontologies and controlled vocabularies is not a common practice. This can lead to situations where a common –*semantically speaking*- data unit can appear with different formats. The HGDB fixes the target format to be used, according to the CSHG where it comes from.
- **Uniqueness** is related to the fact that the database will not have redundant data or duplicate records.

2.4 Exploitation

Once the right data are stored in the database, a software platform to manage them for genome diagnosis purposes is ready to be used. This *"Exploitation"* step is the final one of the SILE method and it is focused on the accurate interpretation of the stored data. The information that has been loaded and stored in the HGDB is transformed into useful knowledge, which can be exploited by a specific tool, i.e. *GenesLove.Me* (GLM) [9]. This tool analyses the information obtained from a patient sample and determines if there are variations associated to a certain disease, according to the data stored in the database.

3 The Colorectal Cancer Case

After having introduced the SILE method, it is time to evaluate how it works in practice when applied to a concrete case –*the Colorectal Cancer use*-. Two main objectives are pursued:

1. to check the value of the method in the selected case,
2. to create a set of practical experiences of the SILE method in particular diseases, in order to be able to compare in the next future commonalities and differences in the application of the method to different diseases.

The paper focuses on the first point, to make viable to face the second one in further works. CRC is the chosen disease to validate the SILE method. The main reasons are associated to the high prevalence of the hereditary form, the knowledge of the genes involved in the disease and its difficulty to be diagnosed. The study of this disease by genome variation analysis focusing on an early detection is fundamental in the diagnosis of new cases. It is necessary because the prognosis[6] of the disease is better as soon as the cancer is detected. All these aspects make the CRC case a very relevant one

---

[6] Prognosis is defined as the likely course of a medical condition.

for our analysis of validity of the SILE method in practice. Our aim is to determine relevant variations related to the disease that could be used for its effective genomic diagnosis in the advanced context of Precision Medicine.

4   Searching Relevant Data Sources

As seen previously in the Search step of the SILE method, the valuable information is dispersed over several data sources. Moreover, data does not always match due to inconsistencies or incongruences, what makes necessary to study each data source and its associated information carefully.

In the context of CRC, the databases DisGeNET [10], ClinVar [11], Ensembl [12], InSiGHT, Phenotype-Genotype integrator [13] and dbSNP [14] have been selected to obtain information from.

From the CSHG, a set of relevant attributes are selected (See Table 1). Each attribute of the CSHG must be linked to a corresponding attribute of the selected data source that will provide its content.

Table 1. An extract of the attributes and information contained in each data source.

Identifier	Description	Data Source
Official symbol	Gene official symbol	HGNC
Allele ID	Allele identifier	ClinVar
Variation ID	Variation identifier	ClinVar
"*rs*" identifier	dbSNP link and identifier	dbSNP
Clinically important	Clinical effect of the variation	ClinVar
Review status	Status of review of the variation	ClinVar
Evaluation	Last evaluation of the variation	ClinVar
Citations	Number of citations	All databases
VAR	Variation-disease association score	DisGeNET
EI	Evidence index	DisGeNET
MAF	Minor Allele Frequency	Ensembl
HPMAF	Highest population MAF	Ensembl
Population	Population of Highest population MAF	Ensembl

Let us close this section remarking that well-known, specific CRC databases as *InSiGHT, the Colon Cancer Gene Variation Databases*[7] and *the Colorectal Cancer Atlas* were not unfortunately be considered because they are outdated, what violates one of our basic quality criteria (i.e., currency).

---

[7] CCGVD, http://chromium.lovd.nl/LOVD2/colon_cancer/home.php.

5 Identifying Significant Variations

The data obtained from the different sources in the previous step were analyzed in order to eliminate redundancies and corroborate the quality. After the interpretation and annotation of the variations from NGS data, they are commonly classified as *benign*, *likely benign*, *pathogenic* or *likely pathogenic*, or as a *variation of unknown significance* (VUS) following the American College of Medical Genetics and Genomics (ACMG) guidelines [15]. As a first applied criteria, only pathogenic or likely pathogenic variations were considered for the final dataset. Then, high Minor Allele Frequencies (MAF) and low Functional Consequence Score (FCS) were used as negative criteria. Variations with a high MAF levels were discarded because it indicates a common variation, and a low FCS indicates a variation without a deleterious effect. A variation with a FCS lower than 0.65 was discarded, variations with a FCS between 0.70 and 0.65 were exhaustively studied and finally, variations with a FCS between 0.9 and 1 were considered pathogenic. Moreover, the prediction of PolyPhen and SIFT was considered. The population genetics is also studied because some variations are common in a population and uncommon in another or can be present in a population and absent in other. Finally, a restrictive p-value was also used in order to select the strength phenotype associated variations. A lower P-value is related to a high level of association, there is an inversely proportional association. Other applied criteria are the status of review, the clinical significance of the references, the absence of discrepancies among data bases, the specificity against the disease -in this case, the CRC-, or pathways to tumorigenesis, and the related bibliography.

The final dataset consists in 34 pathogenic variations. This conforms the most valuable contribution of this work. The application of a sound method leads to the selection of this set of 34 valid variations, ready to be use in clinical contexts for an accurate genomic diagnosis of the disease.

It is important to remark that the result is dependent on the criteria that were applied. But the strength of the method is that we know why the selected variations are these ones and not others. We also know that changing the selected criteria, a different result would be obtained. But the application of the SILE method always let us precisely know why the selected variations are the ones that conform the final dataset. A summary of some of the selected variations and their associated attributes is shown in Table 2.

This result must be open to change and adaptation to the context. Indeed, 105 variations are classified as likely useful in order to be considered in the near future: they are ready to be studied again when more evidence appears. Additionally, there are some variations, as rs555607708, that have potential to become a clinically relevant variation, but they are too recent to be sure.

Table 2. Selection of annotated variations. For each variation, "*rs*" identifier, gene symbol, chromosome, position, HGVS name, reference and alternative allele, variation type, functional consequence score (FC) and minor allele frequency (MAF) are shown. Del: Deletion; SNV: single nucleotide variations; Dup: Duplication; Chr: chromosome.

rs identifier	Gene	Chr	Position	NC identifier	Ref	Alt	Type	FCS	MAF
rs193922376	MSH2	2	47414421	NC_000002.12	A	T	SNV	0,95	0,02
rs267608172	PMS2	7	5982823	NC_000007.14	C	T	SNV	0,95	<0,01
rs63750909	MSH6	2	47799427	NC_000002.12	C	T	SNV	0,95	<0,01
rs267606674	AXIN2	17	65536467	NC_000017.11	C	–	Del	0,95	<0,01
rs587780078	MUTYH	1	45331515	NC_000001.11	–	CC	Dup	0,95	<0,01
rs63751615	MLH1	3	37012098	NC_000003.12	C	T	SNV	0,95	<0,01
rs16892766	EIF3H	8	116618444	NC_000008.11	A	C	SNV	0,6	0,08

6 Data Load and Interpretation: The GenesLove.Me Platform

Once the meaningful information has been identified it is the time to load it in the database. It acts as the main component of the software platform that must perform the genomic analysis by detecting if the relevant variations are present in an analyzed DNA sample of a patient. This is when the genome information system becomes a valuable asset, having the right data for diagnosis purposes thanks to the methodological background that the use of the SILE method provides.

In our case we use the HGDB to store the variations identified related to the CRC, and the GLM platform to present the results to the user according to its specific interest. The load process is performed following three main steps:

1. Extraction of the information from the sources: From each data source the required attributes have been identified so, using the mechanisms provided by each repository, an automatic process has been implemented to extract the associated information.
2. Transformation of the data collected to match the quality criteria established and to solve the possible inconsistencies that can hinder the load process.
3. Load the information into the HGDB.

When the HGDB is populated with the desired information, it can be exploited by proper tools to extract knowledge that can be presented to the final user. In this case we use the GLM platform [5, 9]. This tool allows the user to know the risk of suffering CRC by processing a VCF file that contains the variations detected after sequencing his/her DNA. After the VCF file is processed, the user receives a report with their probability of suffering the disease, according to his genetic characteristics.

7 Conclusions

In a very complex data management domain as the genomic one is, the correct identification of specific DNA variations and the efficient use of Big Data analysis will
enhance our understanding of the molecular basis of diseases. In this context, a
methodological background is essential to guide the process of selecting the right
genomic data sources, identifying the correct variations and interpreting them adequately in the resulting genome analysis. This paper shows how the SILE method has
been applied in a real case associated to the CRC disease.

CRC is the third most prevalent cancer, so the creation of a variation database using
the SILE method helps to improve its diagnosis and comprehension. The method
provides an easy way to perform the analysis of the big amount of available data-bases
and related information. Besides, the use of quality criteria improves the reliability of
the information stored in the final database and confers a sound structure to the whole
process. The complexity of the colorectal study is related to the great variety of syn-
dromes and genes affected and the huge amount of variations. Moreover, as well as in
other types of cancer, the risk of develop the disease depends on specific combinations
of variations. The SILE method was designed having these dynamic aspects in mind. It
means that the application of different quality criteria can generate new results, assuring
the semantic traceability that makes possible to known why the selected variations were
chosen instead of others.

Validating the accuracy of the SILE method in the context of Colorectal Cancer
(CRC) allows scientific community to reproduce it in other complex diseases in order
to create a *useful, curated, actualized* and *promising* database. This is indeed the most
significant further work that we want to do. Our goal is to replicate the use of the
method in different clinical contexts (considering other diseases) in order to accumulate
knowledge on how the application of the method is dependent on the selected disease.
We have results of the application of the method in the case of the Alzheimer's disease,
and we are currently analyzing commonalities and differences that should help to
improve the method application.

Additionally, the SILE method is being applied in two concrete clinical projects. In
the first one, developed together with clinical experts from *"Fundación INCLIVA-
Hospital Clínico"* in València, the selected disease is *breast cancer*. The method is
used to manage data related to micro-RNAs with the aim of extending its application to
other types of genome information. The second application of the SILE method is
being done in the context of an international project with the *"Hospital Nacional del
Cáncer"* and *"Hospital de Clínicas"* in Paraguay. In this case, the analysis focuses on
lung cancer with the aim of extending the conventional Electronic Health Record with
genomic data. The SILE method will be also applied to search and identify the relevant
variations to be considered in this extension, following a similar process in a different
clinical working context.

In all these cases the results of the method application are being successful thanks
to the combination of conceptual modeling and the application of quality metrics,
which ensure the *data quality*, it is for this reason that more applications are planned in
order to extend the HGDB with more information about other diseases.

As a major advantage, even though there is a lot to be discovered and understood, we accumulate evidences that let us to perceive how the SILE method can be used to unravel the variations related to an increased risk of developing a disease.

Acknowledgement. The authors would like to thank members of the PROS Research Centre Genome group for the fruitful discussions regarding the application of CM in the medicine field. This work has been supported by the Spanish Ministry of Science and Innovation through project DataME (ref: TIN2016-80811-P) and the Research and Development Aid Program (PAID-01-16) of the Universitat Politècnica de València under the FPI grant 2137.

References

1. van Dijk, E.L., Auger, H., Jaszczyszyn, Y., Thermes, C.: Ten years of next-generation sequencing technology, Trends Genet. **30**(9), 418–426 (2014). https://doi.org/10.1016/j.tig. 2014.07.001
2. Auffray, C., et al.: Making sense of big data in health research: towards an EU action plan. Genome Med. **8**, 71 (2016). https://doi.org/10.1186/s13073-016-0323-y
3. Wylie, B., Psaty, B.M.: Personalized medicine in the era of genomics. Jama **298**(14), pp. 1682–1684 (2007). https://doi.org/10.1001/jama.298.14.1682
4. Rigden, D.J., Fernández, M.X.: The 2018 Nucleic Acids Research database issue and the online molecular biology database collection. Nucleic Acids Res. **46**(D1), D1–D7 (2017). https://doi.org/10.1093/nar/gkx1235
5. Reyes Román, José F., Iñiguez-Jarrín, Carlos, Pastor, Óscar: Genomic Tools*: web-applications based on conceptual models for the genomic diagnosis. In: Damiani, Ernesto, Spanoudakis, George, Maciaszek, Leszek (eds.) ENASE 2017. CCIS, vol. 866, pp. 48–69. Springer, Cham (2018). https://doi.org/10.1007/978-3-319-94135-6_3
6. Reyes Román, J.F.: Diseño y Desarrollo de un Sistema de Información Genómica basado en un Modelo Conceptual Holístico del Genoma Humano. Universitat Politècnica de València (2018). https://doi.org/10.4995/Thesis/10251/99565
7. Reyes Román, J.F., Pastor, Ó., Casamayor, J.C., Valverde, F.: Applying conceptual modeling to better understand the human genome. In: Comyn-Wattiau, I., Tanaka, K., Song, I-Y., Yamamoto, S., Saeki, M. (eds.) ER 2016. LNCS, vol. 9974, pp. 404–412. Springer, Cham (2016). https://doi.org/10.1007/978-3-319-46397-1_31
8. López, Ó.P., Palacio, A.L., Román, J.F.R., Casamayor, J.C.: Modeling life: a conceptual schema-centric approach to understand the genome. In: Cabot, J., Gómez, C., Pastor, O., Sancho, M., Teniente, E. (eds.) Conceptual Modeling Perspectives. Springer, Cham (2017). https://doi.org/10.1007/978-3-319-67271-7_3
9. Reyes Román, J.F., Iñiguez-Jarrín, C., Pastor López, O.: GenesLove.Me: a model-based-web-Application for direct-To-consumer genetic tests, In: ENASE 2017 - Proceedings of the 12th International Conference on Evaluation of Novel Approaches to Software Engineering, pp. 133–143 (2017). https://doi.org/10.5220/0006340201330143
10. Piñero, J., et al.: DisGeNET: a comprehensive platform integrating information on human disease-associated genes and variants, Nucleic acids research, gkw943 (2016). https://doi.org/10.1093/nar/gkw943
11. Landrum, M.J., et al.: ClinVar: improving access to variant interpretations and supporting evidence. Nucleic Acids Res. **46**(D1), D1062–D1067 (2017). https://doi.org/10.1093/nar/gkx1153

12. Zerbino, D.R., et al.: Ensembl 2018. Nucleic Acids Res. **46**(D1), D754–D761 (2017). https://doi.org/10.1093/nar/gkx1098
13. Ramos, E.M., et al.: Phenotype–Genotype Integrator (PheGenI): synthesizing genome-wide association study (GWAS) data with existing genomic resources. Eur. J. Hum. Genet. **22**(1), 144 (2014). https://doi.org/10.1038/ejhg.2013.96
14. Sherry S. T., Ward M-H, Kholodov M., et al.: dbSNP: the NCBI database of genetic variation. Nucleic Acids Res. **29**(1), 308–311 (2001)
15. Richards, S., et al.: Standards and guidelines for the interpretation of sequence variants: a joint consensus recommendation of the American College of Medical Genetics and Genomics and the Association for Molecular Pathology. Genet. Med. **17**, 405–424 (2015). https://doi.org/10.1038/gim.2015.30

Domain Specific Models as System Links

Vladimir A. Shekhovtsov[1] , Suneth Ranasinghe[1] ,
Heinrich C. Mayr[1(✉)] , and Judith Michael[2]

[1] Alpen-Adria-Universität, Klagenfurt, Austria
{volodymyr.shekhovtsov,suneth.ranasinghe,
heinrich.mayr}@aau.at
[2] Software Engineering, RWTH Aachen University, Aachen, Germany
michael@se-rwth.de

Abstract. Digital Ecosystems consist of a variety of interlinked subsystems. This paper presents a flexible approach to define the links between such subsystems. The idea is to exploit the paradigm of Model Centered Architecture (MCA) and to specify all links/interfaces by means of appropriate Domain Specific Modeling Languages. The approach has been successfully applied and evaluated in several projects. As a proof of concept, we present the model-based interfacing between assistive systems and human activity recognition systems, which showed good performance as needed in real-world applications.

Keywords: Model Centered Architecture · Domain Specific Modeling
Digital ecosystem · Language hierarchies · Interfacing

1 Introduction

From a model centered perspective, any kind of information managed and/or processed by a part of a digital ecosystem (DEC) is an instance of an explicitly specified or implicitly underlying model. The same is true for the processes as well as for the models themselves [6], the latter being instances of metamodels. Consequently, we may see any DEC as a construct consisting of model handlers (consumers and/or producers). This leads to the paradigm of *"Model Centered Architecture (MCA)"* [13], independently from the nature of the handlers that may be digital or living ones.

MCA is a generalization of *Model Driven Architecture (MDA)* and *Model Driven Software Development (MDSD)* [11], as well as *models@runtime* [2]. Like multilevel modeling, MCA advocates, for any system aspect, the use of (possibly recursive) hierarchies of *Domain Specific Modeling Languages (DSML)*, each embedded into a *Domain Specific Modeling Method (DSMM)*. I.e., MCA focuses on models (and their metamodels) in any design and development step up to the running system.

This paper deals with an important aspect of MCA, namely the model-based linking of digital ecosystem parts. We illustrate this by an example taken from the domain of supportive systems in Active and Assisted Living (AAL) [22]. Such systems need as complete information as possible about a person's activities (in the sense of sequences of actions) and the context, in which these activities take place. Humans and so-called Human Activity Recognition (HAR) systems may provide such information. However, although the effectiveness of the latter has significantly improved in recent years [20], most of the accessible HAR systems are either still in project state or have a rather restricted sphere of recognition like fall incidents, anomaly behavior detection in crowded places, traffic monitoring etc. I.e., the stand-alone use of such systems does not meet the needs of comprehensive assistive systems.

Combining several (specialized) HAR systems, each of them covering a particular sphere, seems to be promising, as in combination the HARs could deliver the necessary recognition results. Such a combination requires a smooth coupling with the AAL component to manage whatsoever any output structure and integrate new and upcoming HAR systems. We show that the MCA paradigm provides the means for such a flexible, scalable and transparent integration.

The paper's organization is as follows: Sect. 2 outlines the main MCA concepts and discusses the variety of domain specific modeling and representation languages coming into play. Section 3 discusses the general aspects of model-based component linking. In Sect. 4 we present the proof of concept by means of the integration of AAL ecosystem components. Section 5 sketches related work. The paper concludes with an outlook on future research.

2   MCA: Concepts and Language Hierarchies

MCA treats all processes as well as the data they process (MOF[1] level 0) as instances of models (MOF level 1). These models in turn are instances of metamodels (MOF level 2), described using particular DSMLs, and represented using corresponding domain specific representation languages. Consequently, all system interfaces are defined through models as well, and the system components have the role of model handlers. Figure 1 sketches the related MOF hierarchy (for simplicity we omit here the more complex view when using a multilevel modeling approach). Note that in the case of DECs there might be not only various interface metamodels but also several application metamodels, and consequently several DSMLs.

The M2 interfaces support the management of metamodels and the integration of external metamodels (metamodel exchange). On M1 (model level) the M2 metamodels are instantiated for a concrete application situation. On M0 (instance level), the application itself results from creating extensions of the M1 models.

If the application domain metamodels are comprehensive in the sense of providing concepts for structure, dynamics and functionality, the M0 extensions form the models@runtime, which might be handled by an interpreter (orchestrated by M2, visualized by the arrows from M2 to M0). Alternatively, a MDA tool could generate code from the M1 models (static, dynamic and functional).

In parallel to the model hierarchy, we have to provide languages for representing the semantic artifacts (the models) by appropriate syntactic artifacts. These representation languages again form a hierarchy of three levels: (1) *Grammar definition level* (top level): contains the means of defining the language grammars. In our research, we

---

[1] Meta-Object Facility™ http://www.omg.org/mof/.

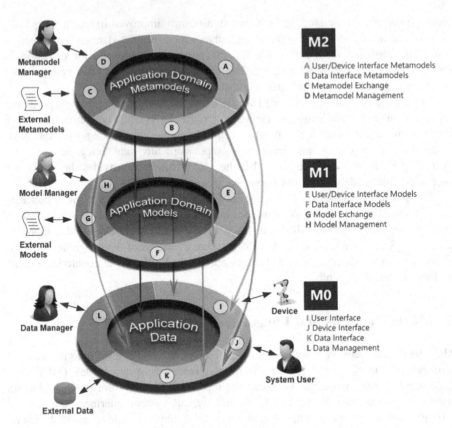

Fig. 1. MCA model hierarchy

use a specific version of EBNF, compatible with the ANTLR grammar definition language. (2) *Language definition level:* defines grammars for the representation languages (RL) related to the defined DSMLs: meta-metamodel RLs, metamodel RLs, model RLs and instance/data RLs. (3) *Language usage level:* representations of the models of all levels.

3   Linking Model-Centered Architectures

MCA proposes the explicit model-based definition for both, user interfaces and interfaces between technical parts of a DEC. This means that appropriate DSMLs and representation languages (mostly subsets of natural language in the case of user interfaces) have to be specified.

There are three general types for linking two systems:

A: Two non-MCA systems ("black boxes"): In this case, no model definitions are known or accessible. Therefore, we have to define (1) a DSML such that the outputs of both systems can be described as instances of models of that DSML, and (2) mappings between the particular models driving the conversion on M0 (if the models are not identical).

B: Black box system and MCA system: If the outputs of the former can be modeled using the DSML of the latter, we could proceed according to A) for the black box system and provide the appropriate model mapping.

C: Two MCA systems: In this case, we only would have to provide the model mapping if the models are not identical.

These types also apply for the other links appearing in a digital ecosystem: E.g., user interfaces and device interfaces are of type B, where the users and devices are seen as black boxes. External data and model interfaces are of type C, as in this case the particular (data) models are expected to be accessible.

The example given in Fig. 2 sketches such links between Systems 1, 2 and 3, all being of type C. The other links (to users and managers) are of type B, as well as the link between System 1 and a device (robot).

Linking may materialize on all MOF levels. We distinguish from top to down:

- M2 level (*metamodel link*): uses the same metamodel but different models. This implies the need of defining appropriate mappings that drive the translation between the particular representation languages. Note that linking based on different meta-models using different DSMLs for one link might be possible, but will not be discussed here.
- M1 level (*model link*): uses the same models.
- M0 level (*data link*): uses the same model instances (possibly represented using different representation languages, which makes conversion necessary).

4 Example: Integrating the Components of an AAL Ecosystem

4.1 Metamodel and Languages

We introduce a metamodel-based HAR (Human Activity Recognition) interface for connecting arbitrary heterogeneous HAR systems in an AAL ecosystem. Such interface has to interpret HAR systems' outputs and deliver appropriately converted data to a given AAL system. Moreover, it has to be adaptable to new HAR semantics in order to be sustainable. As pointed out in the introduction, such 'multi HAR system' approach might increase the recognition realm as needed by an AAL support system.

Our approach is to link a *consumer MCA AAL system* with several *blackbox HAR systems* on M2 (type B). For that purpose, we introduce two link DSMLs that cover the recognition outputs of at least the HAR systems known to us:

- Activity Recognition Environment Modeling Language (AREM-L, metamodel: Fig. 3), a visual conceptual modeling language for describing recognition structures as conceptual models; this language covers both, basic and instrumental (complex) human activities including their context [10, 15].
- Activity Recognition Instance Specification Language (ARIS-L), a textual language for M0 level representations of concrete recognition objects.

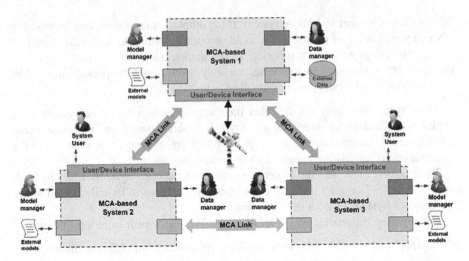

Fig. 2. MCA ecosystem

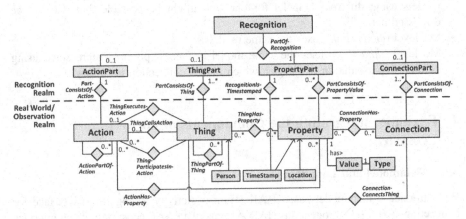

Fig. 3. HAR interface metamodel concepts

The main concept of the AREM-L metamodel is *Recognition* being an aggregate of four sub-concepts: (1) *ActionPart*: contains *Actions* which conceptualize simple or complex activities; (2) *ThingPart*: contains recognized *Things*, i.e. persons or contextual objects; *Person* is defined as a specialization of Thing; (3) *PropertyPart*: contains *Properties*, each having a *Value* of a specific *Type*; (4) *ConnectionPart*: contains *Connections* between Things.

Except from the PropertyPart (containing a mandatory *TimeStamp* Property), the other sub-concepts are optional, i.e., a Recognition may consist of the PropertyPart and none, one, or more of the others.

Things are involved in Actions: they may (1) execute an action, e.g. a coffee machine brews a coffee; this is reflected by the relationship *ThingExecutesAction*, (2) call an action (e.g. a Person pushing the 'brew' button); this is reflected by the relationship *ThingCallsAction*, (3) passively participate in an action (e.g., the coffee in the brewing action); this is reflected by the relationship *ThingParticipatesInAction*.

Actions, Things, and Connections may have Properties. To cover complex situations, Things and Actions can hierarchically be decomposed into structures of Things and Actions, respectively. This is reflected by the corresponding part-of relationships. Location is a specific Property to cover the information necessary for locating Things.

Summing up, an instance of this metamodel, i.e. a model, will describe the concepts underlying the output of a specific HAR system.

AREM-L is a visual modeling language; Fig. 4 shows its basic elements. We waive here a complete specification of the AREM-L syntax, as it should be intuitively understandable based on the example model in Fig. 5a. It represents a type of observation that a HAR system in a smart home environment might make: A person who put objects (e.g. keys) into a container (e.g. a handbag), as step in a "go to shopping" activity.

ARIS-L serves to represent the instances of AREM-L recognition models, i.e., such instances are models of concrete recognitions. Consequently, our goal was to allow for machine-readable representations of recognition data, that are easy to parse and efficient to process. The ARIS-L syntax is inspired by JSON, but includes specific constructs corresponding to the elements of AREM-L. The main rule of ARIS-L is as follows: all data belonging to a specific recognition is encapsulated inside the corresponding rRecognition construct.

Figure 5b shows an ARIS-L fragment for an instance of the recognition model depicted in Fig. 5a. The names of AREM-L model elements precede the representation of their instances' data. These names, in turn, can be preceded optionally by the corresponding concept names (Action, Thing, Property, Connection). Figure 5b illustrates these options by omitting the keyword Thing in the container section, adding this keyword in the object section and providing the keyword Person in the "Observed Person" section. The ConnectionPart shows, as an example, the observed connection "in" between the things "main keys" and "red handbag", which is instance of the model element "in". The links of the ThingPart components to the Action put (i.e., ParticipatesIn, Executes) are instances of the corresponding model element links.

Again, we have to waive presenting the entire grammar.

4.2 Implementation Approach

Our AAL ecosystem consists of (1) a set of HAR systems capturing user's activities (through sensors or by other means), (2) the AAL consumer system providing service to the user based on the captured activities; (3) the HAR interface between them.

According to the classification provided in Sect. 3, all links are of type B, the HAR systems are black boxes and the AAL system is a MCA system. The AREM-L models are used as semantic references for the conversion of HAR output to ARIS-L in the interface, and for parsing ARIS-L in the consumer system. On the data link level, HAR systems' output data is converted into ARIS-L representation.

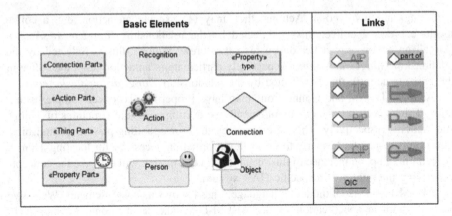

Fig. 4. AREM-L visual modeling elements

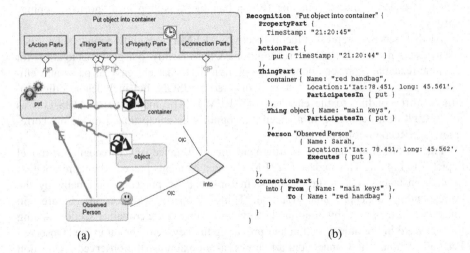

(a)   (b)

Fig. 5. AREM-L (a) and ARIS-L (b) models for the sample recognition

The steps for implementing the HAR interface were as follows:

- *Implementing AREM-L* by deploying the ADOxx meta modeling framework [7] and for generating the AREM-L modeling tool.
- *Implementing the AREM-L parser,* which serves for converting AREM-L models into the internal HAR interface representation.
- *Implementing ARIS-L* by defining the ARIS-L grammar using the ANTLR framework.
- *Implementing the ARIS-L parser* to be included into the consumer system.
- *Implementing the MCA-Links* (see Fig. 2). They consist of (a) a common core, which uses the AREM-L model and the ARIS-L grammar to convert data into ARIS-L representations to be further processed by the model handlers of the consumer system, and (b) a HAR-specific (black box) plug-in converter for each HAR

to be integrated. Such converter transforms HAR output data for being processible in the common core.

- *Configuring the interface.* Creating AREM-L models, making components accessible to each other via the network etc.

The most technically challenging and time-consuming tasks are implementing the AREM-L and the ARIS-L parser, and the MCA-Link common core. Note, that these components are a kind of standard in the sense that they have to be implemented only once and may be reused for any other HAR and consumer system combination. In contrast to that, the problem-specific tasks (e.g., creating the AREM-L model for the given combination) are straightforward and require less technical knowledge except from the concrete HAR setting. I.e., our approach substantially reduces the effort of linking HAR systems.

4.3 Proof of Concept

For a proof of concept, we used an experimental lab, established in the context of a funded research project. As at that time there was no comprehensive HAR system in product stage available, we developed some experimental systems ourselves, namely (1) a sensor-based system using a Nimbits[2] server as a central point for sensor communication, (2) a video-based HAR (V-HAR) system, which uses a technique from [9] to generate semantic descriptions of video frames in text form, and (3) a HAR simulator for providing environmental data.

As a consumer system, we used HBMS [16], an ambient assistance system supporting daily life activities. The HBMS system abstracts, aggregates and integrates the observed behavior data into an individual Human Cognitive Model (HCM) [12], and assists the supported person via a multimodal interface by retrieving knowledge from his/her HCM. The HBMS input consists of sequences of recognized behavioral actions: in our setup, they are coming from the HAR interface. The "observation engine" of the HBMS system integrates these sequences into HCM as behavioral clusters based on reasoning algorithms that deduce the goal of a sequence of actions.

The MCA-Links to the experimental HARs were developed following the approach described in Sect. 4.2. The resulting ecosystem has been tested against various criteria, involving several user groups in different settings. It showed good performance regarding both, response time and recognition accuracy, as needed in real-world applications.

To sum up, following our approach one has to spend significant effort only for implementing standard components like common core and parser parts. Adding new HAR systems causes no additional effort of this kind: only a simple converter has to be implemented for each system. Providing AREM-L models and configuring the link are high-level activities requiring mainly domain knowledge. Generating the parsers' core based on the language grammar specification by means of the ANTLR environment brings further effort reduction. In contrast to that, following the classical approach would require implementing a separate low-level data converter for each new HAR and

---

[2] https://www.nimbits.com/.

embedding it into the consumer system. No semantics would then be intrinsically associated with the incoming data and the conversion process.

5 Related Work

Related Work on Model-Based System Linking. The current body of research on model-based system linking deals with applications of model@runtime and MDA paradigms to this problem. In particular, on-the-fly interoperability for the systems based on the models@runtime paradigm is a topic of [1]. [3] provides a solution for automated synthesis of mediators to support component interoperability. An aspect of the model exchange while coupling systems is treated in [8]. An example of using MDA for establishing system links is [5], which introduces the concept of model-driven domain-specific middleware. The difference to our approach is that the above techniques apply models in a limited way (e.g. only at development time for MDA) without providing an integrated model-centered solution.

Related Work on Universal HAR Interfaces. The need of a generic interface to handle heterogeneous data coming from HAR [4] has been discussed since long. [21] presents a survey of ontological approaches to this problem. [17] shows a specific example of using foundational ontologies underlying the middleware for smart homes. Generic knowledge-based approaches are presented in [14], proposing a knowledge-driven framework for context-aware activity recognition, and in [23], introducing an architectural solution for cognitive sensing of activities. Some research deals with the specific categories of inputs, e.g. [18] describes how domain and contextual knowledge can be utilized for human activity recognition in video streams. Despite the above research, a commonly recognized standard or language for HAR interoperability has not been proposed to date [19] and few practical solutions exist.

6 Future Research

There is still a couple of research questions to answer and challenges to meet. For example, a comprehensive semantic grounding of the languages and models requires an appropriate hierarchy of ontologies. For practical purposes in situations where different kinds of components are to be linked in a digital ecosystem, a modeling/development framework would be advantageous, that allows for the management of several meta-models, their instance models and languages, and provides means for an easy and efficient generation of the necessary parsers and converters. In addition to that, integrating the models@runtime approach or MDA code generation would be a further step to a comprehensive MCA development environment.

References

1. Bencomo, N., Bennaceur, A., Grace, P., Blair, G., Issarny, V.: The role of models@ run. time in supporting on-the-fly interoperability. Computing **95**, 167–190 (2013)
2. Bencomo, N., France, R., Cheng, B.H.C., Aßmann, U. (eds.): Models@run.time: Foundations, Applications, and Roadmaps. LNCS, vol. 8378. Springer, Cham (2014). https://doi.org/10.1007/978-3-319-08915-7
3. Bennaceur, A., Issarny, V.: Automated synthesis of mediators to support component interoperability. IEEE Trans. Softw. Eng. **41**, 221–240 (2015)
4. Chen, L., Khalil, I.: Activity recognition: approaches, practices and trends. In: Chen, L., Nugent, C., Biswas, J., Hoey, J. (eds.) Activity Recognition in Pervasive Intelligent Environments, pp. 1–31. Springer, Paris (2011). https://doi.org/10.2991/978-94-91216-05-3_1
5. Costa, F.M., Morris, K.A., Kon, F., Clarke, P.J.: Model-driven domain-specific middleware. In: Proceedings of ICDCS 2017, pp. 1961–1971. IEEE (2017)
6. Embley, D.W., Liddle, S.W., Pastor, O.: Conceptual-model programming: a manifesto. In: Embley, D., Thalheim, B. (eds.) Handbook of Conceptual Modeling, pp. 3–16. Springer, Heidelberg (2011)
7. Fill, H.-G., Karagiannis, D.: On the conceptualisation of modelling methods using the ADOxx meta modelling platform. EMISA J. **8**, 4–25 (2013)
8. Götz, S., et al.: Adaptive exchange of distributed partial models@run.time for highly dynamic systems. In: Proceedings of SEAMS 2015, pp. 64–70. IEEE Press (2015)
9. Karpathy, A., Fei-Fei, L.: Deep visual-semantic alignments for generating image descriptions. In: Proceedings of IEEE Conference on Computer Vision and Pattern Recognition, pp. 3128–3137 (2015)
10. Kofod-Petersen, A., Cassens, J.: Using activity theory to model context awareness. In: Roth-Berghofer, T.R., Schulz, S., Leake, D.B. (eds.) MRC 2005. LNCS (LNAI), vol. 3946, pp. 1–17. Springer, Heidelberg (2006). https://doi.org/10.1007/11740674_1
11. Liddle, S.W.: Model-driven software development. In: Embley, D.W., Thalheim, B. (eds.) Handbook of Conceptual Modeling, pp. 17–54. Springer, Heidelberg (2011)
12. Mayr, H.C., et al.: HCM-L: domain-specific modeling for active and assisted living. In: Karagiannis, D., Mayr, H., Mylopoulos, J. (eds.) Domain-Specific Conceptual Modeling, pp. 527–552. Springer, Cham (2016)
13. Mayr, H.C., et al.: Model centered architecture. In: Cabot, J., Gómez, C., Pastor, O., Sancho, M., Teniente, E. (eds.) Conceptual Modeling Perspectives, pp. 85–104. Springer, Cham (2017). https://doi.org/10.1007/978-3-319-67271-7_7
14. Meditskos, G., et al.: MetaQ. Perv. Mob. Comput. **25**, 104–124 (2016)
15. Michael, J., Steinberger, C.: Context Modeling for Active Assistance. In: Proceedings of ER Forum 2017. CEUR Workshop Proceedings 1979, CEUR-WS.org, pp. 207–220 (2017)
16. Michael, J., et al.: The HBMS Story. Enterp. Model. Inf. Syst. Arch. **13**, 345–370 (2018)
17. Ni, Q., et al.: A foundational ontology-based model for human activity representation in smart homes. J. Ambient. Intell. Smart Environ. **8**, 47–61 (2016)
18. Onofri, L., et al.: A survey on using domain and contextual knowledge for human activity recognition in video streams. Expert Syst. Appl. **63**, 97–111 (2016)
19. Peláez, M.D., López-Medina, M., Espinilla, M., Medina-Quero, J.: Key factors for innovative developments on health sensor-based system. In: Rojas, I., Ortuño, F. (eds.) IWBBIO 2017,Part II. LNCS, vol. 10209, pp. 665–675. Springer, Cham (2017). https://doi.org/10.1007/978-3-319-56154-7_59

20. Ranasinghe, S., et al.: A review on applications of activity recognition systems with regard to performance and evaluation. Int. J. Distrib. Sens. Netw. **12**, 1550147716665520 (2016)
21. Rodríguez, N.D., Cuéllar, M.P., Lilius, J., Calvo-Flores, M.D.: A survey on ontologies for human behavior recognition (CSUR). ACM Comput. Surv. **46**, 43 (2014)
22. Siegel, C., Dorner, T.E.: Information technologies for active and assisted living—Influences to the quality of life of an ageing society. Int. J. of Med. Inform. **100**, 32–45 (2017)
23. Zgheib, R., et al.: A flexible architecture for cognitive sensing of activities in ambient assisted living. In: Proceedings of WETICE 2017, pp. 284–289. IEEE (2017)

Correction to: An Approach Toward the Economic Assessment of Business Process Compliance

Stephan Kuehnel and Andrea Zasada

Correction to:
Chapter "An Approach Toward the Economic Assessment of Business Process Compliance" in: C. Woo et al. (Eds.): *Advances in Conceptual Modeling*, **LNCS 11158,** **https://doi.org/10.1007/978-3-030-01391-2_28**

In the original version of this chapter, the term "vector product" was used instead of "scalar product" in the first paragraph of page 236. This has now been corrected.

The updated version of this chapter can be found at
https://doi.org/10.1007/978-3-030-01391-2_28

Tutorials 2018

Behavior-Derived Reuse - Conceptual Foundations and Practical Tools for Increasing Software Reuse

Iris Reinhartz-Berger[(✉)] and Anna Zamansky

University of Haifa, Haifa, Israel

Abstract. Software reuse has many known benefits, yet in practice variants are created, challenging their identification. Identification of similar software artifacts can increase reuse in a number of scenarios, such as software product line engineering, organizational knowledge sharing, and acquisitions and mergers. In such scenarios, the similarity analysis needs to take into consideration that the software artifacts may not have been developed by the same teams, and thus may not be similar on the level of implementation.

The goal of this tutorial is to introduce researchers and practitioners to a novel approach for increasing reuse, which looks beyond implementation level similarities by analyzing behavioral similarity and making concrete reuse recommendations based on polymorphism-inspired variability mechanisms. The approach is supported by the VarMeR tool, which enables analysis of multiple large-scale projects developed in Java.

© Springer Nature Switzerland AG 2018
C. Woo et al. (Eds.): ER 2018 Workshops, LNCS 11158, p. 343, 2018.
https://doi.org/10.1007/978-3-030-01391-2

Visual Querying on Graphs: Models and Techniques

Sourav S. Bhowmick[1(✉)] and Byron Choi[2]

[1] Nanyang Technological University, Singapore, Singapore
[2] Hong Kong Baptist University, Kowloon Tong, China

Abstract. Querying graph databases has emerged as an important research problem for real-world applications that center on large graph data. Given the complexity of graph query languages (e.g., SPARQL, Cypher), visual graph query interfaces make it easy for non-expert users to query such graph data repositories. In this tutorial, we survey recent developments in the emerging area of visual graph querying paradigm that bridges traditional graph querying with human computer interaction (HCI). We discuss models and techniques for visual graph query formulation, query processing, and visual exploration of graph query results. In addition, the tutorial suggests open problems and new research directions.

C. Woo et al. (Eds.): ER 2018 Workshops, LNCS 11158, p. 344, 2018.
https://doi.org/10.1007/978-3-030-01391-2

Conceptualizing Analytics

Christoph G. Schuetz[✉] and Michael Schrefl

Johannes Kepler University (JKU) Linz, Linz, Austria

Abstract. Business intelligence and data analytics projects often involve low-level, ad hoc data wrangling and programming, often resulting in high development effort and low usability of analytics solutions. Conceptual modeling allows us to move data analytics onto a higher level of abstraction, facilitating the implementation and use of analytics solutions. This tutorial gives an overview of conceptual modeling methods along the (big) data analysis pipeline, based on literature and experience from cooperative research projects with industry. The tutorial targets practitioners involved in the planning and implementation of analytics projects as well as researchers interested in the state of the art and open research questions of business intelligence and data analytics.

© Springer Nature Switzerland AG 2018
C. Woo et al. (Eds.): ER 2018 Workshops, LNCS 11158, p. 345, 2018.
https://doi.org/10.1007/978-3-030-01391-2

Abstraction in Conceptual Models, Maps and Graphs

Carlo Batin[1]([⊠]) and John Mylopoulos[2,3]

[1] University of Milano Bicocca, Milan, Italy
[2] Universities of Toronto, Toronto, Canada
[3] Universities of Trento, Trento, Canada

Abstract. Models are ubiquitous in Computer and Information Sciences as they help us understand our physical and social environment, and also design systems that facilitate our lives. In this tutorial, we focus on three types of models that are heavily used in the literature: conceptual models, maps, and graphs (for short, CMMGs). To build and make effective use of such models, we use abstractions. An abstraction over a model m removes some detail from m, to make it easier to understand, use, analyze and manage. In this tutorial we focus on conceptual abstractions that are motivated by Cognitive Science and Philosophy, as opposed to mathematical abstractions. (Conceptual) Abstractions that have been extensively studied in the literature include generalization (is-a), aggregation (part-of), classification (instanc`e-of), manifestation (aka materialization), contextualization (context-of) and granularization, but also many others. The tutorial introduces a unifying syntax and semantics for CMMGs, defines the six abstractions mentioned above, and studies the way they have been used for CMMGs.

© Springer Nature Switzerland AG 2018
C. Woo et al. (Eds.): ER 2018 Workshops, LNCS 11158, p. 346, 2018.
https://doi.org/10.1007/978-3-030-01391-2

Author Index

Printed in the United States
by Book masters

Printed in the United States
By Bookmasters